Leaf Development and Canopy Growth

Sheffield Biological Sciences

A series which provides an accessible source of information at research and professional level in chosen sectors of the biological sciences.

Series Editors:

Titles in the Series:

Biology of Farmed Fish
Edited by K.D. Black and A.D. Pickering.

Stress Physiology in Animals
Edited by P.H.M. Balm.

Leaf Development and Canopy Growth
Edited by B. Marshall and J.A. Roberts.

Seed Technology and its Biological Basis
Edited by M. Black and J.D. Bewley.

Leaf Development and Canopy Growth

Edited by

BRUCE MARSHALL
Scottish Crop Research Institute
Dundee, UK

and

JEREMY A. ROBERTS
Plant Sciences Division
School of Biological Sciences
University of Nottingham, UK

Sheffield
Academic Press

CRC Press

First published 2000

Published by
Sheffield Academic Press Ltd
Mansion House, 19 Kingfield Road
Sheffield S11 9AS, England

ISBN 1-84127-044-X

Published in the U.S.A. and Canada (only) by
CRC Press LLC
2000 Corporate Blvd., N.W.
Boca Raton, FL 33431, U.S.A.
Orders from the U.S.A. and Canada (only) to CRC Press LLC

U.S.A. and Canada only:
ISBN 0-8493-9769-3

Printed on acid-free paper in Great Britain by
Bookcraft Ltd, Midsomer Norton, Bath

British Library Cataloguing-in-Publication Data:
A catalogue record for this book is available from the British Library

Library of Congress Cataloging-in-Publication Data:
A catalog record for this book is available from the Library of Congress

Preface

The book takes a fresh, innovative look at the role of leaf development and canopy growth in the survival of the individual and the diversity of the community. It draws together significant advances in a diverse range of disciplines (plant physiology, evolutionary biology, ecology, molecular biology and mathematical biology), answers some questions and, more importantly, stimulates the reader into new conceptual and experimental ways of approaching research. Emergent properties at the plant scale are the result of self-organisation of 'local rules' applied within the cell and the couplings that exist between cells in the plant. In a similar way, emergent properties at the canopy scale depend on properties of the individual plants that are present. The whole is much more than the sum of the constituent parts. The focus in this volume is on the light-intercepting surfaces of the plant and the utilisation of the absorbed energy to fix carbon—the foundation of carbon-based life on earth. The authors consider a diverse range of species throughout, drawing on experience gained with both monocultures and mixed-plant communities. The volume is intended for researchers and professionals at post-doctoral level and above, especially those wishing to cross disciplinary boundaries.

The book opens with an innovative perspective on the evolution of plant form, demonstrating the potential co-existence of alternative forms which are locally optimal on a fitness landscape. It reminds us that plants, and the challenges they have met, have been around for a long time before man's attempted interventions. The traditional, descriptive view of canopy growth—manifest in many crop simulation models with the implicit use of a 'global brain', *deciding* which parts of a plant are deficient in a resource and then *directing* remedial action—is put aside. An alternative view based on 'local rules' (metabolic processes within each cell) is proposed. These rules, together with the interconnections between them (the vascular system in higher plants), provide the means for self-organisation within the plant and emergent global behaviour - the 'apparent brain'. The central chapters (3–6) consider the finer scale processes within which these local rules are implemented and some of the novel cellular and molecular strategies that are used to study them. The book then views the consequences of these local mechanisms in the context of the whole canopy (chapters 7–9), including the latest thinking regarding the self-organisation of nitrogen in relation to the light environment within the canopy, the partitioning of radiative energy between individuals in mixed-communities and the consequences for carbon fixation, evaporative demand

and the partitioning of water flow between individuals. The dynamics of canopy turnover and the ability to exploit transient spatial opportunities are then introduced into the discussion, with special reference to forest gap dynamics. Finally, the book returns to the concepts raised at the beginning, addressing the conditions for the maintenance of individual and community diversity, integrating concepts raised earlier in relation to resource capture and utilisation for the preservation of the individual's genes.

Bruce Marshall
Jeremy A. Roberts

Contributors

Dr N.P.R. Anten	Department of Biological Sciences, Stanford University, Stanford, California 94305-5020, USA
Professor W. Beyschlag	Lehrstuhl für Experimentelle Ökologie, Universität Bielefeld W4-107, Universitätstrasse 25, W4-107, 33615 Bielefeld, Germany
Dr T. Foster	Institute of Molecular Biosciences, Massey University, Palmerston North, New Zealand
Professor J. Grace	Institute of Ecology and Resource, Darwin Building, University of Edinburgh, King's Buildings, Edinburgh EH10 5PB, UK
Dr K. Hikosaka	Biological Institute, Graduate School of Science, Tohoku University, Aoba, Sendai 980-8578, Japan
Professor T. Hirose	Biological Institute, Graduate School of Science, Tohoku University, Aoba, Sendai 980-8578, Japan
Mr F. Magnani	CNR-IMGPF, Via A. Vannucci 13, I-50134 Firenze, Italy
Professor K. Niklas	Department of Plant Biology, Cornell University, Plant Science Building, Ithaca, NY 14853-5908, USA
Dr R.J. Ryel	Department of Rangeland Resources, Utah State University, Logan, UT 84322-5230, USA
Dr U. Schurr	Botanisches Institut der Universität Heidelberg, Im Neuenheimer Feld 360, 69120 Heidelberg, Germany
Professor H. Smith	University of Leicester, University Road, Leicester LE1 7RH, UK
Dr G. Squire	Scottish Crop Research Institute, Invergowrie, Dundee DD2 5DA, UK

Dr B. Veit Institute of Molecular Biosciences, Massey University, Private Bag 11222, Palmerston North, New Zealand

Dr A. Verhoef Institute of Hydrology, Maclean Building, Crowmarsh Gifford, Wallingford, Oxfordshire OX10 8BB, UK

Professor J.S. Wallace Institute of Hydrology, Maclean Building, Crowmarsh Gifford, Wallingford, Oxfordshire OX10 8BB, UK

Mr A. Walter Botanisches Institut der Universität Heidelberg, Im Neuenheimer Feld 360, 69120 Heidelberg, Germany

Professor A.N. Webber Department of Plant Biology, Arizona State University, Box 871601, Tempe, AZ 85287 1601, USA

Contents

1 The evolution of leaf form and function

Karl J. Niklas

1.1 Introduction

This chapter explores the functional divergence that has ultimately resulted in the leaf–stem organographic distinction. This divergence characterizes much of the evolutionary history of the tracheophytes and currently stands as a shared derived feature among the majority of contemporary species. Its origins can be traced back to Ordovician–Silurian times, when the land plants made their first appearance in the form of morphologically unassuming, small organisms, whose highly fragmented fossil remains suggest simple body plans lacking the familiar morphological and anatomical distinctions between leaf, stem and root (Stewart and Rothwell, 1993; Taylor and Taylor, 1993). Indeed, clear evidence for the leaf–stem functional dichotomy is not seen until Devonian times and, even then, only among the members of some plant lineages. Paleobotanical studies continue to shed light on the morphological transformations leading to the separation of leaves from stems. However, the fossil record is largely incomplete for many important lineages, especially during their early evolutionary stages, and rarely, if ever, documents the developmental mechanisms underwriting organographic specialization. This limitation is compensated for by the ability to infer form–function relationships for plants based on a biomechanical perspective—the application of engineering principles and techniques to discover the quantitative relationships between external shape (morphology), internal structure (anatomy) and the performance of functional tasks essential for growth, survival and reproduction (Wainwright *et al.*, 1976; Niklas, 1992)—a perspective that provides the prevailing theme for this chapter.

Unlike zoologists, botanists have the advantage of knowing that all photoautotrophs require the same fundamental resources (oxygen, carbon dioxide, water, minerals and space) and all plants marshal these resources to fabricate their living substance in much the same way, regardless of phyletic ancestry or propinquity of descent (Nobel, 1983; Salisbury and Ross, 1992; Niklas, 1997a). The physiology of photosynthesis and plant respiration has unquestionably remained the same for the past one billion years (Gates, 1980; Nobel, 1983; Niklas, 1997a). By the same token, the physical principles governing the flow of mass and energy between the plant body and the external environment as well as the ways in which plant tissues respond to mechanical forces have not changed throughout evolutionary history. It is possible, therefore, to predict, with a fair degree of accuracy and precision, how specific morphological transformations documented by the fossil record have influenced the ability of plants to perform their fundamental biological obligations.

This chapter outlines how the functional dichotomy between the leaf and stem arose independently, early in the evolutionary history of a number of tracheophyte lineages and why this dichotomy persists today in the majority of their descendants. It is a preamble to the general topic of leaf development and canopy growth, the subject of the present book. It begins with a brief discussion of the traditional definition of the 'leaf'. This discussion is followed by computer simulations that show that the features used to define the leaf are a necessary and sufficient result of selection, satisfying the simultaneous requirements for light-harvesting and mechanical stability as the overall size of the ancient plant body plan increased through evolution. The ancient plant Bauplan consisted of a collection of more or less equally branched, cylindrical, leafless axial elements. The computer simulations specifically indicate that, with increasing size, the plant had to compartmentalize the functions of mechanical support and light-harvesting. As a consequence, the derived body plan acquired a morphology consisting of nearly vertical, mechanically supportive axes ('stems') bearing lateral, horizontally planated structures devoted to harvesting sunlight ('leaves') (Niklas, 1994a).

The computer simulations also show that the angle swept between adjoining lateral structures attached along the lengths of stems must be adjusted to maximize light interception, such that the angle ultimately corresponds to that most frequently seen separating successive leaves in the phyllotactic patterns of many ancient and modern-day plants. Even though this angle may have served as a developmental constraint among early vascular land plants, with the advent of foliage leaves, computer simulations evaluating the effects of different phyllotactic patterns on the ability to intercept sunlight show that a constellation of leaf morphological variables can be varied independently to more than adequately compensate for the potentially negative effects of some patterns of leaf arrangement on light-harvesting (Niklas, 1998a).

The final section of this chapter discusses some of the complex mechanical interactions between leaves and stems, in terms of adaptive responses to static and dynamic mechanical forces. This biomechanical topic is too broad to be dealt with comprehensively and only some of its subject matter is presented, more to inspire further research than to provide synoptic coverage.

1.2 The definition of a 'leaf'

The leaf is traditionally defined as an appendicular organ that typically has a bilateral morphology and anatomy as a consequence of its position, determinate growth in size and developmental subordination with respect to the shoot apical meristem that gives rise to it (Esau, 1967; Bierhorst, 1971; Gifford and Foster, 1988). The appendicular status of the leaf is consistent with the lateral location of the leaf primordium on the shoot apical dome, and by the bilateral external symmetry and vascular supply of the mature leaf, resulting, in part, from the activity of a linear marginal meristem. The generally reduced and determinate growth in the size of the leaf in comparison to that of the stem is taken as further evidence that the shoot apical meristem exerts a controlling influence over leaf development.

Importantly, this definition does not ascribe to the leaf either a unique functional role or a single evolutionary origin. Among the land plants, the leaf has been evolutionarily co-opted to perform a variety of functions (e.g. photosynthetic foliage leaves, protective or defensive cataphylls and spines, and reproductive sporophylls) (Mauseth, 1988), and thus appears to have been the object of intense diversifying selection both within and across tracheophyte lineages. For this reason, leaves are defined in a developmental rather than a functional context. Likewise, even the most conservative estimates of the divergence times among the last common ancestors of some tracheophyte lineages predate the appearance of organographically distinct leaves in the basal members of these lineages (e.g. lycopods, sphenophytes, ferns and seed plants) (Taylor and Taylor, 1993), providing evidence that leaves, as developmentally appendicular organs, have evolved independently, and thus cannot be defined in a strictly taxonomic context.

The independent evolutionary origins and diverse functions of leaves opens the question as to whether the first leaves to appear in phyletically disparate lineages performed the same or different functions. If the former, then the history of leaf evolution is dominated by diversifying selection; if the latter, then convergent evolution resulting from stabilizing selection may be prevalent. It is possible that the ancestral condition of the organs generically called 'leaves' was the foliage leaf, whose original photosynthetic role was co-opted to obtain functionally different roles (e.g. sepals, petals and spines). However, it is equally possible that the ancestral leaf condition for different lineages was functionally diverse, such that the foliage leaf is a derived condition appearing many times in different lineages.

Unfortunately, this issue cannot be resolved unambiguously. The fossil record of many lineages is incomplete and the ancestral condition of the leaves in many taxa is either unknown or doubtful. Likewise, the environmental contexts (and thus the selection forces) attending the appearance of the first leaves in some lineages are difficult to infer, even for some well-preserved fossils. Even though it is attractive to argue that the capacity to produce leaf-like structures is adaptive because it confers a metabolic advantage in terms of harvesting light and exchanging nutrients between the plant body and the external environment, it is evident that the ability to produce appendicular organs capable of serial homology confers many advantages.

1.3 Convergence or divergence?

Leaf-like structures are produced by the algae as well as non-vascular embryophytes (e.g. the blades of kelps and the phyllids of some moss species, respectively) (Bold, 1967). These structures clearly amplify the photosynthetic surface area with respect to the non-photosynthetic tissue-volume of the plant body, and thus unequivocally increase the capacity to intercept light, harvest nutrients from the external medium and dispose of metabolic waste. The evolutionary appearance and retention of foliage leaf-like structures among representatives of the *Rhodophyta*, *Phaeophyta* and *Chlorophyta*, as well as among diverse non-vascular and vascular embryophyte species (Bold, 1967), would appear to support the supposition that foliage leaves or their functional analogs evolved early in the evolutionary history of eukaryotic multicellular photoautotrophs (i.e. 'plants' in the most general sense of the term) by virtue of the metabolic advantage they confer. However, the 'leaves' of many liverwort and moss species with a paucity of rhizoids (e.g. *Porella* and *Funaria*) often serve as the principal organ for the transport and absorption of nutrients or as the main means to hydrate and protect growing apices (Bold, 1967). The early developmental similarity between some moss phyllids and apical paraphyses and 'bracts', which protect and hydrate the gametophyte apex and the reproductive structures it produces, further suggests that developmentally subordinate, multicellular, lateral structures are adaptive regardless of their ability to harvest light (Watson, 1971). There is, therefore, sufficient reason to question whether the current functional correspondence between phyllids and tracheophyte foliage leaves represents an ancestral or a derived condition.

A similar ambiguity exists regarding the selection pressures attending the appearance of leaves among tracheophyte lineages. The colonization of land undoubtedly presented a barrier to the direct carry-over of foliage leaf-like structures, whose elevation above a moist substrate is non-adaptive in the absence of a cuticle to impede uncontrolled evaporation, stomata to regulate the rate of water loss and, in the case of a large organism, of specialized water-conducting tissues, all of which were absent in the aquatic ancestors to the land plants (Raven, 1984). Even if the last common ancestor to the embryophytes had leaf-like structures, it is likely that these structures would have been lost during the initial phase of land colonization and only subsequently re-evolved after the evolutionary appearance of the cuticle, stomata and xylem tissues, which permitted an amplification of surface area relative to tissue volume on land. The fossil record accords with this scenario because the earliest vascular land plant sporophytes consisted of leafless, nearly cylindrical axes, which simultaneously functioned as photosynthetic and anchorage/absorption organs that indirectly or directly provided mechanical support for the reproductive structures.

1.4 Megaphylls and microphylls

Two theories have been advanced to explain the origins of the tracheophyte leaf, the Telome theory and the Enation theory (Stewart and Rothwell, 1993). Both of these theories, however, are more properly said to be 'scenarios', because neither advances a developmental or adaptive explanation for the appearance of leaves. The Telome theory, which originated with Zimmermann (1966 and references therein), states that leaves evolved as the result of the overtopping, planation and webbing of distal and some subtending axes (i.e. 'telomes' and 'mesomes', respectively) of the sporophyte plant body (Figure 1.1). Zimmermann defined 'overtopping' as the differential growth in size of the two axial units resulting from the bifurcation of an apex, such that one unit is subordinate in size and position with respect to the other. Zimmerman defined 'planation' as the change from a three-dimensionally branched truss of telomes and mesomes into a two-dimensionally branched truss, and 'webbing' as the introduction of tissues between adjoining axial elements in a truss. The Telome theory attempts to explain the origin of the megaphyll, the leaf-type traditionally assigned to the fern and seed plant species.

In contrast, the Enation theory was advanced by Bower (1935) to explain the origin of the microphyll (the leaf-type of all lycopods and their ancestors). This theory argues that leaves initially evolved as non-vascularized outgrowths ('enations') on otherwise naked cylindrical axes that, over the course of evolutionary history, increased in size and vascular supply (Figure 1.1). The distinction made between the 'megaphyll' and 'microphyll' by E.C. Jeffrey early in the twentieth century is based on a number of criteria, among which the presence of a leaf gap and of a divergent and branched leaf trace (or traces) is said to be the hallmark of the megaphyll, whereas the absence of these features is said to define the microphyll. The many exceptions to these criteria observed among living as well as fossil species (e.g. Wagner *et al.*, 1982) and the fact that leaf gaps are intrinsically absent in protostelic stems, which occur in some fern species, regardless of whether the leaf is defined as a mega- or microphyll, cast doubt on the synoptic utility of the distinction between these two leaf-types. Likewise, if a distinction between tracheophyte leaf-types exists, as the fossil record suggests, then it rests more firmly on the monophyletic origin of the lycopod leaf as opposed to the multiple origins of leaves in other vascular plant lineages. Under any circumstances, the Telome and Enation 'theories', which are still widely

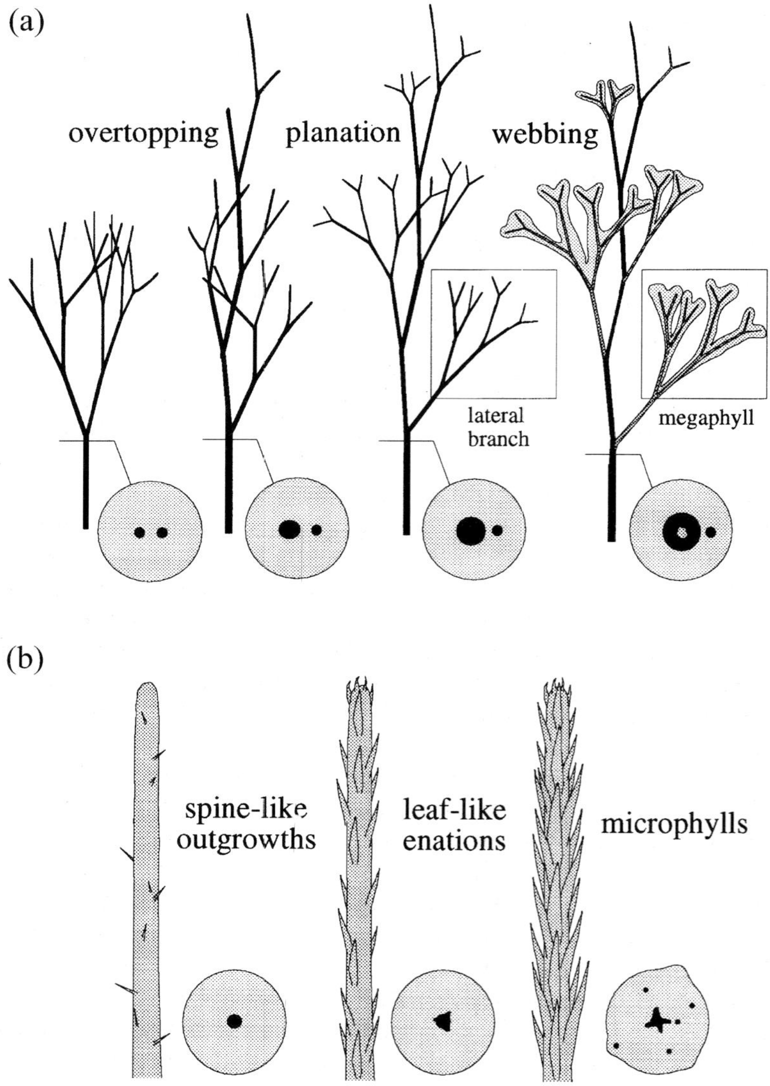

Figure 1.1 Rendering of the Telome and Enation theories for the origin of leaves among early vascular plant lineages. a) The Telome theory assumes that megaphylls evolved as a consequence of the overtopping, planation and webbing of the branched axial members of the early vascular land plants, the most ancient of which were composed of essentially equally branched axes. As a consequence of modifications in the relative size of axes attending overtopping, anatomical differences were established that ultimately led to the distinctions between leaf and stem (see diagrams of representative transverse sections; black denotes vascular tissues). b) The Enation theory assumes that microphylls evolved from small spine-like outgrowths that increased in size and ultimately became vascularized as they assumed the status of leaves (see diagrams of representative cross-sections; black denotes vascular tissues). (See text for further details.)

taught, are far too simplistic conceptually and over-extended in application to merit further discussion other than to say that they draw attention to the polyphyletic origins of tracheophyte leaves.

1.5 The leaf–stem dichotomy

There are perhaps two basic reasons why the functional dichotomy between the 'leaf' and the 'stem' evolved many times in the early history of the vascular land plants and why this distinction has been perpetuated in various lineages ever since. Firstly, the earliest vascular plants grew in size and height by the repeated bifurcation of their apical meristems. This mode of growth, which may be a 'phyletic legacy' in the sense that it was a carry-over from the last common ancestor to all land plants, results in a morphology that is progressively more branched, with increasing apical bifurcation, less efficient at capturing sunlight (as a consequence of increased self-shading) and more prone to mechanical failure (as the weight supported at the base of the plant body increases; see below). The simplest solution to this physiological and mechanical dilemma is to physically partition the functional obligations for mechanical support and light-harvesting; that is, to produce vertical stem-like analogs that physiologically abandon their early historical role as light-capturing devices and now serve exclusively as mechanical support members, and to produce lateral structures devoted exclusively to photosynthesis (Niklas, 1997a,b).

The second reason for the functional dichotomy between 'leaf' and 'stem', which is intimately related to the first, is that the best location for photosynthetic and mechanically supportive tissues is just at the surface of any organ performing these two functions. In terms of photosynthesis, this location facilitates gas exchange between the plant body and the external atmosphere and simultaneously reduces the attenuation in the intensity of light, especially in the blue and red wavelengths, as it passes through photosynthetic cells. This is attested to by the preferential deployment of photosynthetic tissues along the upper surfaces of leaves and in the most peripheral cell layers of a green stem's hypodermis. By the same token, engineering theory shows that the maximum tensile and compressive stresses resulting from any bending force acting on a column or beam invariably occur at the surface of the structure. For extremely small vascular plants, this physiomechanical problem is trivial because small and short photosynthetic stems can be mechanically sustained hydrostatically, especially if they have reliable access to ground water. This follows from the fact that the intensity of any mechanical stress resulting from self-weight increases in proportion to the weight the structure must support. Providing that the weight of a vertical stem results in stress intensities below the stress tolerance level of its most external tissues (epidermis and external hypodermal cell layers), mechanical failure can be avoided. With increasing size and weight, however, bending stresses can approach or exceed the strength and stiffness of photosynthetic tissues, and can thus incur catastrophic mechanical failure.

There are two obvious and biologically common solutions to the desirability of placing the stiffest and strongest tissues in exactly the same location as photosynthetic tissues. One is seen in many grass species, whose vertical stems have vertically aligned strips of sclerenchyma alternating with strips of photosynthetic tissue (Schwendener, 1874). A variant of this is seen among *Equisetum* species, whose vertical stems are fluted in transection and develop their stiffest tissues at locations furthest from the centroid axis of the stem (Bold, 1967; Bierhorst, 1971), where tensile and compressive bending stresses reach their greatest intensities (Niklas, 1992). Both of these configurations permit mechanical support and photosynthesis in the same organ. The second solution, which is not mutually exclusive, is best seen in *Psilotum nudum*, a pteridophyte of uncertain phylogenetic affinity, whose aerial body parts consist of a number of repeatedly bifurcate vertical axes (called trusses) attached to a subterranean rhizome-like system of non-photosynthetic axes (Bierhorst, 1971; Niklas, 1990a). The distal-most elements of the *P. nudum* sporophyte are bright green in color due to the presence of photosynthetic cells just beneath the epidermis (Figure 1.2). However, as the more proximal elements of a truss are progressively examined, these photosynthetic tissues gradually give way to a zone of hypodermal sclerenchyma occupying the same relative location. The most proximal axial element of a mature, fully expanded truss is typically devoid of photosynthetic tissues.

The opposing longitudinal gradients of photosynthetic and mechanical tissues occupying the hypodermal region in *P. nudum* trusses establish leaf- and stem-like analogs (distal and proximal axial elements, respectively) among what are otherwise developmentally homologous elements of the same vertical structure (truss). These gradients reflect a developmental mechanism that obtains functional specialization among axes in what superficially appears to be an extremely ancient plant body plan. This mechanism may have been achieved early in the history of the tracheophytes. The morphology and anatomy of some of the most ancient tracheophytes (e.g. *Rhynia* and *Aglaophyton*) are very similar to those of *P. nudum*, in that there are no obvious organographic distinctions between stem, leaf and root, such that any axial element can function in one or more capacities depending on its anatomy and location within the architecture of the body plan. Some authorities consider *Psilotum* to be a relictual and ancient taxon, while others consider this genus a highly derived and morphologically reduced taxon (Bierhorst, 1971). However, this debate is largely irrelevant to the present argument because, in either case, the axial elements of the *Psilotum* truss are functionally specialized and (serially) homologous portions of the same structure.

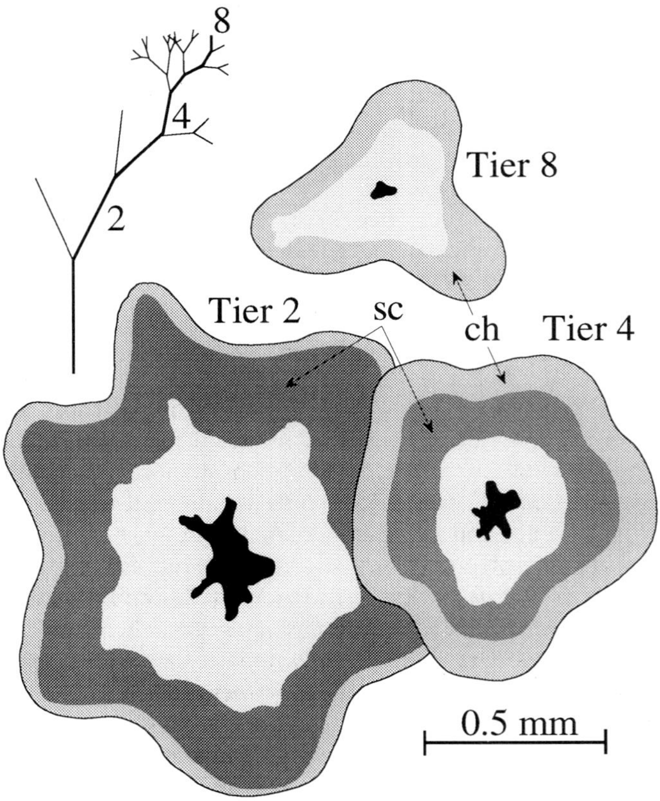

Figure 1.2 Semi-diagrammatic rendition of representative transections through the aerial axes of the pteridophyte, *Psilotum nudum*. The hypodermis of the most distal axial elements in the aerial portions of this species (i.e. axes from branching tier 8 in the 'stick' diagram shown upper left) contains chlorenchyma (ch), which is gradually replaced in successively more proximal axial elements (i.e. axes from tiers 4 and 2) with sclerenchyma (sc), such that the most proximal axis in each aerial truss lacks chlorenchyma. The black region denotes vascular tissues in each transection.

1.6 Computer simulations of the leaf–stem functional dichotomy

The evolutionary trajectories by which the leaf–stem functional dichotomy may have been established can be deduced from the fossil record of early

Paleozoic plants and inferred from computer simulations, which predict the relative fitness of different simple morphologies in terms of their capacity to intercept sunlight and to provide mechanical support for vertical axes. The juxtaposition of the fossil record and the evolutionary trajectories predicted by computer simulations provides a crude test of the hypothesis that selection favoring an increase in plant size and height drove, albeit perhaps indirectly, the evolution of appendicular organs capable of functioning as foliage leaves.

As noted previously, the sporophytes of the early vascular land plants were, on average, small organisms consisting of dichotomously branched, cylindrical, leafless axes bearing sporangia either at their tips or along their sides. This general morphology can be mathematically described by assigning a length to each companion axis resulting from a bifurcation (l_1 and l_2), as well as a diameter (d_1 and d_2). The orientation of each of the companion axes in any dichotomy with respect to the vertical and horizontal planes can be described by means of two angles (Figure 1.3): the bifurcation angle, ϕ, which is the angle swept out between the two adjoining axes, and the rotation angle, γ, which is the angle swept out between the horizontal plane and the apex of an axis (Niklas, 1997b). Since there are two axes in every dichotomy and since the orientations of these axes may not be equivalent, two bifurcation angles (ϕ_1 and ϕ_2) and two rotation angles are required (γ_1 and γ_2). A probability of apical bifurcation must also be assigned to each axis (p_1 and p_2), the numerical value of which ranges from 0 to 1 (resulting in unbranched and branched axes, respectively) (Figure 1.3). With the aid of these five parameters, a 'universe' of mathematically conceivable sporophyte morphologies can be constructed for the early vascular land plants (Niklas, 1997a,b).

This universe can be rendered by as a Cartesian coordinate system if axial length and diameter are held constant (i.e. $l_1 = l_2 = c_1$ and $d_1 = d_2 = c_2$), if the two bifurcation and the two rotation angles for each pair of axes produced by a bifurcation event are set equal to one another (i.e. $\phi_1 = \phi_2$ and $\gamma_1 = \gamma_2$), and if the probabilities of apical bifurcation are likewise equivalent (i.e. $p_1 = p_2$). Under these circumstances, the location of each hypothetical morphology in the universe is specified by the coordinates of three orthogonal axes (i.e. ϕ, γ and p). Naturally, this three-dimensional 'universe' contains only the most simple, isotomously branched morphologies, and is thus incapable of rendering the location of 'overtopped' morphologies bearing 'planated,' unequally branched, lateral axial systems that are made possible when $\phi_1 \sim 0$, $\gamma_1, \gamma_2 \sim 0°$ and $p_1 \neq p_2$. The only 'universe' containing all mathematically conceivable early land plant morphologies is a multidimensional one, which resists simple graphic redaction.

Regardless of how its location is depicted graphically, each hypothetical morphology can be numerically evaluated in terms of its light-harvesting capability and its mechanical ability to sustain its own weight. Light interception, L, is assessed numerically by integrating the area under the curve that

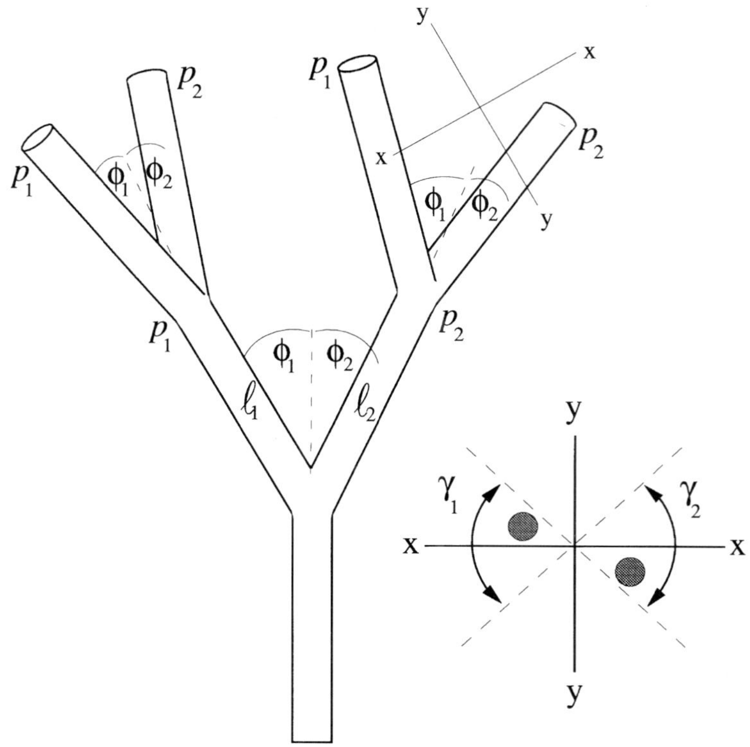

Figure 1.3 Mathematical parameters required to create a 'universe' of morphological variants similar to the morphologies of some early vascular land plants, whose ancestral body plan consisted of a collection of cylindrical bifurcating axes. (For further details, see text.)

results when the dimensionless quotient of projected surface area and total surface area of each morphology (A_p/A) is plotted against the elevation angle of the sun, θ, from dawn to dusk at the Earth's equator ($0° \leqslant \theta \leqslant 180°$). The area under this curve is given by the formula:

$$L = \int_{0°}^{180°} \left(\frac{A_p}{A}\right) i \cdot d\theta \qquad (1.1)$$

where i denotes irradiance intensity (which is assumed to be a constant). It provides a measure of the ability of any hypothetical morphology to intercept sunlight because it accounts for: the self-shading of neighboring axes in terms of their relative size, shape and orientation; and, the daily changes in the total illuminated area that each morphology projects toward the sun. This formula also has the virtue of normalizing the capacity of any morphology to intercept sunlight with respect to the total size of the morphology (here specified in terms of its total surface area, A).

Mechanical stability is also easily evaluated because the length and diameter assigned to each axis specifies the volume of the axis (i.e. $V = \pi d^2 l/4$), and thus the mass-force, F, of the axis if bulk tissue density (ρ) is assumed to be constant throughout the plant body ($F = \pi d^2 l \rho g/4$). This mass-force permits the calculation of the bending moments and stresses acting on each axial element in any hypothetical morphology because the rotation and bifurcation angles assigned to each axis specify the orientation of each axis with respect to the vertical and horizontal planes. In general terms, the bending moment for any axis in a morphology is directly proportional to ϕ and inversely proportional to γ. For example, the bending moment, M, acting on any terminal axis in any hypothetical morphology is given by the formula:

$$M = \left(\frac{\pi}{\gamma}\right) d^2 l^2 \rho g \left(\frac{\phi}{2}\right) \tag{1.2}$$

The maximum bending stress for this axis, σ, is given by the formula:

$$\sigma = \left(\frac{d}{2}\right)\left(\frac{M}{I}\right) \tag{1.3}$$

where, I is the axial second moment of area. Since, for any circular cylinder, $I = 0.049d^4$, it follows that $\sigma \sim 10(M/d^3)$ for a terminal axis. Naturally, bending moments are additive and increase in a basipetal direction toward the base of any vertical structure. The 'running sum' of all the bending moments acting on all the axes in any hypothetical morphology can be easily calculated with the aid of a computer.

The relative fitness, ω, of any hypothetical morphology can be evaluated if it is assumed that plants should maximize their capacity to intercept sunlight and simultaneously minimize the bending moment acting on their base. Simplistically, therefore, it may be assumed that $\omega = L/M$. This measure of relative fitness can be used to evaluate the benefits that different morphological expressions of leaf- and stem-like analogs confer to a plant in terms of growth and survival, and thus, indirectly, to sexual reproduction. For example, simultaneously optimizing light interception and mechanical stability is a physical impossibility for most branched structures because the horizontal orientation represents the optimal one for light-harvesting, whereas this orientation incurs the largest bending moment (Niklas, 1992). Conversely, a vertical axis experiences no bending moment as a consequence of its own weight, yet is far less efficient at intercepting sunlight. Consequently, a morphological trade-off exists whenever relative fitness is defined in terms of maximizing the numerical value of L/M.

The morphological constraints imposed by the formula, $\omega = L/M$, are more fully appreciated when they are evaluated in terms of an 'adaptive walk' through the universe containing all mathematically conceivable morphologies predicated on maximizing L/M. This walk begins with the most ancient vascular land plant morphology, which currently belongs to the form genus *Cooksonia*, a Siluro–Devonian taxon representing a comparatively small, isobifurcate organism bearing terminal globose sporangia on some of its axes. Assuming that this morphology reflects the ancestral body plan for the majority of tracheophyte lineages, it provides the starting point for any 'adaptive walk' to locate variants in the universe of morphologies that are progressively more fit.

Using the *Cooksonia*-mimic as the starting point, the 'walk' is propelled by a simple search-and-find algorithm that locates the morphological variant closest to the mimic that has a higher relative fitness, at which point it then seeks the next nearest morphology with a higher fitness. This search protocol, which is permitted to branch whenever two or more morphologies with equivalent and higher relative fitnesses are located, is repeated indefinitely until one or more morphologies are located that have a relative fitness higher than that of all the previously identified morphologies and higher than any of those 'surrounding' them, at which point the 'walk' is terminated. By this means, the morphological solutions for maximizing L/M can be identified and compared to the major trends seen in the fossil record.

The results of one such adaptive walk indicate that L/M can be increasingly maximized, up to a point, by simply reducing the bifurcation angle and adding paired axes to the tips of the *Cooksonia*-mimic's axes (Figure 1.4). The result is a larger morphology vaguely reminiscent of that of *Rhynia* or *Aglaophyton*, both of which are found in geological strata younger than those containing *Cooksonia* fossils. Further addition of axes to this *Rhynia*-like morphology, however, results in significant self-shading and noncompensating, larger bending moments. The morphological solution to this functional impasse is to produce axes with unequal probabilities of branching and unequal bifurcation and rotation angles, such that some portions of the morphology 'grow' above others to yield a pseudomonopodial growth form (Figure 1.4). The more vertical axes in this growth form have smaller bending moments than their more horizontal companions, which, in turn, are more efficient at gathering sunlight. The result is a global reconciliation between the potentially antagonistic design requirements for mechanical support and light interception.

This solution works for a time, such that larger, equally fit morphological variants are possible. However, with additional 'growth' in size, the pseudo-

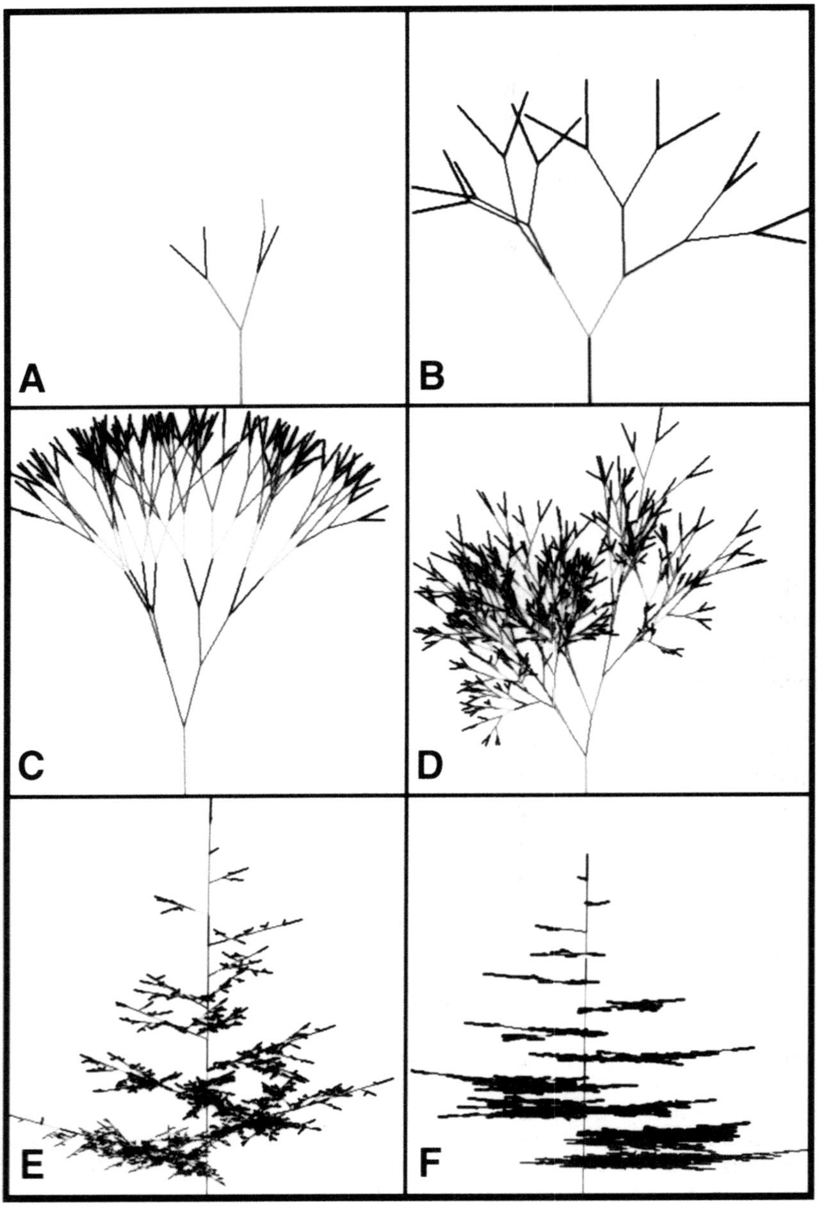
A
B
C
D
E
F

monopodial growth habit becomes inefficient and the relative fitness of more branched morphologies conforming to this theme starts to decrease. The solution next identified by the adaptive walk is to produce a main, vertical 'stem' (consisting of a series of axes for which $\phi = 0°$) bearing smaller lateral branching systems consisting of axes with low probabilities of branching and small angles of rotation. This morphological variant is vaguely reminiscent of those of the trimerophytes, such as *Psilophyton*, a taxon that occurs in geological strata younger than those containing the remains of *Rhynia* or *Aglaophyton*. Nonetheless, with further increases in overall size, the fitness of this morphological theme begins to decline because the rate at which the magnitude of the bending moments acting on the lateral branching systems increases exceeds the rate at which light-harvesting increases. This erosion in relative fitness is temporarily forestalled if the divergence angle, δ, swept between consecutively formed lateral branches is set at 137.5° (this angle is a new morphological parameter that must be arbitrarily introduced at this point in the 'adaptive walk' to temporarily stop the reduction in relative fitness; the significance of this parameter is discussed in Section 1.7 in the context of phyllotaxy). Further increases in size are attended by a reduction in relative fitness, unless the axial elements in all of the lateral branching systems are planated (i.e. $\gamma = 0°$). At this point, the 'adaptive walk' stops because all of the morphological variants with horizontally flattened lateral branching systems are 'surrounded' by variants with less relative fitness (Niklas, 1997b).

Taken at face value, the morphological trends identified by this particular adaptive walk, which are capable of reconciling the antagonistic design specifications for light interception and mechanical stability, are in reasonable agreement with the fossil record of early Paleozoic vascular plants. This record demonstrates a general increase in the maximum height among progressively younger contemporaneous species, attended by the transition

---

Figure 1.4 Computer-generated, three-dimensional branching morphologies mimicking the appearance of some early vascular land plants arranged in a sequence of increasing capacity to harvest sunlight, while simultaneously reducing the total bending moment acting on the most basal axial element in each morphology (i.e. arranged in an 'adaptive walk,' from A to F): A) 'adaptive walk' is initiated with a short sparsely branched *Cooksonia*-like morphology consisting of more or less equally branched axial members; B) the addition of branched units at the tips of the *Cooksonia*-like morphology and a reduction in the bifurcation angle results in a *Rhynia*-like morphology that is more efficient at intercepting sunlight and maintaining mechanical stability; C) *Psilophyton*-like morphology results from overtopping (i.e. the unequal growth of adjoining axes), which is required to maintain light-harvesting and mechanical efficiency with further increases in size; D) confining the divergence angle between successively formed lateral branching systems to 137.5° is required to maintain light-harvesting and mechanical efficiency; E) incipient planation of lateral branching systems is required to maintain light-harvesting and mechanical efficiency; and F) closure of 'adaptive walk' with an *Archeopteris*-like morphology bearing completely planated lateral branching systems attached to a vertical stem.

from an open dichotomous growth habit to a pseudomonopodial growth habit, culminating in the appearance of arborescent growth forms with a main, vertical stem bearing lateral, more or less planated branching systems (e.g. *Archaeopteris*). This does not justify the claim that early vascular land plant evolution was driven by selection forces favoring phenotypes capable of reconciling the functional obligations of light interception and mechanical support. The results from computer simulations can mimic a biological phenomenon for all the wrong reasons and, under any circumstances, the early history of the tracheophytes is not subject to experimental manipulation. What can be said is that this adaptive walk demonstrates that a functional dichotomy between leaf- and stem-like organs or organ systems is functionally beneficial for light interception and mechanical support, and that this dichotomy can be achieved by subordinating the growth and development of some parts of the plant body to produce lateral, planated ('appendicular') structures.

In passing, it is worth noting that the morphological transformations of 'overtopping' and 'planation' predicted by the adaptive walk to facilitate photosynthesis and mechanical stability are also identified by the Telome theory as phenomenologies attending the origin and evolution of the megaphyll (see Figure 1.1). Nevertheless, the Telome theory was previously rejected as a viable method for understanding the early land plant evolution, despite the emergence of overtopping and planation as important morphological trends, because it fails to establish the sequence of morphological transformations that obtains functional benefits. In this sense, the Telome theory provides a useful lexicon for morphological transformations but it does not provide information regarding which transformations are adaptive.

1.7 Phyllotactic patterns

The previously described adaptive walk for light interception and mechanical stability shows that the reconciliation between the design specifications for these two biological functions reaches an impasse that can be breached at one stage in the walk provided that the divergence angle swept out between successively formed 'lateral branching systems' is set at 137.5°. This angle was not selected at random. It figures prominently in the literature dealing with phyllotaxy, which shows that leaves separated by this angle and arranged in a spiral along the length of a stem will never overlap one another (Wright, 1873).

For reasons as yet unknown, the different spiral phyllotactic patterns observed among real plants can be represented by a procession of fractions called the Fibonacci series—1/2, 1/3, 2/5, 3/8, 5/13, 8/21, 13/34, 21/55, etc.—in which the numerator of each fraction is the sum of the previous two numerators and the denominator is likewise the sum of the two denominators in the previous two fractions. In terms of patterns of leaf arrangements, the numerator in each fraction represents the number of circuits around a stem that separate two superimposed leaves, whereas the denominator of each fraction equals the number of leaves separating the superimposed leaves. For example, the distichous pattern of leaf arrangement can be represented by the fraction $1/2$ because one circuit around the stem is required to reach two superimposed leaves. By the same token, the tristichous pattern of leaf arrangement is represented by $1/3$ because one circuit around the stem produces the superimposition of every third leaf.

Note that the angle swept out by one gyration around a stem for any particular spiral phyllotactic pattern is always equal to 360° multiplied by the corresponding fraction. Thus, 180° separates successively formed leaves in the distichous pattern of leaf arrangement ($1/2 \times 360° = 180°$), whilst 120° separates successive leaves in the tristichous pattern of leaf arrangement ($1/3 \times 360° = 120°$). It can be seen that the significance of 137.5° lies in the fact that the Fibonacci series converges on the decimal fraction 0.38197, which, when multiplied by 360°, gives 137.5°. As a consequence, whenever successively formed leaves (or leaf-like analogs) are formed at just this angular distance around the axis of a stem, no two will ever lie precisely above or below one another. This observation was taken as evidence that phyllotactic patterns corresponding to the higher fractions represented in the Fibonacci series reflect adaptations to avoiding self-shading (Wright, 1873). As will be seen, any computer simulation attempting to identify the leaf divergence angle that reduces the extent to which neighboring leaves shade one another will invariably identify 137.5° as the 'optimal angle'. But this result is misleading and sheds no light on the adaptiveness or lack thereof of different phyllotactic patterns because it ignores the simple fact that many other morphological variables influence the ability of leafy shoots to intercept sunlight (Niklas, 1998a). Indeed, the real question is not whether phyllotactic patterns produced by divergence angles near 137.5° are efficient—they will invariably be found to be so—but rather whether other morphological features are capable of compensating for the effects of 'inefficient divergence angles' on light interception.

It is possible to construct a simple universe of all mathematically conceivable phyllotactic patterns and to evaluate their capacity to intercept sunlight in much the same way that a universe of all mathematically conceivable early vascular land plant morphologies was constructed and evaluated. For present purposes, each 'leafy shoot' is initially confined to a single vertical 'stem' bearing elliptically-shaped leaves with length L and width W, such that the area of each leaf is given by the formula $\pi LW/4$ (i.e. L and W denote the major and minor axes of the elliptically-shaped leaf). Each shoot in this phyllotactic universe must also bear the same number of leaves and have the same total leaf surface area to assure legitimate comparisons across different morphologies in terms of their ability to intercept sunlight. In addition to the divergence angle, δ, which essentially defines the phyllotactic pattern of leaf arrangement for any morphology, only five other variables are required to describe the morphology of each variant: the leaf rotation angle, γ, which here defines the orientation of the leaf lamina with respect to the horizontal plane; the leaf deflection angle, ϕ, which is now the angle between the axis of the leaf lamina and the longitudinal axis of the vertical stem; petiole length, l; the height of the first leaf from the base of the stem, h_o; and the distance between successive leaves—the 'internodal distance'—denoted as i, such that the total height of the shoot H equals $h_o + ni$, where n is the number of leaves.

Low values of ϕ produce leaves that are almost vertical in orientation, with upper surfaces nearly appressed to the stem's surface; horizontally cantilevered leaves result when $\phi = 90°$. The leaf lamina parallels the Earth's surface when $\gamma = 0°$ and lies perpendicular to it when $\gamma = 90°$, much like the surface of an iris leaf blade. Comparatively complex morphologies can be produced by varying some or only a few of these variables. For example, morphological variants with a 'flattened' rosette growth habit are produced as the deflection angle is increased to 90° and as the internodal distance, i, approaches zero. Tall phenotypes with grass-like leaves are produced by increasing the internodal distance and leaf length with respect to width and by reducing petiole length and the deflection angle. Nonetheless, only three variables are required to specify the orientation of leaves with respect to the direction of the sun: the deflection, the rotation and the divergence angles (ϕ, γ and δ, respectively).

As was the case for the early vascular land plant universe of all mathematically conceivable morphologies, the universe of all conceivable phyllotactic patterns is multidimensional and difficult to render graphically in a synoptic way. However, the effect of any one morphological variable on the ability of variants—with the same or different leaf divergence angle—to intercept sunlight can be easily graphed. Each such 'adaptive walk' shows that divergence angles converging on 137.5° maximize light interception, particularly for variants with a rosette growth habit, whose leaves tend, on average, to overlap significantly. However, numerous 'walks' also show that the consequence of the divergence angle on the ability to intercept sunlight can be significantly reduced or all but eliminated if leaf shape and orientation are free to vary in a fashion independent of the leaf divergence angle. For example, rosette morphologies with long and very slender grass-like leaves are far more efficient than those with circular or near circular leaf outlines because they minimize the overlap of leaf laminae (Figure. 1.5).

Alterations in the angle between leaves and the vertical stem (i.e. variance in ϕ) likewise all but eliminate the 'superiority' of the 137.5° divergence

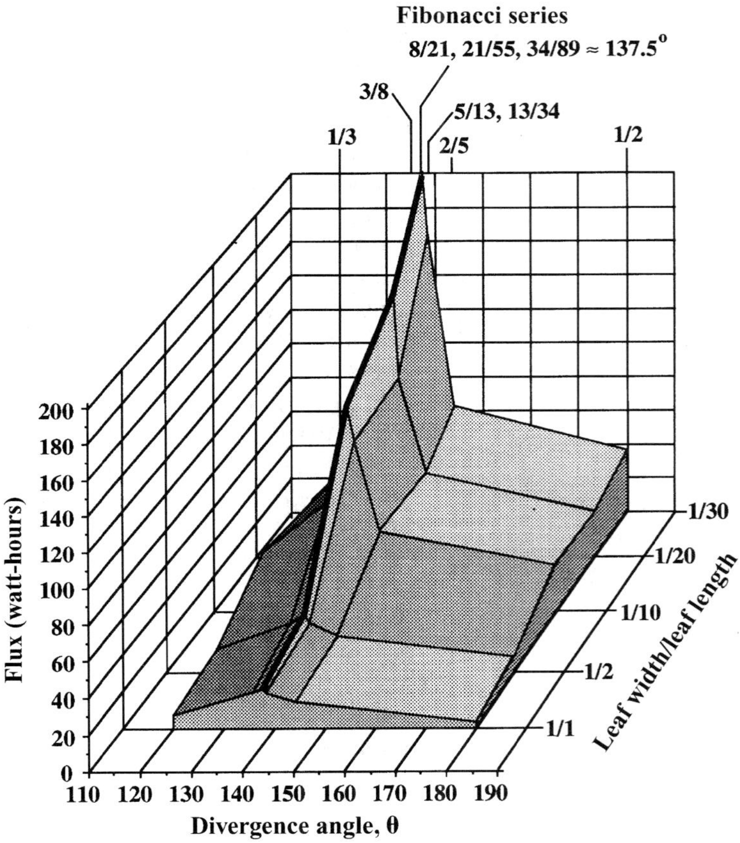

Figure 1.5 Three-dimensional portion of a more complex hypothetical 'phyllotactic' universe containing rosette-like morphologies bearing leaves with equivalent total surface area depicted by plotting light-harvesting efficiency (flux in watt-hours) against the leaf divergence angle, θ, and the ratio of leaf width and leaf length (Rosette: $i = 0.05$ cm, $l = 3$ cm, and $\phi = 45°$). The leaf divergence angle is translated into the corresponding fractions of the Fibonacci series toward the rear of this rendering. The resulting three-dimensional surface has a 'ridge' (shown by a thick black line) corresponding to the angle 137.5°, which is approximated by the higher fractions in the Fibonacci series.

angle (Figure 1.6). Moreover, large internodal distances maximize the ability to intercept sunlight regardless of the leaf divergence angle because leaves are more widely spaced and, once again, avoid shading one another (Figure 1.7). Long petioles have a similar affect on light-harvesting (Figure 1.8). Finally, although they require additional variables, morphological variants with 'compound leaves' can be easily constructed. Comparisons among these variants reveal that 'pinnated leaves' further foster light interception (Figure

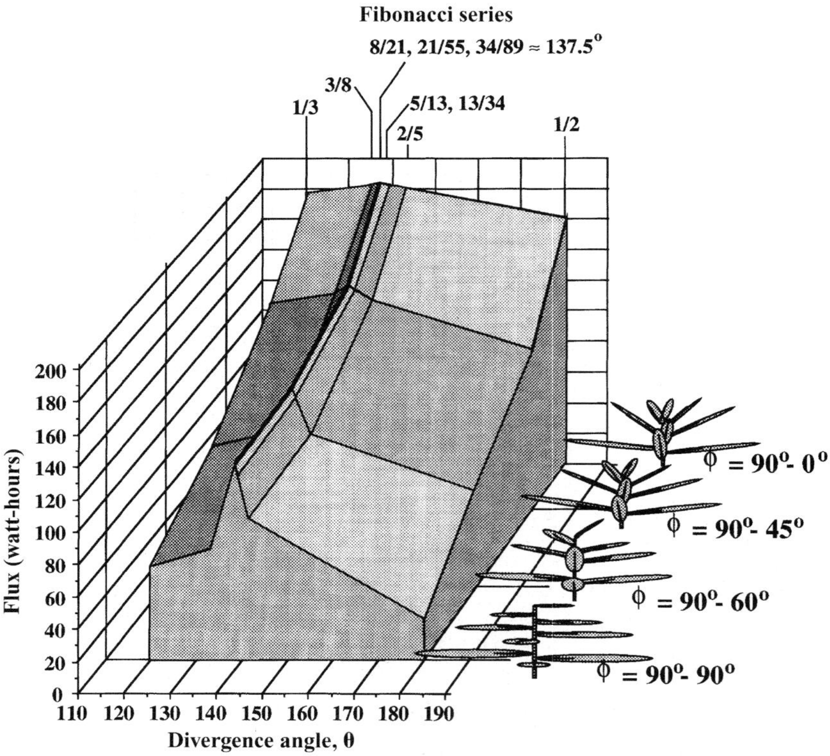

Figure 1.6 Three-dimensional portion of a more complex hypothetical 'phyllotactic' universe containing rosette-like morphologies bearing leaves with equivalent total surface area depicted by plotting light-harvesting efficiency (flux in watt-hours) against the leaf divergence angle, θ, and the variance in the leaf deflection angle, ϕ (depicted graphically on axis shown to the right).

For further details see legend to Figure 1.5. Rosette: $W/L = 1/20$, $i = 0.05$ and $l = 0.5$ cm.

1.9), especially if the deflection angles among the leaves on a shoot can be varied in an independent fashion (not shown).

It should be noted that these simulations serve as an object lesson regarding the nature of 'developmental constraints' in evolution, which has received considerable attention, although insufficiently documented in terms of plants (see Gould, 1980; Alberch, 1980, 1981; Arthur, 1984). The notion of a developmental constraint is that development, which must be highly organized and coordinated to achieve the final size and form of an organism, necessarily imposes limits on how far morphology or anatomy can be varied in an adaptive way. However, a subtle distinction must be made when dealing with 'constraints' because a developmental system can limit adaptive solutions in one of two ways: it can set strict boundary conditions on the

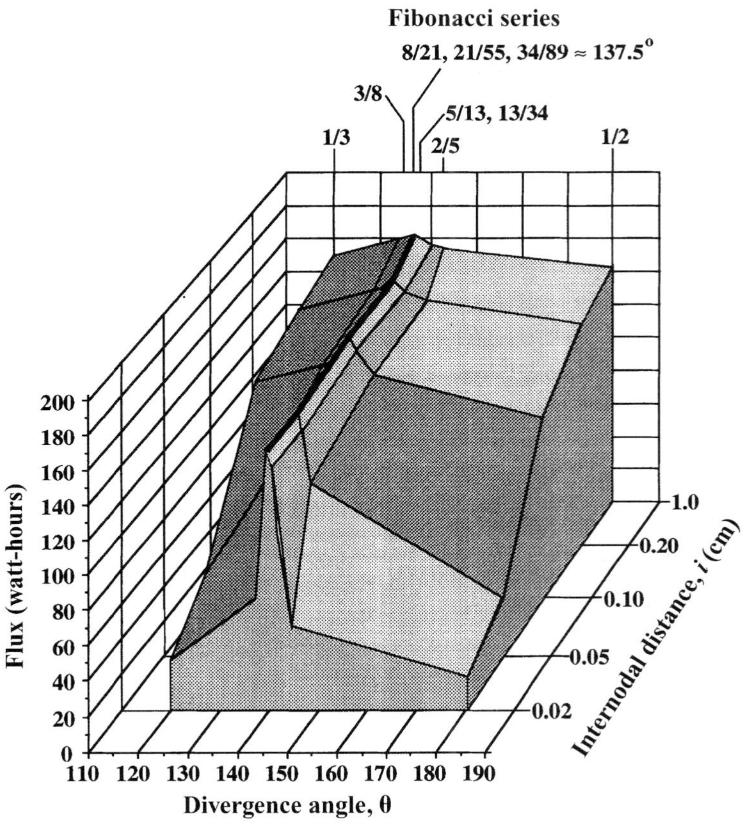

Figure 1.7 Three-dimensional portion of a more complex hypothetical 'phyllotactic' universe containing rosette- to herb-like morphologies bearing leaves with equivalent total surface area depicted by plotting light-harvesting efficiency (flux in watt-hours) against the internodal distance, *i*.
For further details see legend to Figure 1.5. Rosette → Herb: $W/L = 1/20$, $l = 3$ cm and $\phi = 45°$.

expression of form, shape and size, which an organism cannot transgress under any circumstances; or, it can necessitate compensatory changes in other features not directly under the control of development, thereby giving an organism some adaptive latitude.

The 137.5° divergence angle is a case in point. All other things being equal, it might be imagined that phyllotactic patterns must converge onto patterns of leaf arrangement with this divergence angle because it provides the optimal arrangement for light-harvesting. However, computer simulations show that virtually any leaf divergence angle can achieve the same light-harvesting capacity provided that other morphological variables are free to

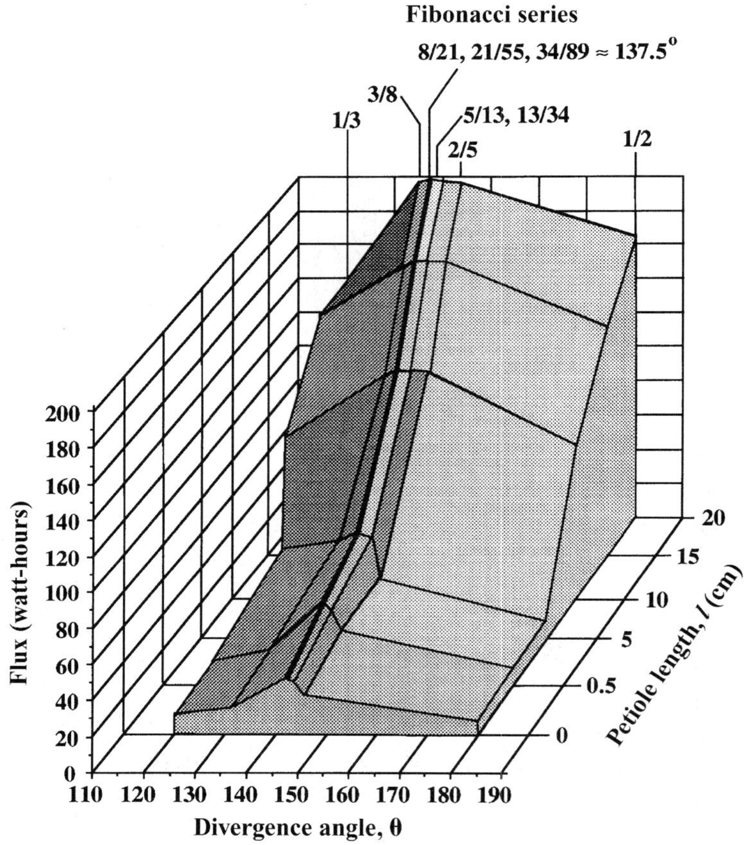

Figure 1.8 Three-dimensional portion of a more complex hypothetical 'phyllotactic' universe containing rosette-like morphologies bearing leaves with equivalent total surface area depicted by plotting light-harvesting efficiency (flux in watt-hours) against the leaf divergence angle, θ, and petiole length, *l*.

For further details see legend to Figure 1.5. Rosette: $W/L = 1/5$, $i = 0.05$ cm and $\phi = 90°$.

vary independently. Numerous studies have shown that the unyielding precision with which new leaves are produced by a shoot apical meristem is only part of a broader developmental repertoire, which involves the ability to interact with the external environment in such a way that a number of phenotypic features can be modified in a fashion that compensates for ineffective leaf arrangements. Features such as leaf size and shape and petiole length can clearly be varied in accordance with ambient light conditions to achieve extremely efficient light-harvesting morphologies, such that genetically proscribed leaf divergence angles do not represent a developmental constraint as traditionally defined.

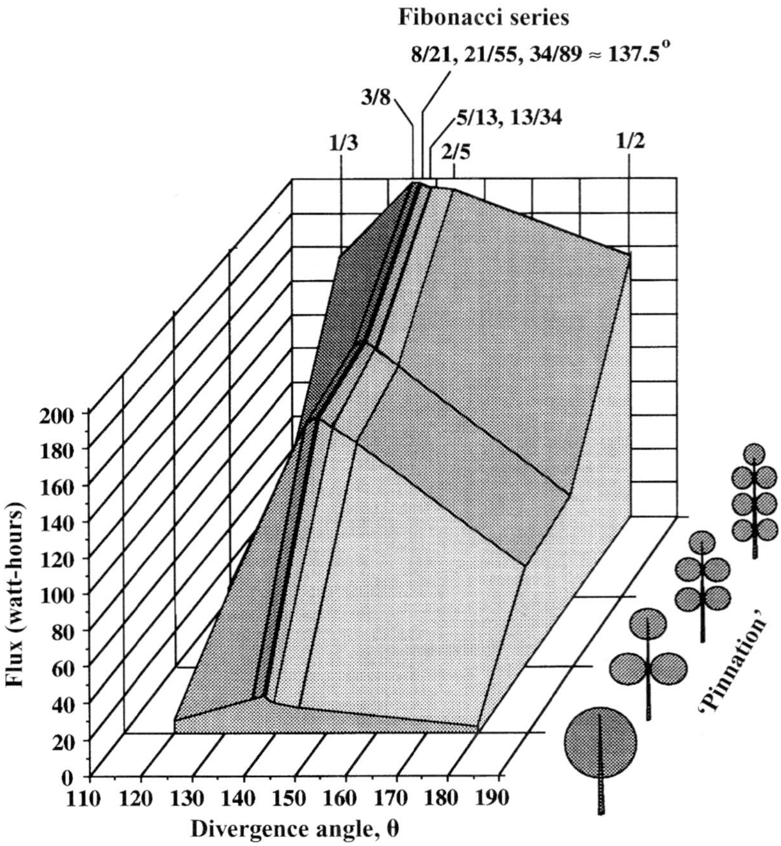

Figure 1.9 Three-dimensional portion of a more complex hypothetical 'phyllotactic' universe containing rosette-like morphologies bearing leaves with equivalent total surface area depicted by plotting light-harvesting efficiency (flux in watt-hours) against 'pinnation' (i.e. the degree to which the total surface area of each leaf is represented by separate circular 'leaflets').

For further details see legend to Figure 1.5. Rosette: $W/L = 1/1$, $i = 0.05$ cm, $l = 3$ cm, and $\phi = 45°$.

1.8 Mechanical interactions between leaves and stems

Leaves generate mechanical forces in stems as a result of gravity acting on their biomass (static loads) and as a consequence of wind-induced drag (dynamic loads). Over the course of land plant evolution, the morphology and anatomy of leaves and stems have been adapted to deal with these mechanical forces. One obvious expression of this is the size-dependent relationship between the length and stiffness of the petiole and the mass and surface area of the leaf lamina. Size-dependent relationships become evident when it is appreciated that most leaves are cantilevered structures and are thus subject to all the allometric (size-dependent) relationships that are proscribed for cantilevers.

For example, a petiole of length, L, bending under a force, P, acting on its free end (resulting from the weight of the lamina of a simple or palmately compound leaf) will deflect from the vertical, δ, measured at any distance, χ, from the free end of the petiole ($\chi = 0$, where the lamina is attached) according to the formula:

$$\delta_{\chi} = \left(\frac{-PL^3}{3EI}\right)\left[1 - \left(\frac{3\chi}{2L}\right) + \left(\frac{\chi^3}{2L^3}\right)\right]. \tag{1.4}$$

In this formula, E is Young's elastic modulus, a measure of a material's stiffness, which is a size-independent property. In contrast, I is the axial second moment of area, a mathematical device that provides a numerical index of the contribution made by both the size and shape of the cross-sectional geometry of any beam to the ability of the beam to resist bending (Bisshop and Drucker, 1945; Timoshenko and Gere, 1961). The term EI, which is called flexural rigidity, is thus a measure of the ability of any beam or column to resist a bending force by virtue of its material properties and the amount and shape of the materials used to fabricate its cross-section. For very small elastic deflections resulting from self-loading, such that $\delta \leqslant 0.10\ L$, the tip deflection of a petiole is approximated by the formula:

$$\delta_{\chi=0} = \frac{PL^3}{3EI}. \tag{1.5}$$

Thus, regardless of the materials used in its construction (i.e. regardless of the value of E), the tip-deflection of any petiole is expected to be proportional to the third-power of petiole length and to be directly proportional to the mass of the leaf's lamina. It is also apparent that, for equivalent lengths, petioles composed of stiff materials will have smaller tip-deflections than those composed of less stiff materials. Likewise, thicker petioles, which have larger axial second moments of area, will deflect less under their own weight and that of the lamina they must support than thinner petioles.

These relationships were reasonably well borne out by allometric analyses of 193 leaves, differing in size, drawn from 19 dicotyledonous and monocotyledonous species. The analyses revealed that petiole flexural rigidity scales with respect to petiole length, such that $EI \propto L^{3.24}$ ($r^2 = 0.78$) for simple and palmately compound leaves and $EI \propto L^{2.45}$ ($r^2 = 0.74$) for pinnately compound leaves (Figure 1.10). These relationships comply reasonably well with the mechanical expectation that $EI \propto L^3$. Likewise, petiole rigidity scales with respect to the load, P, generated by laminae, such that $EI \propto P^{1.63}$ ($r^2 = 0.84$) for simple and palmately compound leaves and

$EI \propto P^{2.34}$ ($r^2 = 0.73$) for pinnately compound leaves, whereas the relationship governing the performance of cantilevered beams predicts that, at minimum, $EI \propto P$ (Figure 1.10). The relative constancy of tip-deflections of leaves differing dramatically in size and drawn from very different taxa by virtue of proportionate changes in the petiole length, tissue stiffness, transverse geometry or shape with respect to lamina weight suggests that leaf biomechanical properties have evolved in accordance with inviolable physical principles. From a functional perspective, it is reasonable to argue that this evolution, permitting nearly uniform tip-deflections across taxa differing in leaf size, is likely to be adaptively beneficial in terms of orienting leaves with respect to sunlight.

There is also ample evidence that leaf development adaptively responds to ambient wind conditions (Venning, 1949; Jacobs, 1973; Niklas, 1996). For example, the petioles of *Acer saccharum* leaves removed from trees growing in windy sites are found to be shorter but significantly less rigid than their counterparts removed from trees growing in sites sheltered from the wind (Figure 1.11). Leaves with shorter petioles are expected to resist bending more than leaves with longer petioles but less rigid petioles are expected to bend and twist more easily in the wind.

The data from *A. saccharum* illustrate the phenomenon known as thigmomorphogenesis, a developmental response to chronic mechanical perturbation (Knight, 1811; Jacobs, 1973; Telewski and Jaffe, 1986; Telewski, 1995). The typical response of plants to persistent mechanical disturbance is to produce shorter and broader stems and petioles. Unless this 'geometric response' is compensated for by means of a reduction in the stiffness of tissues, thigmomorphogenesis would result in mechanically unyielding organs that would experience extremely high drag forces and would potentially break. In the case of *A. saccharum* leaves, the petioles produced in windy sites are shorter but their tissue stiffness is much less than those removed from trees growing in sheltered sites. Thus, the petioles produced in windy sites are highly flexible and bend and twist easily. Analyses of the volume fractions of *A. saccharum* petiole tissues show that the reduction in the stiffness of petioles growing in windy sites is a result of a reduction in the percentage of vascular tissue and a corresponding increase in the amount of collenchyma, which is a viscoelastic 'material' capable of considerable extension before it breaks. In passing, it is noteworthy that the allometry of *A. saccharum* leaves does not differ between windy and protected sites; the differences seen in the physical and material properties of leaves are merely the extremes of a continuous scaling relationship that reflects a conservative yet 'plastic' developmental repertoire.

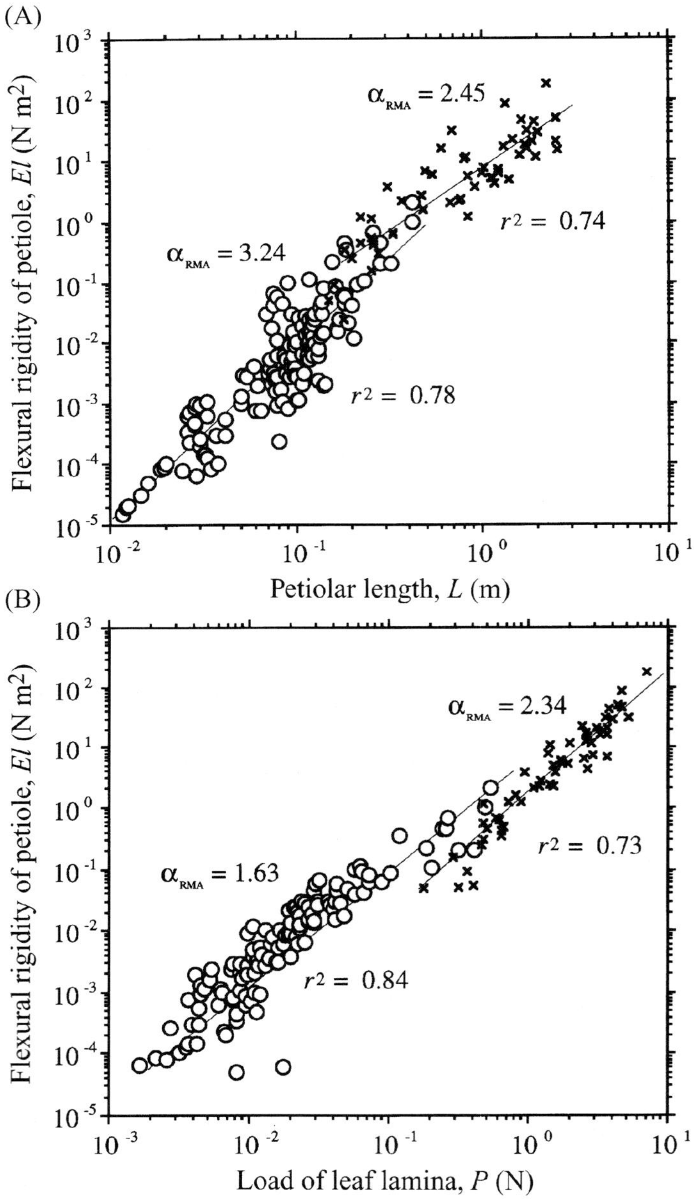

Figure 1.10 Allometric relationships between flexural rigidity of petiole, *EI*, and: A) petiole length, *L*, and B) lamina load, *P*, for simple and palmately compound leaves (○) and for pinnately compound leaves (x) (n = 119) drawn from different species. Solid lines denote reduced major axis (RMA) regression curves with slopes, α_{RMA}.

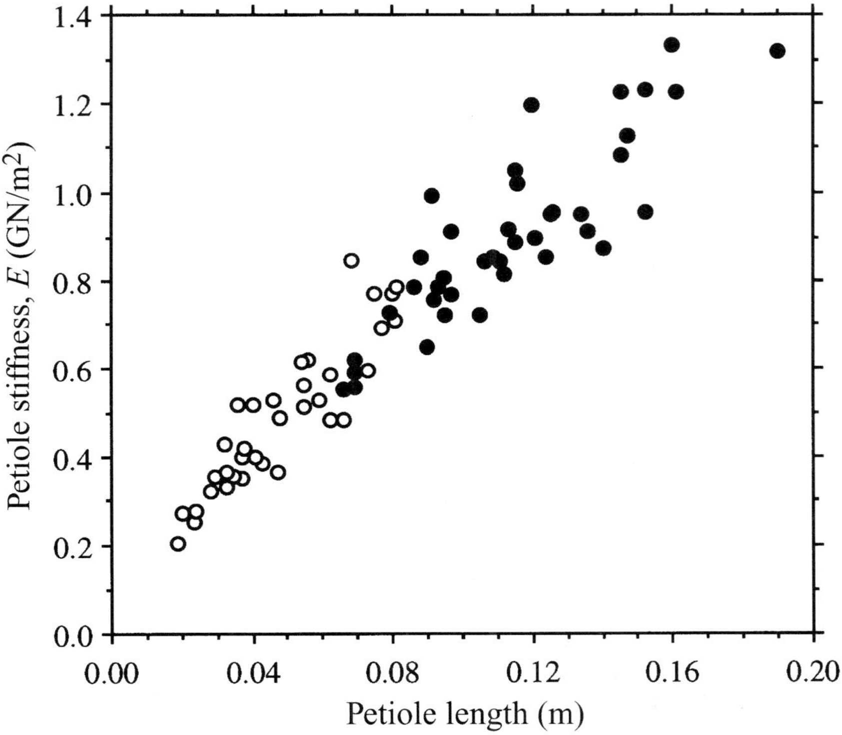

Figure 1.11 Petiole stiffness *E* of *Acer saccharum* petioles plotted against petiole length, *L*, for leaves drawn from trees grown in an exposed, windy site (○) and in a sheltered, calm site (●). (See text for further details.)

1.9 Drag-reduction of the canopy

The display of a large canopy of leaves to sunlight unavoidably results in the projection of a large surface area to moving air. Unless this surface area can be temporarily reduced when wind speeds increase dramatically, the dynamic mechanical forces exerted on subtending stems by wind acting on leaves can exceed the capacity of stem tissues to cope with bending or torsional stresses and stems can fail catastrophically. Wind-induced drag forces are the most conspicuous cause of mechanical failure, since healthy woody stems are typically capable of supporting their own weight and that of the organs they normally support (Metzger, 1893; Esser, 1946a,b; Alexander, 1971; Banks, 1973; King and Loucks, 1978; Niklas, 1992, 1994b; Vogel, 1996). Prior studies have shown that the leaves and young stems of many woody species either elastically flex backward or fold upon themselves in high winds, thereby decreasing their projected surface areas with concomitant reductions in the drag forces they would otherwise impart to older subtending stems (Fraser, 1962; Hutte, 1968; Mayhead, 1973; Vogel, 1996).

The manner in which leaf laminae deform and reduce drag in response to wind movement varies across species. A dorsiventral morphology with a thin cross-section is easily modelled as a stress-skin panel or a polylaminated sandwich board (Allen, 1969; Gibson *et al.*, 1988) with facings (lined above and below with epidermis) rigidly bonded to interconnecting stringers (the vascular bundles running throughout the lamina). This structural configuration can be shown to spread any force acting normal to its surface, such as wind drag, over a large surface area, resulting in small lateral deflections that are well accommodated by the stringers holding the webbing together (Niklas, 1992). The advantage of this behavior is that lateral loadings are resisted primarily by in-plane membrane forces that develop within the webbing or 'skin' rather than by the usual bending forces seen typically in structures made up of columns or beams.

Curved folded plates have the additional mechanical advantage of supporting lateral loads by arch compression—a curvilinear deformation spanning the width and breadth of the whole structure, in which all components share a common load by distributing strains as equally as possible. In this way, no single portion of the construction bears a larger than average load per surface area and the whole structure can be rolled upon itself when the force acting normal to its projected surface area exceeds a certain intensity. Indeed, stress-skin panels and thin polylaminated sandwich boards are remarkably flexible when bent or twisted and they can rapidly deform elastically without damage, especially if the stringers radiate outward from (and converge on) a single point. The thin laminae of many leaves fold easily in high winds, thereby reducing their drag and the probability of mechanical damage (Vogel, 1989).

The disadvantage of very thin stress-skin panels is that they are prone to ripping at their edges when excessively folded. This risk is much like that experienced by fabrics that are stretched or pulled. For woven structures, the solution is a selvedge or woven border, or hem or gusset. The hem is used to strengthen a continuous margin by doubling it up; the gusset is inserted as a triangular patch at the base of some indentation to reduce the risk of tearing inward from a weakened point. These familiar mechanical solutions coincide with the marginal strengthening seen in the laminae of many leaves. Arched selvedges in leaves are common among dicotyledonous species. A network consisting of stronger and weaker vascular strands, which describe broad intramarginal archings, succeeded by others of smaller curvature nearer the margin. The curve of each arch runs for some distance nearly parallel to the leaf margin. The parallel venation of monocotyledonous leaves acts in much the same way, since the most marginal strand and the sclerified leaf margin act as a single, long, strengthening arch. Another effective design is to place a strong band of stiff tissue running along the entire margin of the lamina. This is seen in the leaves of *Ulmus* and *Platanus*. Finally, the mechanical equivalent of gussets is seen in the form of hardened patches of tissue located at the base of deep indentations at the margins of leaves, like those of many ferns (e.g. *Asplenium horridum*).

The presence of a polylaminated construction with very stiff and much less stiff layers also reduces the risk of a crack or tear passing through the whole structure because the compliant (less stiff) layer can deform extensively and, thereby, reduce the stress concentration at the tip of the advancing crack (Vincent, 1990). The sharp end of a crack acts as a concentrator of force and can be stopped by running either into a void or into a material that deforms rapidly and easily. The greater the difference between the material properties of the adjoining layers of material, the less likely that a crack will propagate through the two layers. A reduction in the size of the components in the stiffer of the two layers is also helpful because every crack requires a minimum size before it will propagate. Increasing the surface contact made between the stiff and less stiff layers also increases the probability that a crack will run into the less stiff layer, and thus run its course. The parallel venation of grass leaves is a case in point. If a grass leaf is slit at its margin, it does tear more easily but only in direct proportion to its reduced cross-section. When it does tear, it does so in a jagged way as the crack passes up and down between the parallel veins and intervening softer tissues of the leaf (Vincent, 1982, 1983).

The leaves of grasses and other monocotyledons provide a special case because their sheaths clasp stem internodes and thus contribute to the mechanical support of the shoot as a whole (Tomlinson, 1962; Niklas, 1990b, 1992, 1998b). Developmental studies show that the clasping leaf sheaths are stiffer and more rigid than the internodes they surround during the early development of the stem, when internodes are not fully extended. The internode tissues stiffen gradually over the course of their development and become as stiff as the sheaths once they extend beyond their enclosing leaves. The sheaths continue to provide mechanical support for the fully extended stem internodes. In this case, they externally rigidify internodes against local (Brazier) buckling when shoots bend. The leaf sheath also helps internodes to resist torsional deformations.

In terms of the adaptive value of reducing the projected surface area of leaves in response to wind movement, it is clear that the drag force, F_D, experienced by any object obstructing the flow of air is equal to the product of the fluid's density, ρ, the square of the fluid velocity, U, the area the object projects in the direction of flow, S_p, and the drag coefficient, C_D (Vogel, 1996). Mathematically, it can be seen that $F_D = 0.5\rho U^2 S_p C_D$. One important point is that, regardless of reports in the literature, the drag coefficient is not constant (Mayhead, 1973), especially for objects that can twist, bend and fold onto themselves and so reconfigure their shape in response to a moving fluid (Vogel, 1989, 1996). Wind tunnel experiments using portions of intact shoots show that the laminae of leaves with very different morphologies curl and fold when subjected to high winds, and thereby reduce their projected surface areas and their drag coefficients (Vogel, 1989). Since the drag force is proportional to the product of S_P and C_D and since the stems of plants experience the cumulative drag force exerted by the wind on all their leaves, elastic changes in lamina shape confer obvious benefits (Figure 1.12). When combined with the effects of the bending of stems, this configuration greatly reduces the drag forces acting on entire shoots. Clearly, the leaves of many species are incapable of reconfiguring in the wind and thereby reducing their drag; the pinnate leaves of cycads, such as *Dioon* and *Zamia*, do not

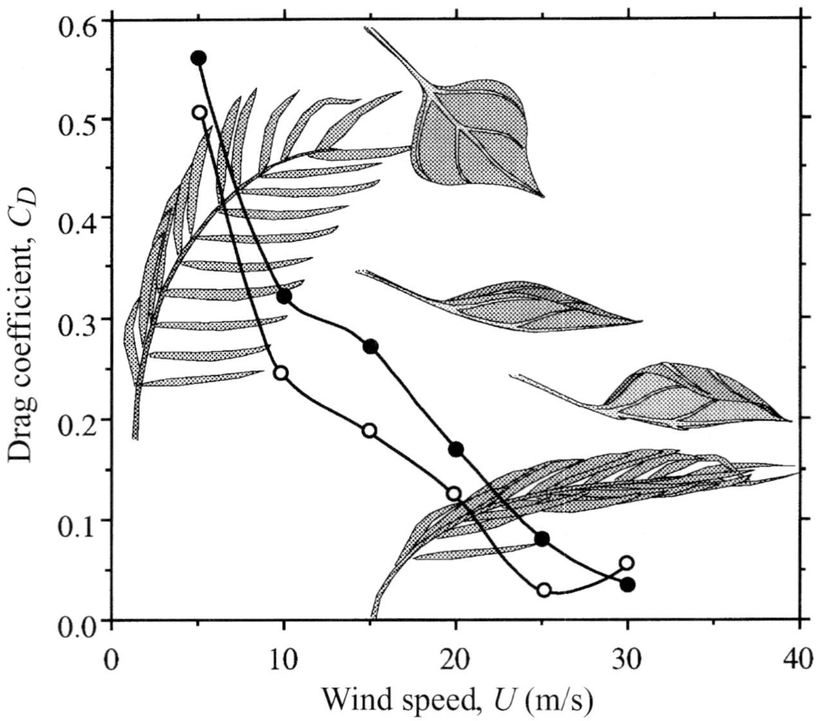

Figure 1.12 Drag coefficient, C_D plotted against wind speed, U, for a *Catalpa* (●) and a *Chamaedorea* (○) leaf with equivalent surface areas (227 cm^2) placed in a wind tunnel. Cartooned reconfigurations of leaves for progressively higher wind speeds are based on photographs taken orthogonal to the direction of airflow.

reconfigure to any appreciable extent. However, for many species, the pinnate leaf operates mechanically as a flexible branch bearing numerous simple leaves with elastic hinges at their bases.

Branches and tree trunks nonetheless frequently snap somewhere along their length, such that trees typically do not uproot when subjected to very large wind-induced drag forces (Jones, 1956; Hutte, 1968; Wood, 1970; Runkle, 1982; Harcombe and Marks, 1983; Putz *et al.*, 1983). This phenomenon indicates that the susceptibility to stem mechanical failure is not uniform throughout the infrastructure of large woody plants but rather varies, such that some stems or portions of stems are more prone to dynamic wind-failure than others. Although limited in scope, a few studies have revealed complex patterns of variation in the mechanical reliability of the stems in tree canopies and provide evidence for preferential stem failure. That is, some stems are more prone to fail than others. In the majority of cases, these failure-prone stems are peripheral branches, the mechanical failure of which will necessarily reduce the drag-induced bending moments and stresses experienced by the rest of the plant body. In severe wind storms, the forfeiture of these stems and the leaves they bear can obviously reduce the probability that the trunks of large trees will break. Since the older portions of a tree represent a large investment in biomass and have the capacity to regenerate new shoots, the mechanical failure of a constellation of comparatively young stems is arguably functionally adaptive. As a consequence of their normal growth and development, trees may thus establish a class of stems prone to mechanical failure during windstorms, thereby providing a margin of safety against the catastrophic failure of the tree as a whole.

This hypothesis is consistent with surveys of tree damage and death and with a limited number of studies estimating the mechanical reliability of stems based on static rather than dynamic loads. For example, in their study of Barro Colorado Island, Putz and co-workers (1983) reported that 70% of a total of 310 damaged trees ($>$ 10 cm diameter at breast height) snapped near but not at their base (of which 217 subsequently developed new shoots), whereas 25% uprooted (of which five produced viable shoots) and only 5% broke at their base. Likewise, among mesic eastern USA forest tree species, Runkle (1982) reported that the typical mode of failure is the snapping of trunks (67%). Importantly, these and other similar surveys of tree failure (e.g. Behre, 1921; Curtis, 1943) excluded from consideration trees that had sustained comparatively minor damage to their lateral branches, and did not exclude trees whose trunks snapped as a result of disease (e.g. 13% of the trees with snapped trunks in the Barro Colorado Island study had heartrot; Putz *et al.*, 1983), suggesting that the typical form of tree failure may be the loss of branches or the failure of the trunk some distance from its base.

1.10 Concluding remarks

This treatment of leaf form–function relationships and the evolution of the canopy is admittedly idiosyncratic in its pointed neglect of a large body of physiological literature in favor of perhaps the less familiar and still inadequately appreciated perspective gained by biomechanics. This bias was intentional, not because physiological principles are unimportant but because they are incomplete without a biomechanical perspective, which draws attention to the many physical forces that leaves normally experience. With the aid of computer simulations, this perspective shows that the organographic distinction between the foliage leaf and the stem, which has evolved independently among the various major tracheophyte lineages, follows from first principles, in the sense that it is a predictable outcome of selection acting on a plant body that by evolution increases in size and must simultaneously support itself mechanically and intercept sunlight. The selective advantages of increasing plant body size were not discussed but they are numerous in terms of acquiring, retaining and competing for nutrients, enhancing the capacity for long-distance dispersal of reproductive structures and producing supernumerary organs that afford redundancy when some are damaged or lost. It is thus reasonable to suppose that selection favoring an increase in the plant body was prevalent during the long course of land plant evolution. Under any circumstances, the fossil record shows that the maximum size of the plant body rapidly increased during the formative evolutionary stages of terrestrial colonization and that the leaf–stem distinction attended this trend in body size with equal alacrity.

A biomechanical perspective on phyllotaxy likewise shows that manifold morphological parameters affect the capacity of leaves to intercept sunlight, suggesting that there is no single 'optimal' leaf shape or size but rather a constellation of equally effective morphologies. This may account for the tremendous diversity in leaf size and shape observed within, as well as across, individual lineages in comparison to the morphological and anatomical conservatism observed for corresponding reproductive structures.

Considerable research is still required to fully appreciate the numerous adaptive responses of leaves to the various mechanical forces they normally experience. We are only just beginning to appreciate the complexity of these responses and how they may affect physiological processes. Clearly, future research agendas must address leaf form–function relationships from a biomechanical as well as a physiological perspective.

References

Alberch, P. (1980) Ontogenesis and morphological differentiation. *American Zoologist*, **20** 653–667.

Alberch, P. (1981) Convergence and parallelism in foot morphology in the neotropical salamander, *Bolitoglossa*. 1. Function. *Evolution*, **35** 84–100.
Alexander, R.M. (1971) *Size and Shape*, Edward Arnold, London, UK.
Allen, H.G. (1969) *Analysis and Design of Structural Sandwich Panels*, Pergamon, Oxford, UK.
Arthur, W. (1984) *Mechanism of Morphological Evolution*, John Wiley and Sons, Chichester, UK.
Banks, C.C. (1973) The strength of trees. *Journal of the Institute of Wood Science*, **6** 44–70.
Behre, C.E. (1921) A study of windfall in the Adirondacks. *Journal of Forestry*, **19** 632–637.
Bierhorst, D. (1971) *Morphology of Vascular Plants*, Macmillan, New York, USA.
Bisshop, K.E. and Drucker, D.C. (1945) Large deflection of cantilever beams. *Quarterly Applied Mathematics*, **3** 272–275.
Bold, H.C. (1967) *Morphology of Plants*, Harper and Row, New York, USA.
Bower, F.O. (1935) *Primitive Land Plants*, Macmillan, London, UK.
Curtis, J.D. (1943) Some observations of wind damage. *Journal of Forestry*, **41** 877–882.
Esau, K. (1967) *Plant Anatomy*, John Wiley, New York, USA.
Esser, M.H.M. (1946a) Tree trunks and branches as optimal mechanical supports of the crown. I. The trunk. *Bulletin of Mathematical Biophysics*, **8** 65–74.
Esser, M.H.M. (1946b) Tree trunks and branches as optimal mechanical supports of the crown. II. The branches. *Bulletin of Mathematical Biophysics*, **8** 95–100.
Fraser, G.F. (1962) Wind tunnel studies of the forces acting on the crowns of small trees, in *Report on Forest Research for the Year Ended 1962*, Forestry Commission, UK, pp. 178–183.
Gates, D.M. (1980) *Biophysical Ecology*, Springer-Verlag, Heidelberg, Germany.
Gibson, L.J., Ashby, M.F. and Easterling, K.E. (1988) Structure and mechanics of the iris leaf. *Journal of Materials Science*, **23** 3041–3048.
Gifford, E.M. and Foster, A.S. (1988) *Morphology and Evolution of Vascular Plants*, W.H. Freeman and Co., New York, USA.
Gould, S.J. (1980) The evolutionary biology of constraint. *Daedalus*, **109** 39–52.
Harcombe, P.A. and Marks, P.L. (1983) Five years of tree death in a *Fagus-Magnolia* forest, southeast Texas, USA. *Oecologia*, **57** 49–54.
Hutte, P. (1968) Experiments on windflow and wind damage in Germany: site and susceptibility of spruce forests to storm damage. *Forestry Supplement*, **41** 20–27.
Jacobs, M.R. (1973) Thigmomorphogenesis: the response of plant growth and development to mechanical stress. *Planta*, **114** 143–157.
Jones, E.W. (1956) Ecological studies on the rain forest of southern Nigeria. IV. The reproduction and history of the Okuma Forest Reserve. Part II. The reproduction and history of the forest. *Journal of Ecology*, **44** 83–117.
King, D. and Loucks, O.L. (1978) The theory of tree bole and branch form. *Radiation and Environmental Biophysics*, **15** 141–165.
Knight, T.A. (1811) On the causes which influence the direction of the growth of roots. *Philosophical Transactions of the Royal Society of London*, **1811** 209–219.
Mauseth, J.D. (1988) *Plant Anatomy*. Benjamin Cummings Publishing Co., Menlo Park, CA, USA.
Mayhead, G.J. (1973) Some drag coefficients for British forest trees derived from wind tunnel studies. *Agricultural Meteorology*, **12** 123–130.
Metzger, K. (1893) Der Wind als massgebender Faktor für das Wachtsum der Bäume. *Mundener Forstliche Hefte*, **3** 35–86.
Niklas, K.J. (1990a) Biomechanics of *Psilotum nudum* and some early Paleozoic vascular sporophytes. *American Journal of Botany*, **72** 590–606.
Niklas, K.J. (1990b) The mechanical significance of clasping leaf sheaths in grasses: evidence from two cultivars of *Avena sativa*. *Annals of Botany*, **65** 505–512.
Niklas, K.J. (1992) *Plant Biomechanics: an Engineering Approach to Plant Form and Function*, University of Chicago Press, Chicago, IL, USA.

Niklas, K.J. (1994a) Morphological evolution through complex domains of fitness. *Proceedings of the National Academy of Sciences*, **91** 6772–6779.

Niklas, K.J. (1994b) Interspecific allometries of critical buckling height and actual plant height. *American Journal of Botany*, **81** 1275–1279.

Niklas, K.J. (1996) Differences between *Acer saccharum* leaves from open- and wind-protected sites. *Annals of Botany*, **78** 61–66.

Niklas, K.J. (1997a) *The Evolutionary Biology of Plants*, The University of Chicago Press, Chicago, IL, USA.

Niklas, K.J. (1997b) Adaptive walks through fitness landscapes for early vascular land plants. *American Journal of Botany*, **84** 16–25.

Niklas, K.J. (1998a) Light harvesting 'fitness landscapes' for vertical shoots with different phyllotactic patterns, in *Symmetry in Plants*, Vol. 4 (eds. R.V. Jean and D. Barabé), World Scientific Publishing Co. Ltd, Singapore, pp. 759–773.

Niklas, K.J. (1998b) The mechanical roles of clasping leaf sheaths: evidence from *Arundinaria técta* (Poaceae) shoots subjected to bending and twisting forces. *Annals of Botany*, **81** 23–34.

Nobel, P.S. (1983) *Biophysical Plant Physiology and Ecology*, W.H Freeman and Co., San Francisco, CA, USA.

Putz, F.E., Coley, P.D., Lu, K., Montalvo, A. and Aiello, A. (1983) Uprooting and snapping of trees: structural determinants and ecological consequences. *Canadian Journal of Forest Research*, **13** 1011–1020.

Raven, J.A. (1984) Physiological correlates of the morphology of early vascular plants. *Botanical Journal of the Linnean Society*, **88** 105–126.

Runkle, J.R. (1982) Patterns of disturbance in some old growth mesic forests of eastern North America. *Ecology*, **63** 1533–1546.

Salisbury, F.B. and Ross, C.W. (1992) *Plant Physiology*, Wadsworth Publishing Co., Belmont, NY, USA.

Stewart, W.N. and Rothwell, G.W. (1993) *Paleobotany and the Evolution of Plants*, Cambridge University Press, Cambridge, UK.

Schwendener, S. (1874) *Das Mechanische Prinzip im anatomischen Bau der Monocotylen*, W. Engelmann, Leipzig, Germany.

Taylor, T.N. and Taylor, E.L. (1993) *The Biology and Evolution of Fossil Plants*, Prentice-Hall, Englewood Cliff, NJ, USA.

Telewski, F.W. (1995) Wind-induced physiological and developmental responses in trees, in *Wind and trees* (eds. M.P. Coutts and J. Grace), Cambridge University Press, Cambridge, UK.

Telewski, F.W. and Jaffe, M.J. (1986) Thigmomorphogenesis field and laboratory studies of *Abies frazei* in response to wind or mechanical perturbation. *Physiologia Plantarum*, **66** 211–218.

Timoshenko, S.P. and Gere, J.M. (1961) *Theory of Elastic Stability*, McGraw-Hill, New York, USA.

Tomlinson, P.B. (1962) The leaf base of palms: its morphology and mechanical biology. *Journal of the Arnold Arboretum*, **43**, 23–50.

Venning, F.D. (1949) Stimulation by wind motion of collenchyma formation in celery petioles. *International Journal of Plant Sciences*, **110** 511–514.

Vincent, J. (1990) *Structural Biomaterials*, Princeton University Press, Princeton, NJ, USA.

Vincent, J.F.V. (1982) The mechanical design of grass. *Journal of Materials Science*, **17** 856–860.

Vincent, J.F.V. (1983) The influence of water content on the stiffness and fracture properties of grass leaves. *Grass and Forage Science*, **38** 107–114.

Vogel, S. (1989) Drag and reconfiguration of broad leaves in high winds. *Journal of Experimental Botany*, **40** 941–948.

Vogel, S. (1996) Blowing in the wind: storm-resisting features of the design of trees. *Journal of Arboriculture*, **22** 92–98.

Wainwright, S.A., Biggs, W.D., Currey, J.D. and Gosline, J.M. (eds.) (1976) *Mechanical Design in Organisms*, Princeton University Press, Princeton, NJ, USA.

Wagner, W.H., Jr, Beitel, J.M. and Wagner, F.S. (1982) Complex venation patterns in the leaves of *Selaginella*: megaphyll-like leaves in lycophytes. *Science*, **218** 793–794.

Watson, E.V. (1971) *The Structure and Life of Bryophytes*, Hutchinson University Library, London, UK.

Wood, T.W.W. (1970) Wind damage in the forest of Western Samoa. *Malaysia Forestry*, **33** 92–99.

Wright, C. (1873) On the uses and origin of arrangements of leaves in plants. *Memoirs of the American Academy of Arts and Sciences*, **9** 379–415.

Zimmermann, W. (1966) Kritische Beiträge zu einigen biologischen Problemen. VII. Die Hologenie. *Zeitschrift Pflanzenphysiologie*, **54** 125–144.

2 Plants as self-organising systems

Federico Magnani and John Grace

2.1 Introduction

Where does order come from? We know from the laws of physics that dynamic, irreversible processes should progress towards ever greater entropy until they reach a state of maximum disorder. Yet we can see plenty of systems and organisms all around us that maintain a high internal order, in apparent defiance of the laws of thermodynamics. In these so-called dissipative systems, internal organisation can be maintained thanks to a continuous flow of energy, be it heat applied to a boiling pan, electricity supplied to a laser or solar energy harnessed by green plants through photosynthesis (Nicolis and Prigogine, 1989).

What is most fascinating is that order seems to arise spontaneously from the interaction of those parts of which complex systems are composed. In stark contrast with what is generally observed in artificial systems (Dorf and Bishop, 1995), natural systems are not generally governed by an external controller or by any top-down rules. Nor is their ordered behaviour the mere sum of the dynamics of their components; rather, it arises as an emergent property from the action and interaction of all the parts of which the system is composed (Salt, 1979). The system is said to be self-organising (Prigogine and Stengers, 1985).

Self-organising systems are often characterised by distinctive emergent features. The behaviour of individual components, for example, is often coordinated, i.e. they appear to act in phase, like individual gas particles in a laser cell, which all emit at a single wavelength (Häken, 1979). Self-organisation, moreover, typically results in the emergence of characteristic patterns, often of amazing beauty, that are maintained by homeostatic mechanisms despite large variations in external conditions. These patterns frequently correspond to an optimal system behaviour, i.e. the one that maximises system order and performance for a given amount of available energy. Laser itself, however, is a nice example of another property of self-organising systems: the loss of coordination and pattern and the transition to a chaotic behaviour when a critical range of external conditions (input energy in the case of laser) is exceeded. In this case, several organised states are equally possible (a condition referred to as bifurcating behaviour) and the continuous transition of the system between them results in the chaotic pattern observed.

Self-organising systems are widespread in the non-living world. They include crystals, lasers, convection cells and cloud formations, atmospheric circulation patterns and river systems. These systems have little in common, except that they seem to defy the natural law of entropy, whereby the universe tends to a state of disorder, and that their ordered patterns on the macroscopic scale arise like magic from a sea of microscopic disorder.

At a global level, plant life is a self-organising system par excellence, self-replicating and continually evolving. It has been on the planet for almost four billion years, and has transformed the atmosphere remarkably, from anaerobic to aerobic, creating and maintaining the conditions for its own survival (Lovelock, 1982). From time to time the genetic code constraining the behaviour of each individual mutates. In a few cases, this can result in offspring of greater fitness and hence evolution occurs, which is the basis of life self-organisation on a global scale (Kaufmann, 1993; see also Chapters 1 and 10).

At the individual level, internal organisation and coordination are key features of plant life. The solar energy harnessed through photosynthesis is used by the plant to capture scarce resources, to grow and to reproduce itself. This is made possible by the coordinated action of specialised structures (organs) and a hierarchy of repetitive and semi-independent modules within each organ, such as branches, leaves, cells and chloroplasts within the photosynthetising canopy (Sprugel *et al.*, 1991). Therefore it is imperative to ask again the initial question: where does order come from? How can all these organs and modules work together in a coordinated way, producing the emergent properties that are observed at the whole plant level?

This is no trivial question. Coordination can be expected to arise not only from self-organisation, whereby the mutual interaction of system parts acts as an internal controller, but also as a result of an external control, as in artificial cybernetic systems (Dorf and Bishop, 1995). In this case, coordination results from the fact that every part in the system is playing by the same rules. This is largely the case in living organisms. The genetic information shared by all the cells in an individual (and, to a lesser extent, in a population) could be likened, in a way, to an external controller that dictates their coordinated response to the environment. From this perspective, plant function could be seen as a collective property, the simple summation of the properties of largely independent units, rather than an emergent property of the system.

Little attention has been paid so far to these intriguing questions, in contrast to the lively debate on self-organisation in plant population dynamics and ecological processes (Solé and Manrubia, 1995; Odum, 1983). In this chapter, some examples of coordination on different scales are proposed, from the single leaf to the canopy to the whole plant in an attempt to determine the role of plant self-organisation. The implications of self-organisation and homeostasis for the representation of plant response to the environment are then discussed.

2.2 Coordination of stomatal function at the leaf level

The dimensions of individual stomatal pores are known to vary greatly within a leaf (Willmer and Fricker, 1996). However, stomata have been reported to respond together with each other to external stimuli, such as humidity and light (Weyers and Lawson, 1997); even under extreme conditions, when patchy stomatal closure is often reported, the behaviour of individual pores within a patch appears to be largely coordinated (Cardon *et al.*, 1994). How can this coordinated pattern arise? The fact that stomatal function appears to be less variable than pore dimensions suggests that self-organisation and internal control are at work, possibly as a result of the same stabilising feedback structure proposed by Jarvis and McNaughton (1986) at the canopy level (Figure 2.1a). At the level of the individual pore, stomatal opening will result in a proportional increase in transpiration. Higher transpiration rates,

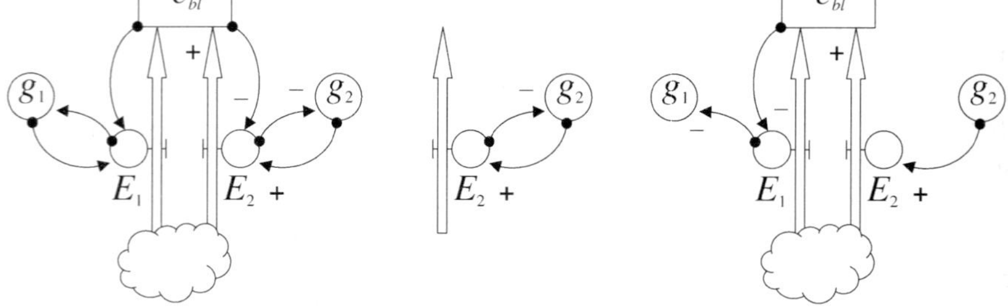

Figure 2.1 Self-organisation at the leaf level: coordination of stomatal function (Jarvis and McNaughton, 1986). a) Mutual interaction between two adjoining stomata: stomatal conductance (g_1 and g_2) controls pore transpiration (E_1 and E_2), which in turn determines the absolute humidity of air within the boundary layer (c_{bl}); water eventually diffuses from the boundary layer to the atmosphere. b) Negative feedback of stomatal opening onto itself, mediated by greater pore transpiration. c) Positive effect of stomatal opening on the behaviour of neighbouring stomata: increased transpiration results in higher boundary layer humidity, which in turn depresses transpiration, inducing stomatal opening in adjoining pores. The overall effect is a partial coordination of stomatal behaviour.

however, are known to trigger stomatal closure (Mott and Parkhurst, 1991; Monteith, 1995); in the specific case, this negative feedback loop will result in a reduction of pore opening (Figure 2.1b). Greater pore transpiration, on the other hand, will marginally moisten the air within the leaf boundary layer, so reducing the transpiration rate and forcing the partial opening of surrounding stomata (Figure 2.1c). The combination of a negative feedback of the stoma onto itself and a positive effect on its neighbours will eventually minimise differences between stomata, promoting their coordinated behaviour.

It is interesting to note that several other features of stomatal function are also typical of complex self-organised systems. Bifurcation for example (i.e. the ability of a system to respond in two completely different ways to the same environmental conditions, making it impossible for an observer to predict the path the system will take) was first predicted theoretically (Rand *et al.*, 1981), and later observed experimentally (Shababa *et al.*, 1997). Oscillatory stomatal behaviour has also been often observed (Barrs, 1971; Cardon *et al.*, 1994). Both process models and experimental evidence, therefore, seem to point to a central role of self-organisation in stomatal function.

2.3 Nitrogen acclimation to light at the canopy level

Let us now move one step up the hierarchical scale and consider the population of leaves that makes up a plant's canopy. In this case, rather than homogeneity, a clear differentiation is observed among leaves, which closely matches the environmental heterogeneity inside the crown. As an example, leaf nitrogen content (on a leaf area basis) has often been reported to decline exponentially inside the canopy, mirroring the vertical profile of available light (Kull and Jarvis, 1995). This has been suggested to maximise the efficiency of nitrogen use and productivity, by concentrating the photosynthetic machinery preferentially where higher radiative levels provide the substrate for assimilation (Hikosaka and Terashima, 1995; see also Chapters 6 and 7). How can such an optimal distribution be achieved?

It has been suggested that leaves and branches within a crown can be considered largely independent of each other (Sprugel *et al.*, 1991); yet, they rely on the same rooting system for nutrients and water. It is therefore reasonable to think of a population of independent modules competing for limited resources. Following this approach, Kull and Kruijt (1999) have proposed that leaves competing for a common nitrogen pool will exploit it at a rate that is proportional to the local availability of carbohydrates, hence to leaf photosynthesis and, ultimately, to the local light environment. The simulation diagram, Figure 2.2, represents two such leaves, positioned at different levels in the canopy and, therefore, exposed to contrasting radiation

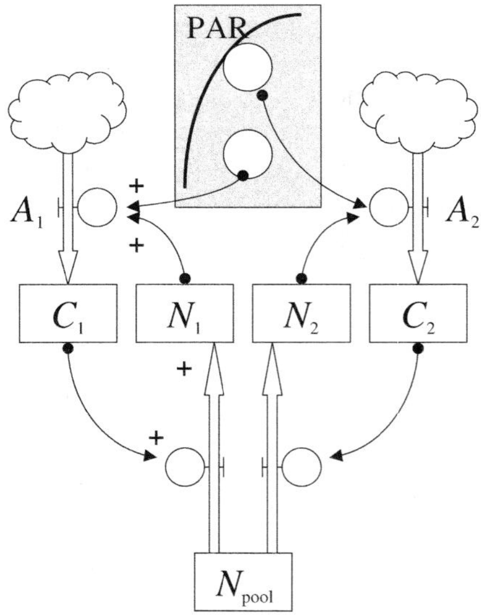

Figure 2.2 Self-organisation at the canopy level: nitrogen acclimation to light (Kull and Kruijt, 1999). Two individual leaves, exposed to different radiation levels, compete for a common nitrogen pool (N_{pool}). Import of nitrogen into the leaf labile pool (N_1 and N_2) is assumed to be proportional to local carbohydrate availability (C_1 and C_2), itself a function of leaf assimilation (A_1 and A_2). Assimilation, in turn, is positively affected both by light and by leaf nitrogen content. The resulting positive feedback is supposed to establish a vertical gradient in nitrogen concentration across the canopy. Abbreviation: PAR, photosynthetically active radiation.

levels (as captured by the vertical gradient of photosynthetically active radiation in the insert). Higher photosynthetically active radiation (PAR) will enable the second leaf to photosynthesise more actively, so increasing its availability of carbohydrates and enabling it to tap the common nitrogen pool more effectively. In a positive feedback loop, faster protein synthesis will, in turn, increase the photosynthetic potential of the leaf, which as a result will be able to exploit solar energy even more effectively. The opposite is true, of course, for the shade leaf, which will be out-competed in the race for nitrogen; in the game of competition for limited resources, the success of one leaf is detrimental to all the others.

It is interesting to compare the influence structures of Figures 2.1 and 2.2 and to observe how they result in completely different patterns: at the leaf level, a homogeneous behaviour was achieved through the self-stabilisation of each pore (negative feedback) and the stimulation of neighbouring stomata (positive effect). At the canopy level, on the contrary, differentiation is the result of the self-stimulation of each leaf (positive feedback) and the repression of competing neighbours (negative effect). Both are the result of system self-organisation.

The model of Kull and Kruijt (1999), however, fails to explain why the same pattern of nitrogen distribution in response to light is also observed when the comparison is extended to different plants within a heterogeneous stand (Ellsworth and Reich, 1992), or to seedlings grown under contrasting light regimes (Liang *et al.*, 1995). In this case, plants would not be expected to compete for a common pool of resources; lacking any interaction, self-organisation cannot be invoked to explain the pattern observed.

An alternative explanation, suggesting a central role for genetic control rather than self-organisation, has been proposed by Lam and co-workers (1996): light appears to exert a positive effect on the expression of genes involved in assimilation of ammonia into the amino-acids, the basic building blocks for protein synthesis, whilst repressing the genes involved in the opposite process. A similar effect has also been demonstrated on the genes encoding for nitrate reductase and nitrite reductase (Migge *et al.*, 1997; Foyer *et al.*, 1995). Nitrogen assimilation into amino-acids and proteins would, therefore, appear to be regulated by light in each individual leaf, independently. The observation that phytochrome is involved in such a regulation would explain why acclimation to light seems to respond to the red:far-red ratio of radiation rather than to PAR itself (Kwesiga and Grace, 1986; Ballaré *et al.*, 1995). Coordination could, therefore, result from the application of the same genetic rules over the whole plant (and the entire plant population, to a large extent) even in the absence of self-organisation. Tissue response would be based on some sensing of the external environment (Gleeson, 1993), possibly by phytochrome in this case, rather than on the interaction with other plant components.

2.4 Resource allocation: response to nutrient availability

Quite early on, the idea of the plant as a wise investor, able to allocate resources to above- and below-ground organs in an optimal manner, was proposed in the literature (Brenchley, 1916; Blackman, 1919). For example, Brenchley wrote 'the plant makes every endeavor to supply itself with adequate nutrients and... when the food supply is low, it strives to make as much root as possible'. The idea of internal coordination resulting in a functional balance between absorbing roots and assimilating foliage, later formalised by Davidson (1969), has since been a central paradigm in the study of allocation response to nutrient availability (Wilson, 1988; Cannell and Dewar, 1994).

An even higher degree of homeostasis is often reported in marine biomass: there tends to be a constant ratio of C:N:P, the so-called Redfield Ratio (Redfield, 1958), such that marine biomass can be represented as $C_{106}N_{16}P$. A similar stoichiometric constancy has been suggested in terrestrial plants by

McDonald and Davies (1996), based on a review of experimental evidence. On a global scale, Schulze and co-workers (1994) indeed demonstrated that foliar nitrogen content is rather conservative within plant functional groups under field conditions, despite large differences in other leaf characteristics. The reason for the conservative nature of the ratio C:N is that cytoplasm has a rather constant composition of protein, whilst the relative contribution of cytoplasm, storage carbohydrates, vacuolar sap and carbon-rich structural materials, such as cellulose and lignin, does not change profoundly. According to Schulze and co-workers (1994), differences between functional groups are related mainly to leaf longevity and to the amounts of (carbon-rich) structures and defences that are associated with long-lived foliage.

Detailed experimental studies, on the other hand, have proved that nitrogen concentrations can vary up to fivefold in a single species in response to artificial changes in nutrient availability (Ingestad and Ägren, 1992), mainly as a result of a variable balance between structural and labile pools.

Functional balance in nutrient uptake, on the other hand, is expected to result in a constant concentration of nutrients in a plant's reserves (Reynolds and Thornley, 1982). Experimental evidence, however, is not conclusive. In response to a change in nutrient availability, for example, Buysse and Merckx (1995) did indeed observe a large effect on growth allocation and overall relative growth rate in plants, but only minor changes in the ratio between sugars, starch and amino-acids in the leaf and in the root. A build-up in carbohydrate levels under nutrient deficiency, on the contrary, has been reported by McDonald and co-workers (1986).

Finally, it has been proposed that coordination of foliage and root function could lead neither to stoichiometric homeostasis nor to functional balance but to optimal plant growth (Johnson and Thornley, 1987; Gleeson, 1993): under conditions of high nutrient availability, the plant would not only allocate more resources towards foliage production but would also increase leaf protein content, so as to maximise light interception and the efficiency of radiation use and, eventually, growth. The hypothesis still lacks experimental confirmation.

In conclusion, all available experimental evidence suggests a high degree of internal coordination, possibly the outcome of plant self-organisation, resulting in partial chemical and functional homeostasis. This is in agreement with the view that the growth of individual organs cannot be understood in isolation from the rest of the plant, i.e. that sink strength is not an absolute entity but is variously affected by the activity of the source and of other competing sinks (Farrar, 1993).

The mechanisms of carbon and nitrogen translocation, and of the underlying processes of phloem loading, Münch pressure flow and phloem unloading, are rather complex and still not fully understood (Komor, 1994; Van Bel, 1993; Patrick, 1997). Over the years, however, several mechanistic models have been proposed to capture the key features of growth allocation (Thornley, 1972; Christy and Ferrier, 1973; Minchin *et al.*, 1993; Cheeseman, 1993; Dewar, 1993; Komor, 1994). These models differ in many respects, a reminder of how little is known of the mechanisms involved. A good example is the way the process of translocation itself is represented.

The basic structure of the Thornley (1972) model is sketched in Figure 2.3, as a starting point for model analysis. Here, nitrogen uptake and carbon assimilation are assumed to be a function of fine root and leaf biomass, respectively. Hence, nitrogen and carbon are independently translocated to the rest of the plant by a diffusion-like process, which is supposed to simulate phloem flow. In Dewar's (1993) model, on the contrary, some of the nutrients are supposed to be transported from the root to the shoot via the xylem and then to be circulated back by phloem flow, driven by a standing gradient in carbohydrate concentration. The idea that resources are continuously recirculated around the plant (Komor, 1994) is even more explicit in Cheeseman's (1993) model: here, N and C are assumed to be exported independently in

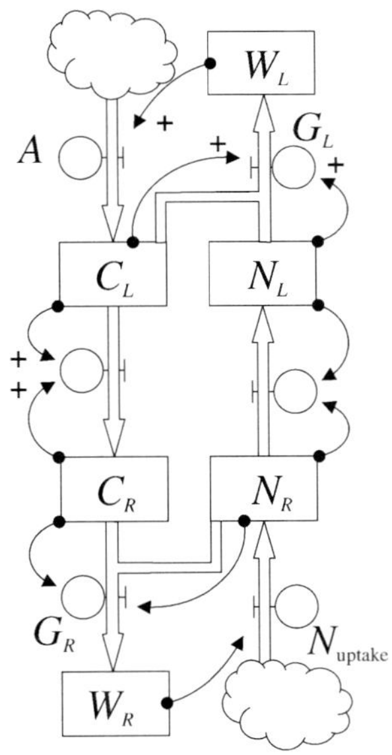

Figure 2.3 Self-organisation at the whole plant level: resource allocation and functional homeostasis (Thornley, 1972; Cheeseman, 1993). The growth of leaves (G_L) and absorbing roots (G_R) is assumed to be a function of the carbon and nitrogen concentration in the local labile pool (C_L and C_R and N_L and N_R, respectively). Nitrogen uptake (N_{uptake}) and assimilation (A) are assumed to be proportional to the biomass of absorbing roots and leaves (W_R and W_L), respectively. Carbon and nitrogen are assumed to diffuse independently between foliage and roots. Both homeostatic and optimal behaviour can be predicted by the model, depending on the degree of compensation of C and N in driving growth.

both directions at a rate that is determined solely by the dynamics of active phloem loading. In contrast with this view, phloem flow is dominated by the rate of unloading of competing sinks in the model by Minchin and co-workers (1993), as the assumption is made that phloem concentration at the loading site is kept constant by some homeostatic mechanism.

What pattern does emerge from such a complex web of root–shoot interactions? As discussed by Mäkelä and Sievänen (1987), the outcome strongly depends upon the internal dynamics of tissue growth. If growth is assumed to be limited at any one time by the factor in shortest supply (Liebig's rule), a functional balance between nutrient uptake and photosynthesis and a constant C:N ratio is predicted. If, on the other hand, growth is assumed to be co-limited by carbon and nitrogen availability, following bi-substrate enzyme kinetics (Thornley, 1972; Dewar, 1993), coordination is expected to result in optimal plant growth. The outcome of plant self-organisation appears to depend upon the inner functioning of its parts as much as on their mutual interaction.

2.5 Resource allocation: response to water relations

A central role in plant self-organisation has often been attributed to plant water relations. Leonardo da Vinci observed that 'all the branches of a tree at every stage of its height, when put together, are equal in thickness to the trunk below them' (Richter, 1989). Such a structural constancy has been suggested to result in a fractal crown geometry (Long, 1994), and to stem from hydraulic requirements (West *et al.*, 1997). Leonardo da Vinci's rule has also been extended to the balance between foliage and the area of conducting sapwood (Shinozaki *et al.*, 1964a), and attributed to the hydraulic demands of water transport. According to the pipe model (Shinozaki *et al.*, 1964b), the sapwood can be viewed as a bundle of conducting pipes, extending from the roots to the canopy, each supplying a unit amount of transpiring foliage with water and nutrients. In fact, a rather constant foliage-to-sapwood area ratio has indeed been observed in many tree species (Margolis *et al.*, 1995).

Allocation between transpiring foliage, absorbing roots and conductive xylem has also been suggested to result in functional homeostasis in water transport. In reviewing plant water relations, Passioura (1982) noted that 'the hydraulic resistance of a whole plant has often been found to be variable, with a tendency to keep leaf water potential constant over a wide range of transpiration rates'. Boyer (1985) noted that such a pattern is probably associated with growth processes. Homeostasis in water relations had long been observed in herbaceous plants in response to changes both in air humidity (Tinklin and Weatherley, 1966; Stoker and Weatherley, 1971) and in root water potential (Macklon and Weatherley, 1965). Whitehead and co-workers (1983) extended this observation to *Pinus radiata* seedlings acclimated to different conditions of air humidity, coming to the conclusion that minimum leaf water potential is indeed conserved to a large extent irrespective of transpiration rates. The same homeostatic behaviour is also observed over the lifetime of the plant. Meinzer and Grantz (1990), for example, noted that the minimum leaf water potential of sugar cane plants does not vary appreciably with plant development, as the increase in leaf area and plant transpiration rates is exactly balanced by parallel changes in hydraulic conductance resulting from root and shoot growth.

A review of literature data concerning minimum water potential in *Pinus sylvestris* seedlings and mature trees seems to confirm this tendency (Figure 2.4; Magnani *et al.*, 1999a): despite large ontogenetic and environmental differences, data are grouped around a mean value of −1.4 MPa, with a standard deviation of 0.2 MPa. This contrasts with the more than threefold variation observed in specific conductance of the leaf, between 2.0 and 6.3×10^{-8} m MPa^{-1} s^{-1}. As already noted by Tyree and Sperry (1988), minimum leaf water potentials under field conditions closely approach the critical value that would trigger extensive xylem embolism (Figure 2.4). This led to the suggestion (Magnani *et al.*, 1999a,b) that self-organisation could maximise plant growth within the limits imposed by hydraulic architecture and by the risk of diffuse xylem embolism.

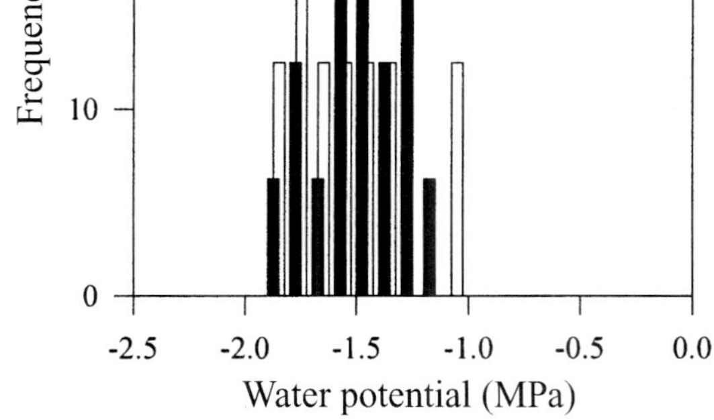

Figure 2.4 Evaluation of the hypothesis of functional homeostasis in water transport. Review of literature data of minimum leaf water potential (shaded bars; -1.4 ± 0.2 MPa, mean ± standard deviation; $n = 16$) and critical water potential for cavitation (white bars; -1.5 ± 0.2 MPa, mean ± standard deviation; $n = 8$) in *P. sylvestris*. (From Magnani *et al.* 1999a.)

Two mechanisms that could explain the functional homeostasis observed in water transport are reviewed in Figure 2.5, where only a root and a leaf compartment are considered for the sake of simplicity. Carbon and nitrogen translocation is now treated in a simplified way: the focus is on the implications of root and leaf growth on shoot water relations, on the one hand, and on the effects of leaf and root water potential on tissue growth, on the other. Foliage growth is assumed to increase plant transpiration rates and, as a result, to induce more negative leaf water potentials; fine root production, on the other hand, by reducing plant hydraulic conductance would have the opposite effect.

Tissue water potential, in turn, could affect either sink activity or translocation rates. According to Dewar (1993), the main effect of tissue water potential would be on sink activity (continuous arrows in Figure 2.5). Under conditions of high transpirational demand, frictional potential losses would drive leaf water potential well below root water potential, and thus strongly reduce foliage growth. This would have a stabilising effect on leaf water potentials, as already discussed. It is interesting to note that, by curbing

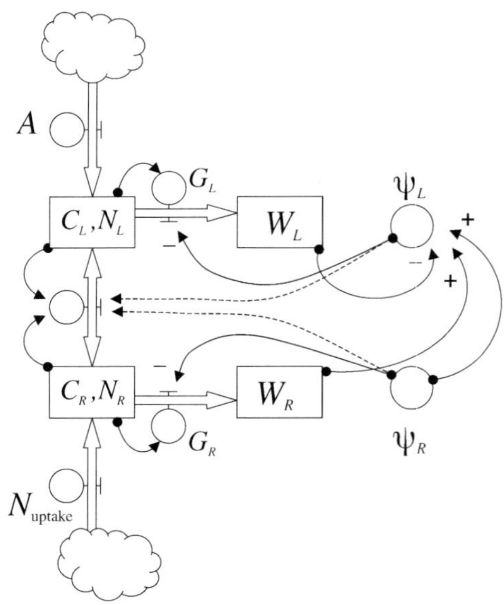

Figure 2.5 Self-organisation at the whole plant level: allocation and homeostasis in water transport (Dewar, 1993; Komor, 1994). Leaf water potential (Ψ_L) is positively affected by root water potential (Ψ_R) and root biomass (assumed to be proportional to plant hydraulic conductance) and negatively by foliage biomass (proportional to plant transpiration). Tissue water potentials are supposed to directly affect either local growth (Dewar, 1993) or phloem pressure gradients and resource allocation (Komor, 1994). In both cases, homeostasis in water transport is expected. For explanation of symbols see legend to Figure 2.3.

foliage growth more than gas exchange, a mild water stress would be predicted to result in a build-up of carbohydrates in the leaf, as often observed (Wilson, 1988).

To what extent water relations directly affect Münch phloem flow and resource translocation (dotted arrows in Figure 2.5) has long been debated (Milburn, 1975; Lang, 1983). To address this point, one can refer to the detailed representation of Münch pressure flow by Christy and Ferrier (1973), as later integrated by Komor (1994). A steep water potential gradient between roots and foliage, as would be induced by active transpiration, would reduce phloem turgor at the loading site. This, in turn, would result in diminished mass flow rates towards the root, solute build-up in the leaf phloem and greater uptake by the growing leaf. Eventually, Komor's (1994) model would therefore predict an increased allocation to foliage under water stress conditions, contrary to the experimental evidence available.

As already noted when discussing nutrient effects, the outcome of self-organisation seems, therefore, to be determined by the local behaviour of system parts as much as by the exchange of information (in this case carbon, nitrogen and water) at the system level. Local actions and global interactions both contribute to the process.

2.6 Sensory apparatus

The preceeding sections have presented examples of how self-organisation may be achieved at the leaf, plant and canopy level, and have indicated how development of the plant may proceed in such a way as to buffer the internal environment in relation to the availability of resources in the external environment, by developing shoot as opposed to root, or vice versa. In most cases, to achieve the fine control that these homeostatic adjustments require, there is an *a priori* need for sensory apparatus to acquire information from the external and internal environment, as well as a transduction method, which conveys signals from the sensors to activate gene expression, cell division and growth. This section briefly reviews the state-of-knowledge of the sensory apparatus of higher plants.

2.6.1 Light

Much work has been carried out on light sensing, following the observation in the 1960s that morphogenetic responses, such as seed germination, flowering and stem elongation, are controlled by the ratio of red (R) to far-red (FR) light, and the discovery that these responses are triggered by the photoexcitation state of the pigment phytochrome (Smith, 1995 and Chapter 5). Work on phytochrome has aided the interpretation of many of the classical plant responses to shade, such as the increase in the specific leaf area, leaf area ratio and the decrease of the chlorophyll *a*:*b* ratio. It has also helped in understanding the nature of plant response to competition, as many plants detect their neighbours by sensing a shift in R:FR reflected from them, and respond by increasing stem elongation (Ballaré *et al.*, 1990). Several other photoreceptors are believed to have roles analogous to phytochrome. They include a family of UV-A and blue light sensors, known as cryptochromes (Cashmore, 1998), and a UV-B receptor (Ballaré *et al.*, 1995). Root elongation as well as shoot elongation may be directly controlled by light, despite the rapid attenuation of light by soil (Kurata and Yamamoto, 1997).

It is, however, still not clear how far all the known responses to shade are triggered by the phytochrome response, as changes in photon flux density will cause changes in many other parameters: electron flow in the thylakoid membranes; a lower rate of production of adenosine triphosphate (ATP) and the reduced form of nicotine adenine dinucleotide phosphate (NADPH); and a reduction in the pools of glucose and quantity of starch grains (Tanada, 1997). Any or all of these could be the basis of a developmental response or could bring about fine adjustments, such as leaf or chloroplast movement (Haupt and Scheuerlein, 1990).

2.6.2 Gravity and mechanical sensing

Gravity sensing enables the development of polarity, whereby the roots grow downward and shoots grow upward (even in the dark). According to the starch-statolith model of gravity sensing, rather large sedimenting particles, generally amyloplasts contained in specific cells, stimulate microfilaments in the cell membrane (Bjorkman, 1992; Slack, 1997; Sinclair and Trewavas, 1997), eliciting the characteristic gravitropic response. There is extensive literature to support this model, particularly from the columella of root caps (Sievers *et al.*, 1996), in *Arabidopsis* (Slack, 1997) and in giant algal cells (Staves, 1997); however, other sensing mechanisms would seem to play some role as starch-free *Arabidopsis* still shows some sensitivity to gravity (Trewavas and Knight, 1994). Mechanical sensing, and touch sensitivity in general, is an important part of the observed response to wind (Grace, 1977), whereby mechanical tissues develop to enable the plant to withstand mechanical forces. Some plants have a naturally high sensitivity to mechanical stimuli. Tendrils and the leaves of so-called 'sensitive plants', such as *Mimulus*, move in response to very small mechanical stimuli. It has been possible to show that mechanical stimulus causes massive transfer of calcium to the cytosol from storage in the vacuoles, and its combination with the small protein molecule, calmodulin. Cellular processes are then initiated by the action of protein kinase enzymes (Trewavas and Knight, 1994). Presumably, the mechanical stimulus is sensed as a change in pressure within the cell, and may be related to the sensing of changes in turgor during drought (see below).

2.6.3 Drought

Stomata close in response to soil water deficits. It has been postulated that when the root encounters dry soil it sends a chemical signal in the transpiration stream, which causes the stomata to shut (Gollan *et al.*, 1986; Tardieu and Davies, 1992). The suggested identity of the signal is abscisic acid (ABA), which is known to be transported in the xylem sap and to make stomata shut by regulating ion transporters in guard cells (Leung and Giraudat, 1998). The strongest evidence for a chemical signal comes from split-root experiments. When part of the root system is experimentally diverted to a container, where it can be droughted, the act of droughting the root makes the stomata close, even though the leaf is well supplied with water from the remaining, well-watered part of the root system, and the water potential is not reduced by the droughting treatment (Gollan *et al.*, 1986; Tardieu and Davies, 1992). Currently, however, it is evident that the situation is not so simple as was first thought, and it is not clear whether there really is a chemical signal in all droughted plants. In trees, for example, there is a problem in the length of the transport path. In tall conifers, any chemical signal would take 5–20 days to reach the leaves, as the sap velocity is only 1–2 m per day. Such a signal would arrive too late to be useful. A hydraulic signal, on the other hand, would be immediate, as can be demonstrated in an experimental system where the soil and roots are pressurised (e.g. Comstock and Mencuccini, 1998). Hydraulic signals are rapidly translated into changes in turgor pressure within cells, which in turn might activate pressure-sensitive calcium channels; or they could lead to an adjustment in the permeability of plasmodesmata, which may have a controlling influence on the transmission of developmental signals (Epel, 1994).

2.6.4 Sensors and models

Our overall understanding of the mechanisms whereby plants sense their external environment and internal states, and respond developmentally to signals received from their sensors, remains primitive. Until more is known, most models of plant responses to their environment will continue to make implicit assumptions about sensing, thus providing sets of hypotheses which physiologists will take much longer to test. Similarly, that plants will exercise an economy of effort in their growth and development is an underlying assumption of many models, but is really an untested, and difficult-to-test, hypothesis based on expectations arising from evolutionary theory.

2.7 Emergent properties and top-down system analysis

In the preceding paragraphs, an attempt was made to understand how self-organisation could explain much of the structural and functional coordination that is often observed in a plant's life. The aim now is to see what the consequences of internal self-organisation are for the response of the whole plant to the environment. Here, because of the uncertainty still surrounding many physiological processes, instead of a bottom-up approach, a top-down perspective will be adopted, i.e. an attempt will be made to explore the implications of structural and functional homeostasis, the emergent properties that have been seen to be the final outcome of internal self-organisation.

A top-down model based on the homeostatic principle of a constant nitrogen concentration in the mobile pool is demonstrated in Figure 2.6. The model is equivalent to that presented by Reynolds and Thornley (1982), assuming an on-off control of resource allocation: whenever nitrogen concentration falls below a set threshold, growth is allocated entirely to fine roots, so increasing nitrogen absorption and re-establishing the homeostatic concentration. When nitrogen is in excess, on the contrary, there is no requirement for root growth, but shoot growth will permit more surface area for photo-assimilation, thus matching the nitrogen with carbon and speeding up the rate of growth. Grace (1997) applied this simple model to explore the effects of nutrition on plant response to atmospheric CO_2 and shading, and suggested that a similar top-down perspective could be effectively used in individual models to simulate plant competition, geographical spread and succession.

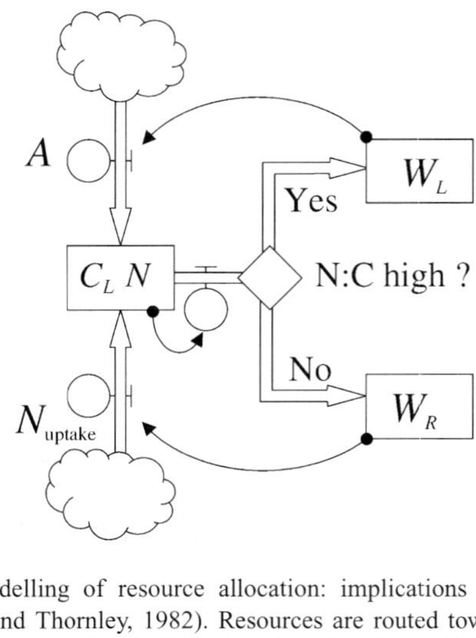

Figure 2.6 Top-down modelling of resource allocation: implications of homeostasis in nitrogen concentration (Reynolds and Thornley, 1982). Resources are routed towards assimilating foliage or absorbing roots depending on a threshold partitioning function, which keeps a constant nutrient concentration in the labile pool. For explanation of symbols see legend to Figure 2.3.

A similar approach can also be applied to study the implications of homeostasis in water transport (Figure 2.7a). Whenever leaf water potential falls below a critical threshold, no more leaves can be safely supported by the plant. Growth is then fully allocated to feeder roots, so reducing plant hydraulic resistance and re-establishing the homeostatic value of leaf water potential. In trees and other woody plants, however, xylem contributes a significant proportion of plant hydraulic resistance (Meinzer *et al.*, 1992; Magnani and Borghetti, 1995), and stem growth accounts for up to 60% of total available carbon (Santantonio, 1989).

The approach has, therefore, been extended to take xylem characteristics and growth into account (Magnani *et al.*, 1999a,b), making the further assumption that evolution has resulted in an allocation strategy that maximises plant growth and fitness (Parker and Maynard Smith, 1990). The resulting model is exemplified in Figure 2.7b. The production of either sapwood or fine roots results in a reduction of plant hydraulic resistance, but at a different cost. The return from sapwood production, in particular, will sharply decline with plant height: if one unit of invested carbon can be used to produce ten conductive pipes working in parallel in a seedling of 1 m, it will suffice to produce only one conductive pipe along the length of a tree

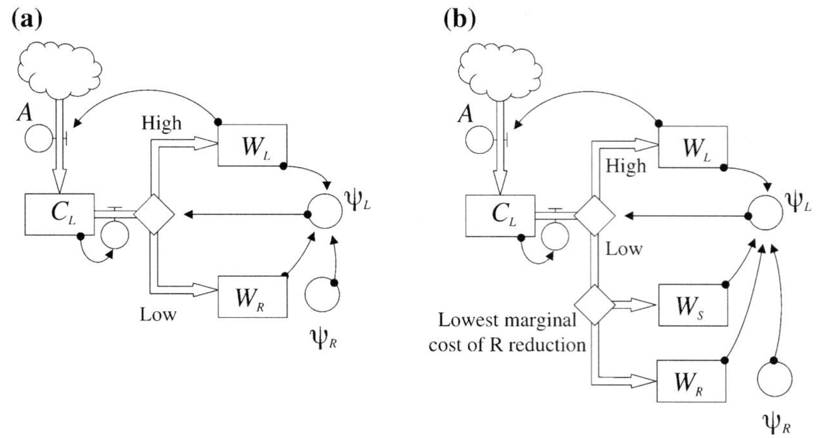

Figure 2.7 Top-down modelling of resource allocation: implications of homeostasis in water transport (Magnani *et al.*, 1999a,b). a) Resources are routed towards foliage (responsible for both assimilation and transpiration) or absorbing roots depending on a threshold partitioning function, which keeps leaf water potential at a constant critical level. b) When optimal growth under hydraulic constraints is assumed, the approach can be expanded to model allocation among sapwood (W_S) and fine roots (W_R), which are both involved in water transport. Total plant hydraulic resistance, R, is assumed to be lowered at the minimum possible cost; this implies preferential allocation to roots or sapwood, depending on tree height and environmental conditions. For explanation of symbols see legends to Figures 2.3 and 2.5.

10 m tall, with a total gain in hydraulic conductance that is only 1% of that gained in the case of the seedling. Despite the greater longevity of the sapwood, which would favour it as a long-lasting investment, allocation to fine roots will thus increase with age, at the expense of xylem and foliage production. As a result, above-ground net primary production would be expected to decline with age (Magnani *et al.*, 1999b), in good agreement with experimental reports (Gower *et al.*, 1996; Ryan *et al.*, 1997). The amount of sapwood cross-sectional area that is required to feed one unit of foliage (A_s/A_f) is also predicted to increase with height, as sapwood becomes less effective in water conduction because of the increasing length of the hydraulic path. Experimental data from a *P. sylvestris* chronosequence (Mencuccini and Grace, 1996) confirmed this trend (Figure 2.8A), as well as other model predictions.

The model can also be applied to explore the response of allocation and growth to climate. If a constant value of leaf water potential is to be maintained, for example, high transpiration rates should force a reduction in plant hydraulic resistance and induce an increase in A_s/A_f, despite the greater allocation to fine roots (Magnani *et al.*, 1999a). This, again, is consistent with experimental observations in two *P. sylvestris* plantations in a wet and a dry area of Britain (Mencuccini and Grace, 1995), as shown in Figure 2.8B.

The age-related decline in above-ground net primary productivity and the structural acclimation to climatic conditions both represent important properties of the system, which emerge as direct corollaries of plant functional homeostasis. Here is the answer to the question posed concerning the ecological relevance of plant self-organisation.

2.8 Conclusions

Where does order come from? In this chapter, an attempt has been made not so much to answer this basic question but to generate some ideas for further research in such a fascinating area.

Coordination is pervasive in plant life, and self-organisation appears to play a central role, from the molecular to the global scale. As we have seen, the emergent properties that result from self-organisation are evidently a characteristic of the system as a whole, yet they cannot be explained unless the functioning of the parts one level down the hierarchical scale are understood. Their analysis thus requires that the holistic and the reductionistic approach are combined.

However, self-organisation is not the only process responsible for the emergence of order at the individual level, and it could be misleading to invoke it whenever coordination arises in a complex system. It has been

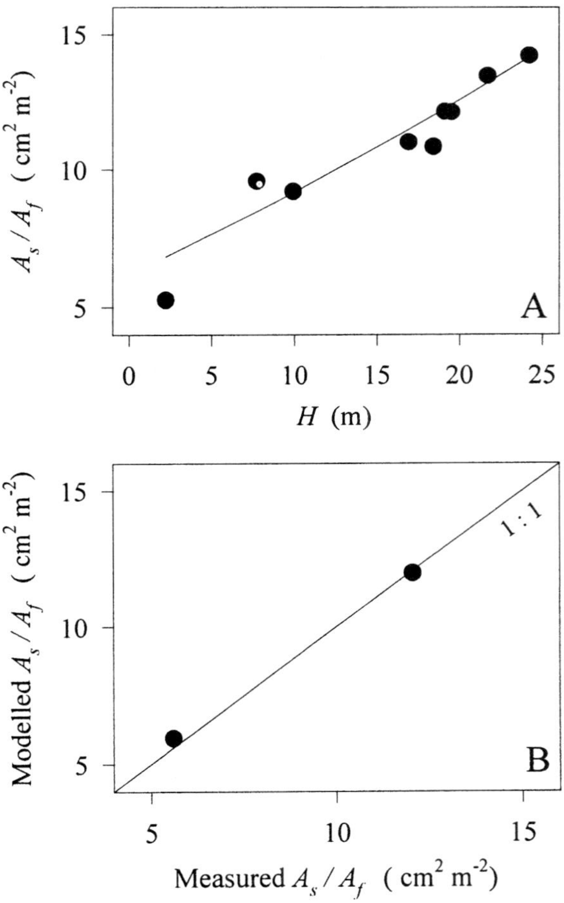

Figure 2.8 Implications of the hypothesis of functional homeostasis in water transport: structural acclimation in *P. sylvestris*. A) Height-related changes in sapwood-to-foliage area ratio (A_S/A_f) in a Scots pine chronosequence in Thetford, UK. Experimental data are compared with predictions based on the hypothesis of optimal growth under hydraulic constraints (from Magnani *et al.*, 1999b). A constant ratio would be expected from the pipe model theory. B) Modelled *vs* measured A_S/A_f values for two mature *P. sylvestris* stands growing in a wet and a dry area of Britain (from Magnani *et al.*, 1999a).

hinted that genetic control could also result in orderly patterns, acting as an external controller for a homogeneous population of cells. This view, however, could also be overly simplistic: not only is the genetic code the result of evolution and plant self-organisation at the global level, it is increasingly apparent that, rather than a fixed set of rules, it is an ever-changing book, different portions of which are switched on and off depending on the physiological status of the cell (Maddox, 1998). At some level of the hierarchical scale, therefore, order could always be associated with self-organisation.

References

Ballaré, C.L., Scopel, A.L. and Sánchez, R.A. (1990) Far-red radiation reflected from adjacent leaves: an early signal of competition in plant canopies. *Science*, **247** 329–332.

Ballaré, C.L., Scopel, A.L., Roush, M.L. and Radosevich, S.R. (1995) How plants find light in patchy environments: a comparison between wild-type and phytochrome-B deficient mutant plants of cucumber. *Functional Ecology*, **9** 859–868.

Barrs, H.D. (1971) Cyclic variations in stomatal aperture, transpiration and leaf water potential under constant environmental conditions. *Annual Review of Plant Physiology*, **22** 223–236.

Bjorkman, T. (1992) Perception of gravity by plants. *Advances in Space Research*, **12** 195–201.

Blackman, V.H. (1919) The compound interest law and plant growth. *Annals of Botany*, **33** 353–360.

Boyer, J.S. (1985) Water transport. *Annual Review of Plant Physiology and Plant Molecular Biology*, **36** 473–516.

Brenchley, W.E. (1916) The effect of the concentration of the nutrient solution on the growth of barley and wheat in water cultures. *Annals of Botany*, **30** 77–90.

Buysse, J. and Merckx, R. (1995) Diurnal variations in growth rate and growth substrate levels of spinach (*Spinacia oleracea* L.) under nitrogen-limiting conditions. *Plant, Cell and Environment*, **18** 1419–1425.

Cannell, M.G.R. and Dewar, R.C. (1994) Carbon allocation in trees: a review of concepts for modelling. *Advances in Ecological Research*, **25** 50–104.

Cardon, Z.G., Mott, K.A. and Berry, J.A. (1994) Dynamics of patchy stomatal movements, and their contribution to steady-state and oscillating stomatal conductance calculated using gas-exchange techniques. *Plant, Cell and Environment*, **17** 995–1007.

Cashmore, A.R. (1998) The cryptochrome family of blue/UV-A photoreceptors. *Journal of Plant Research*, **111** 267–270.

Cheeseman, J.M. (1993) Plant growth modelling without integrating mechanisms. *Plant, Cell and Environment*, **16** 137–147.

Christy, A.L. and Ferrier, J.M. (1973) A mathematical treatment of Münchs pressure-flow hypothesis of phloem translocation. *Plant Physiology*, **52** 531–538.

Comstock, J. and Mencuccini, M. (1998) Control of stomatal conductance by leaf water potential in *Hymenoclea salsola* (T. and G.), a desert subshrub. *Plant, Cell and Environment*, **21** 1029–1038.

Davidson, R.L. (1969) Effects of soil nutrients and moisture on root/shoot ratios in *Lolium perenne* L. and *Trifolium repens* L. *Annals of Botany*, **33** 571–577.

Dewar, R.C. (1993) A root-shoot partitioning model based on carbon-nitrogen-water interactions and Münch phloem flow. *Functional Ecology*, **7** 356–368.

Dorf, R.C. and Bishop, R.H. (1995) *Modern Control Systems*, Addison-Wesley, Reading, UK.

Ellsworth, D.S. and Reich, P.B. (1992) Leaf mass per area, nitrogen content and photosynthetic carbon gain in *Acer saccharum* seedlings in contrasting forest light environments. *Functional Ecology*, **6** 423–435.

Epel, B.L. (1994) Plasmodesmata: composition, structure and trafficking. *Plant Molecular Biology*, **26** 1343–1356.

Farrar, J.F. (1993) Sink strength: what is it and how do we measure it? *Plant, Cell and Environment*, **16** 1015–1018.

Foyer, C.H., Valadier, M.H. and Ferrario, S. (1995) Co-regulation of nitrogen and carbon assimilation in leaves, in *Environment and Plant Metabolism* (ed. N. Smirnoff), BIOS Scientific Publishers, Oxford, UK, pp. 17–33.

Gleeson, S.K. (1993) Optimization of tissue nitrogen and root-shoot allocation. *Annals of Botany*, **71** 23–31.

Gollan, T., Passioura, J.B. and Munns, R. (1986) Soil water status affects the stomatal conductance of fully turgid wheat and sunflower leaves. *Australian Journal of Plant Physiology*, **13** 459–464.

Gower, S.T., McMurtrie, R.E. and Murty, D. (1996) Above-ground net primary production decline with stand age: potential causes. *Trends in Ecology and Evolution*, **11** 378–382.

Grace, J. (1977) *Plant Response to Wind*, Academic Press, London, UK.

Grace, J. (1997) Toward models of resource allocation by plants, in *Plant Resource Allocation*, (eds. F.A. Bazzaz and J. Grace), Academic Press, New York, USA, pp. 279–291.

Häken, H. (1979) *Synergetics*. Springer-Verlag, Berlin, Germany.

Haupt, W. and Scheuerlein, R. (1990) Chloroplast movement. *Plant, Cell and Environment*, **13** 595–614.

Hikosaka, K. and Terashima, I. (1995) A model of the acclimation of photosynthesis in the leaves of C3 plants to sun and shade with respect to nitrogen use. *Plant, Cell and Environment*, **18** 605–618.

Ingestad, T. and Ägren, G.I. (1992) Theories and methods on plant nutrition and growth. *Physiologia Plantarum*, **84** 177–184.

Jarvis, P.G. and McNaughton, K.G. (1986) Stomatal control of transpiration: scaling up from leaf to region. *Advances in Ecological Research*, **15** 1–49.

Johnson, I.R. and Thornley, J.H.M. (1987) A model of shoot:root partitioning with optimal growth. *Annals of Botany*, **60** 133–142.

Kaufmann, S.A. (1993) *The Origins of Order, Self-Organization and Selection in Evolution*, Oxford University Press, Oxford, UK.

Komor, E. (1994) Regulation by futile cycles: the transport of carbon and nitrogen in plants, in *Flux Control in Biological Systems* (ed. E.-D. Schulze), Academic Press, San Diego, USA, pp. 153–201.

Kull, O. and Jarvis, P.G. (1995) The role of nitrogen in a simple scheme to scale-up photosynthesis from leaf to canopy. *Plant, Cell and Environment*, **18** 1174–1182.

Kull, O. and Kruijt, B. (1999) Acclimation of photosynthesis to light: a mechanistic approach. *Functional Ecology*, **13** 24–36.

Kurata, T. and Yamamoto, K.T. (1997) Light-stimulated root elongation in *Arabidopsis thaliana*. *Journal of Plant Physiology*, **151** 346–355.

Kwesiga, F. and Grace, J. (1986) The role of the red/far-red ratio in the response of tropical tree seedlings to shade. *Annals of Botany*, **57** 283–290.

Lam, H.M., Coschigano, K.T., Oliveira, I.C., Melo-Oliveira, R. and Coruzzi, G.M. (1996) The molecular genetics of nitrogen assimilation into amino-acids in higher plants. *Annual Review of Plant Physiology and Plant Molecular Biology*, **47** 569–593.

Lang, A. (1983) Turgor-regulated translocation. *Plant, Cell and Environment*, **6** 683–689.

Leung, J. and Giraudat, J. (1998) Abscisic acid signal transduction. *Annual Review of Plant Physiology and Molecular Biology*, **49** 199–122.

Liang, N., Nagayama, M., Nakata, M. and Maruyama, K. (1995) Growth, photosynthesis and nitrogen content in Japanese beech (*Fagus crenata* Bl.) seedlings grown under five irradiances. *Photosynthetica*, **31** 257–268.

Long, C.A. (1994) Leonardo da Vinci's rule and fractal complexity in dichotomous trees. *Journal of Theoretical Biology*, **167** 107–113.

Lovelock, J. (1982) *Gaia: A New Look at Life on Earth*, Oxford University Press, Oxford, UK.

Macklon, A.E.S. and Weatherley, P.E. (1965) Controlled environment studies of the nature and origins of water deficits in plants. *New Phytologist*, **64** 414–427.

Maddox, J. (1998) *What Remains to be Discovered?* MacMillan, London, UK.

Magnani, F. and Borghetti, M. (1995) Interpretation of seasonal changes of xylem embolism and plant hydraulic resistance in *Fagus sylvatica*. *Plant, Cell and Environment*, **18** 689–696.

Magnani, F., Borghetti, M. and Grace, J. (1999a) Acclimation of coniferous tree structure to the environment under hydraulic constraints (submitted).

Magnani, F., Mencuccini, M. and Grace, J. (1999b) Age-related decline in stand productivity: the role of structural acclimation under hydraulic constraints. *Plant, Cell and Environment* (in press).

Mäkelä, A. and Sievänen, R.P. (1987) Comparison of two shoot–root partitioning models with respect to substrate utilization and functional balance. *Annals of Botany*, **59** 129–140.

Margolis, H., Oren, R., Whitehead, D. and Kaufmann, M.R. (1995) Leaf area dynamics of conifer forests, in *Ecophysiology of Coniferous Forests* (eds. W.K. Smith and T.M. Hinckley), Academic Press, San Diego, USA, pp. 181–223.

McDonald, A.J.S. and Davies, W.J. (1996) Keeping in touch: responses of the whole plant to deficits in water and nitrogen supply. *Advances in Botanical Research*, **22** 229–300.

McDonald, A.J.S., Lohammar, T. and Ericsson, A. (1986) Growth response to step-decrease in nutrient availability in small birch (*Betula pendula* Roth.). *Plant, Cell and Environment*, **9** 427–432.

Meinzer, F.C. and Grantz, D.A. (1990) Stomatal and hydraulic conductance in growing sugar-cane: stomatal adjustment to water transport capacity. *Plant, Cell and Environment*, **13** 383–388.

Meinzer, F.C., Goldstein, G., Neufeld, H.S., Grantz, D.A. and Crisosto, G.M. (1992) Hydraulic architecture of sugar-cane in relation to patterns of water use during plant development. *Plant, Cell and Environment*, **15** 471–477.

Mencuccini, M. and Grace, J. (1995) Climate influences the leaf area-sapwood area ratio in Scots pine. *Tree Physiology*, **15** 1–10.

Mencuccini, M. and Grace, J. (1996) Developmental patterns of above-ground xylem conductance in a Scots pine (*Pinus sylvestris* L.) age sequence. *Plant, Cell and Environment*, **19** 939–948.

Migge, A., Meya, G., Carryol, E., Hirel, B. and Becker, T.W. (1997) Co-action of light and the nitrogen substrate in controlling the expression of the tomato genes encoding nitrite reductase and nitrate reductase. *Journal of Plant Physiology*, **151** 151–158.

Milburn, J.A. (1975) Pressure flow, in *Encyclopedia of Plant Physiology New Series*, Vol. 1, Transport in Plants, I, Phloem Transport (eds. M.H. Zimmermann and J.A. Milburn), Springer-Verlag, Berlin, Germany, pp. 328–353.

Minchin, P.E.H., Thorpe, M.R. and Farrar, J.F. (1993) A simple mechanistic model of phloem transport which explains sink priority. *Journal of Experimental Botany*, **44** 947–955.

Monteith, J.L. (1995) A reinterpretation of stomatal responses to humidity. *Plant, Cell and Environment*, **18** 357–364.

Mott, K.A. and Parkhurst, D.F. (1991) Stomatal responses to humidity in air and helox. *Plant, Cell and Environment*, **14** 509–515.

Nicolis, G. and Prigogine, I. (1989) *Exploring Complexity*, Freeman, New York, USA.

Odum E.P. (1983) *Basic Ecology*, CBS College, Philadelphia, USA.

Parker, G.A. and Maynard Smith, J. (1990) Optimality theory in evolutionary biology. *Nature*, **348** 27–33.

Passioura, J.B. (1982) Water in the soil-plant-atmosphere continuum, in *Encyclopedia of Plant Physiology*, Vol. 12B (eds. O.L. Lange, P.S. Nobel, C.B. Osmond and H. Ziegler), Springer-Verlag, Berlin, Germany, pp. 5–33.

Patrick, J.W. (1997) Phloem unloading: sieve element unloading and post-sieve element transport. *Annual Review of Plant Physiology and Plant Molecular Biology*, **48** 191–222.

Prigogine, I. and Stengers, I. (1985) *Order Out of Chaos*, Flamingo, London, UK.

Rand, R.H., Upadhyaya, S.K., Cooke, J.R. and Storti, D.W. (1981) Hopf bifurcation in a stomatal oscillator. *Journal of Mathematical Biology*, **12** 1–11.

Redfield, A.C. (1958) The biological control of the chemical factors in the environment. *American Scientist*, **46** 1–18.

Reynolds, J.F. and Thornley, J.H.M. (1982) A shoot–root partitioning model. *Annals of Botany*, **49** 585–597.

Richter, J.P. (1989) *The Notebooks of Leonardo da Vinci (1452–1519)*, compiled and edited from the original manuscripts, Dover, New York, USA.

Ryan, M.G., Binkley, D. and Fownes, J.H. (1997) Age-related decline in forest productivity: pattern and processes. *Advances in Ecological Research*, **27** 213–262.

Salt, G.W. (1979) A comment on the use of the term emergent properties. *American Naturalist*, **113** 145–148.

Santantonio, D. (1989) Dry-matter partitioning and fine root production in forests: new approaches to a difficult problem, in *Biomass Production by Fast-Growing Trees* (eds. J.S. Pereira and J.J. Landsberg), Kluwer Academic, Dordrecht, The Netherlands, pp. 57–72.

Schulze, E-D., Kelliher, F.M., Korner, C., Lloyd, J. and Leuning, R. (1994) Relationships among maximum stomatal conductance, ecosystem surface conductance, carbon assimilation rate and plant nitrogen nutrition: a global ecology scaling exercise. *Annual Review Ecology Systematics*, **25** 629–660.

Shababa, S., Delbourgo, R. and Newman, I. (1997) Observations of bifurcation and chaos in plant physiological responses to light. *Australian Journal of Plant Physiology*, **24** 91–96.

Shinozaki, K., Yoda, K., Hozumi, K. and Kira, T. (1964a) A quantitative analysis of plant form: the pipe model theory. I. Basic analyses. *Japanese Journal of Ecology*, **14** 97–105.

Shinozaki, K., Yoda, K., Hozumi, K. and Kira, T. (1964b) A quantitative analysis of plant form: the pipe model theory. II. Further evidence of the theory and its application in forest ecology. *Japanese Journal of Ecology*, **14** 133–139.

Sievers, A., Buchen, B. and Hodick, D. (1996) Gravity sensing in tip-growing cells. *Trends in Plant Science*, **1** 273–279.

Sinclair, W. and Trewavas, A.J. (1997) Calcium in gravitropism: a re-examination. *Planta*, **203** s85–s90.

Slack, F.D. (1997) Plastids and gravitropic sensing. *Planta*, **203** s63–s68.

Smith, H. (1995) Physiological and ecological function within the phytochrome family. *Annual Review of Plant Physiology and Plant Molecular Biology*, **46** 289–315.

Solé, R.V. and Manrubia, S.C. (1995) Are rainforests self-organized in a critical state? *Journal of Theoretical Biology*, **173** 31–40.

Sprugel, D.G., Hinckley, T.M. and Schaap, W. (1991) The theory and practice of branch autonomy. *Annual Review, Ecology Systematics*, **22** 309–334.

Staves, M.P. (1997) Cytoplasmic streaming and gravity sensing in *Chara* internodal cells. *Planta*, **203** s79–s84.

Stoker, R. and Weatherley, P.E. (1971) The influence of the root system on the relationship between the rate of transpiration and depression of leaf water potential. *New Phytologist*, **70** 547–554.

Tanada, T. (1997) The photoreceptors in the high irradiance response of plants. *Physiologia Plantarum*, **101** 451–454.

Tardieu, F. and Davies, W.J. (1992) Integration of hydraulic and chemical signalling in the control of stomatal conductance and water status of droughted plants. *Plant, Cell and Environment*, **98** 540–545.

Thornley, J.H.M. (1972) A balanced quantitative model for root:shoot ratios in vegetative plants. *Annals of Botany*, **36** 431–441.

Tinklin, R. and Weatherley, P.E. (1966) On the relationship between transpiration rate and leaf water potential. *New Phytologist*, **65** 509–517.

Trewavas, A. and Knight, M.R. (1994) Mechanical signalling, calcium and plant form. *Plant Molecular Biology*, **26** 1329–1341.

Tyree, M.T. and Sperry, J.S. (1988) Do woody plants operate near the point of catastrophic xylem dysfunction caused by dynamic water stress? Answers from a model. *Plant Physiology*, **88** 574–580.

Van Bel, A.J.E. (1993) Strategies of phloem loading. *Annual Review of Plant Physiology and Plant Molecular Biology*, **44** 253–281.

West, G.B., Brown, J.H. and Enquist, B.J. (1997) A general model for the origin of allometric scaling laws in biology. *Science*, **276** 122–126.

Weyers, J.D.B. and Lawson, T. (1997) Heterogeneity in stomatal characteristics. *Advances in Botanical Research*, **26** 317–352.

Whitehead, D., Sheriff, D.W. and Greer, D.H. (1983) The relationship between stomatal conductance, transpiration rate and tracheid structure in *Pinus radiata* clones grown at different water vapour saturation deficits. *Plant, Cell and Environment*, **6** 703–710.

Willmer, C. and Fricker, M. (1996) *Stomata*, Chapman and Hall, London, UK.

Wilson, J.B. (1988) A review of evidence on the control of shoot:root ratio, in relation to models. *Annals of Botany*, **61** 433–449.

3 Genetic analysis of leaf development and differentiation

Toshi Foster and Bruce Veit

3.1 Introduction

Evolution has fashioned a multitude of leaf characters to best exploit particular environments. Certain characters, such as absolute size, orientation in space or pigmentation may be seen as influenced by prevailing growth conditions. By contrast, characters such as leaf shape or phyllotaxy reflect less flexible programmes of development. In both cases, however, genetic factors clearly constrain the underlying patterns of development to yield forms that are characteristic of a particular species. That such genes might be inherited as discrete entities, and therefore be analysed as such, was first appreciated by Mendel. In the last century, this view has been further elaborated in three ways: firstly, classical methods of genetic analysis have been refined, which enable gene function to be assigned in a general way. In its most basic form, individuals in which the activity of a particular gene has been altered through mutation are compared to normal siblings; any differences observed can be attributed to changes in the activity of that gene. A second development has grown out of recent advances in molecular cloning techniques that make it possible to isolate physically specific genes, thus providing insight into the molecular mechanism by which they function. Finally, transgenic approaches by which gene activity can be artificially modified enable incisive tests of models for gene function. Together, these approaches offer a useful framework for describing complex developmental processes.

This chapter offers a review of the process of leaf development and differentiation. Because space limitations preclude a comprehensive review of this topic, the goal was to review the rapidly developing conceptual framework that has grown out of genetic and molecular studies, and to present them in a way that is accessible to the interested non-specialist audience. Much of the discussion focuses on leaf development in angiosperms, though many of the principles apply equally well to lower plants. As the description of these genetic analyses given here is by no means complete, wherever possible more focused reviews on particular topics are referenced.

Where appropriate, accounts are given of both descriptive and experimental studies that provide the foundation for these genetically based models.

3.2 Essential aspects of leaf organisation and development

A general description of leaf development necessarily begins with some agreement as to what constitutes a leaf. Taking the tobacco leaf as an example (Avery, 1933), primordia are first seen appearing as bulges in predictable positions on the flanks of the shoot apex (Figure 3.1). As each primordium is displaced away from the tip by continued growth of the apical dome, it enlarges through anisotropic patterns of growth into a flattened, bilaterally symmetrical structure. As the primordium enlarges, cellular differentiation begins with the formation of the hair-like trichome cells on superficial layers, while vascular cell types form more internally. Once these differentiation processes conclude, the leaf functions for a time before it senesces.

The example given might suggest that a purely geometrical definition of the leaf is adequate, especially given the widely accepted meaning of the lay term 'leaf-like'. However, a thin, flattened structure, so well suited to the

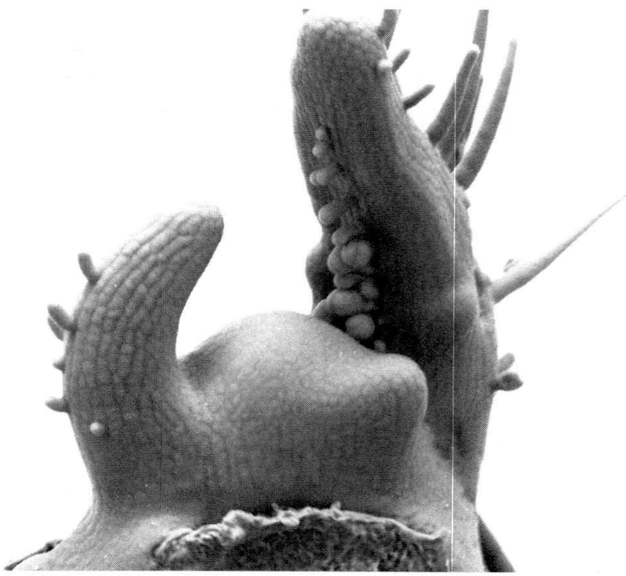

Figure 3.1 Scanning electron micrograph (SEM) image of the vegetative shoot apex of a tomato plant illustrating a spiral pattern of initiation of three-leaf primordia with trichome formation evident (the fourth and oldest primordium removed to expose shoot apex). Image kindly provided by D. Reinhardt and C. Kuhlemeier.

capture of light and the exchange of gas, could be confounded with many examples of flattened stems known within the plant kingdom and would fail to recognise a myriad of non-planar leaf shapes. For similar reasons, a narrowly cast, functional definition also falls short. The developmentally inclined might propose defining leaves in terms of their mode of origin and developmental potential. Once again, such a definition would appear less than universal by failing to recognise leaf-like structures that form without the benefit of a shoot tip (for discussion see Poethig, 1997).

Despite the limitations of these criteria, they have considerable heuristic value by focusing attention on individual factors upon which commonly occurring patterns of development depend, thereby revealing essential elements of the underlying mechanisms. The sequence of development described for tobacco approximates the pattern for most other vascular plants, though the basic pattern would be modified in specific ways to obtain various leaf forms. For simplicity, in the present discussion the process of leaf development has been divided into three phases. This scheme first considers the physical and genetic context in which the cells that make up the leaf become defined, then addresses developmental processes by which basic growth axes of the leaf become established, and finally assesses factors that influence terminal differentiation of cells within the leaf. Within this framework, an attempt has been made to elicit consistent themes that perhaps reflect conserved developmental mechanisms. The topic concludes with a discussion on how these basic programmes of development may be modified to obtain various leaf forms manifest within a single individual as well as among different members of the plant kingdom.

3.3 The initiation of leaves

'*Meristems make the plant*', Ian Sussex

For the vast majority of higher plants, leaves are derived from groups of undetermined, embryonic-like cells that are produced at the tips of growing shoots (Steeves and Sussex, 1989). Unlike animals, for which a basic body plan is established during a brief embryogenesis, plants elaborate their architecture gradually. This emphasis on post-embryonic development permits plants to elaborate a form that is best suited to prevailing growth conditions. Thus, a seedling might take on a spindly form when grown in crowded, low-light conditions but adopt a more branched, compact habit in full sunlight. At least some of this adaptation may reflect specific tropisms. Just as significant, however, are the idiosyncratic patterns of growth initiated at the tips of growing shoots, in which specific phyllotaxies (the arrangement of leaves on the stem) are established. Moreover, the coupling together of specific patterns of axillary branching and internode development with leaf initiation can exert a profound effect on plant architecture.

3.3.1 The role of the shoot apical meristem

Studies to determine what factors influence the position of leaf initiation are logically focused on the shoot apex. Shoot apices continue to initiate leaves relatively normally when removed from older adjacent tissue (Wetmore, 1954; Ball, 1960; Smith and Murashige, 1970), suggesting that the factors that guide the initiation process are likely to act over relatively short distances within the shoot apex itself. The arena in which these factors are thought to act has become embodied in the concept of the 'shoot apical meristem' (SAM) (Sussex, 1989). As well as providing a venue for a leaf's debut, the SAM also functions to maintain a source of undifferentiated cells from which leaves become organised. Cells at the apex of the SAM remain undifferentiated, while cells that are displaced from the tip become organised into the various structures that comprise the plant body.

While the size and shape of SAMs varies greatly between species and even within different growth phases of a single plant, certain features are consistent. The SAM exhibits a so-called 'zonation', in which different regions can be distinguished by histological criteria (reviewed by Steeves and Sussex, 1989). In lower plants, a single large, prism-shaped cell that occupies a central apical position divides infrequently, but in highly oriented planes, to produce smaller derivatives. These derivatives, in turn, divide more frequently as they are displaced further away from the apex. A similar gradient of mitotic activity is observed in higher plants, though the apical architecture is more complex (Figure 3.2). Instead of a single apical cell, several cells appear to function as initials, dividing relatively infrequently to give rise to more rapidly dividing derivatives. Cells within this 'central zone' tend to be larger and more vacuolated than their more mitotically active and cytoplasmically dense derivatives, which lie within the surrounding organogenic 'peripheral zone'.

Ultimately, cells displaced into the peripheral zone become determined to form various structures of the growing shoot, while those that remain within the central zone are somehow maintained in an undetermined embryonic-like state. The unique behaviour of each cell population appears to be determined largely by its position within the shoot apex. Surgical ablation of initial cells within the central zone leads to their functional replacement by adjacent cells from the meristem or in the regeneration of one or more new meristems from undamaged parts of the original (Pilkington, 1929; Sussex, 1964; Wardlaw, 1949). Thus, the SAM appears not only perpetually embryonic but self-organising as well.

In higher plants, the position-dependent, polarised patterns of cell division

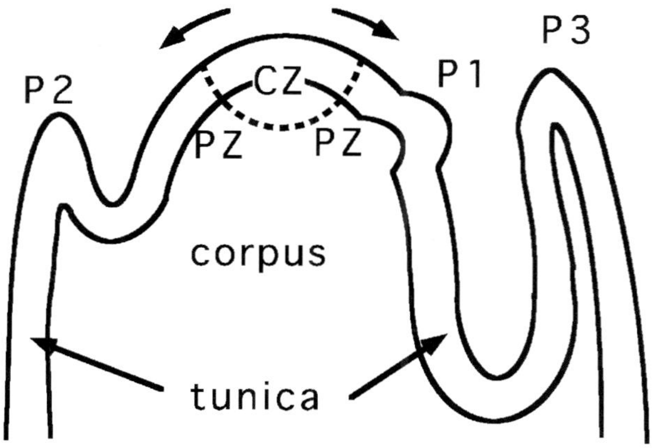

Figure 3.2 Schematic representation of a median longitudinal cross-section of a shoot apical meristem (SAM) showing positions of the central zone (CZ), peripheral zone (PZ), first and second leaf primordia (P_1 and P_2) and position of the corpus and tunica. Arrows indicate polarity of cell division.

within the apex provide not only a steady stream of undifferentiated cells, but also yield a characteristic, layered tissue organisation made up of concentric, clonally distinct cell populations. The polyclonal nature of the shoot apex is suggested by its appearance in median longitudinal sections, in which files of cells appear to emanate from a small number of apical initials. This view has received confirmation from classic studies of cytochimeras, in which various clonal subpopulations within the shoot apex were marked by an increased ploidy number that could be recognised in terms of an increased nuclear diameter (Satina *et al.*, 1940). In cases where an increase in ploidy occurs in a cell that functions as an initial for a particular layer, a stable cytochimera is formed.

Analyses of such cytochimeras suggest the existence of two or more clonally distinct layers, each originating from a separate set of apical initials. This organisation appears typical among angiosperms, as suggested by the many variegated forms in which the identity of specific layers is marked by differences in the capacity to make or accumulate pigment (Tilney-Bassett, 1986). The most superficial of these, the L1, contains cells that divide predominantly in an anticlinal plane (the new wall is perpendicular to the tissue surface), thus creating an epidermal layer one cell thick. One or more internal cell lineages (designated L2, ... etc.) may also divide anticlinally to produce another more internal layer or, alternatively, in a number of planes to form a solid tissue more than one cell diameter thick.

Differences in the general plane of division between different layers are embodied in the tunica/corpus concept (Schmidt, 1924). The tunica is composed of one or more superficial clonal layers, in which predominantly anticlinal divisions yield a tissue of uniform thickness. More internal layers divide in various planes to produce the corpus. It is unclear how the distinct behaviours of the tunica and corpus are specified, although the contribution of specific cell layers to each within a species and even within a single plant over its normal growth cycle is variable. This observation suggests that regulation relies predominantly on positional rather than lineage-dependent factors (Szymkowiak and Sussex, 1996).

A number of factors have been described that appear to influence the geometry of the SAM. Since the maintenance of the SAM can be viewed as a balance between the rate at which undifferentiated cells are produced *vs* the rate at which they are consumed by organogenesis, factors influencing either process would be expected to have an effect (Lyndon, 1998). Indeed, the fluctuation in size of the SAM during each leaf initiation cycle may play a critical role in the mechanism that regulates this periodic phenomenon. In a larger context, the SAM may undergo seasonal changes in size relating to bursts of rapid leaf initiation which may be followed by relatively quiescent periods. It is unclear how these cycles are regulated, though in some cases specific plant growth regulators have been implicated.

The genetic analysis of the SAM has progressed rapidly, especially in the model plant *Arabidopsis*, where a number of genes have been defined that influence the formation of the SAM or its maintenance. Perhaps the most basic of these is the *SHOOT MERISTEMLESS* (*STM*) gene, which appears to be essential for the formation of the SAM in the developing embryo (Long *et al.*, 1996). Loss of *STM* activity results in the generation of an embryo that forms relatively normal cotyledons, but is completely lacking cells that constitute the normal SAM (Figure 3.3). Molecular analysis reveals that the *STM* gene product belongs to the homeobox family of transcription factors and is normally highly expressed in the SAM region, suggesting that it somehow specifies the undetermined nature of these cells.

The establishment of *STM* expression in the developing embryo depends on at least two genes, *CUC1* and *CUC2* (cup-shaped cotyledon). In mutants that lack *CUC1* and *CUC2* activity, a failure of a normal SAM to form is associated with a lack of *STM* expression (Aida *et al.*, 1999). Plants which are mutant for *CUC1* and/or *CUC2* instead form abnormal cotyledons, which are fused into a cup-shaped structure (Aida *et al.*, 1999). Expression of *CUC2* appears to occur initially in the region separating the sites where the two cotyledons will eventually form, suggesting that this expression somehow prevents a cotyledonary pattern of outgrowth in these regions. A

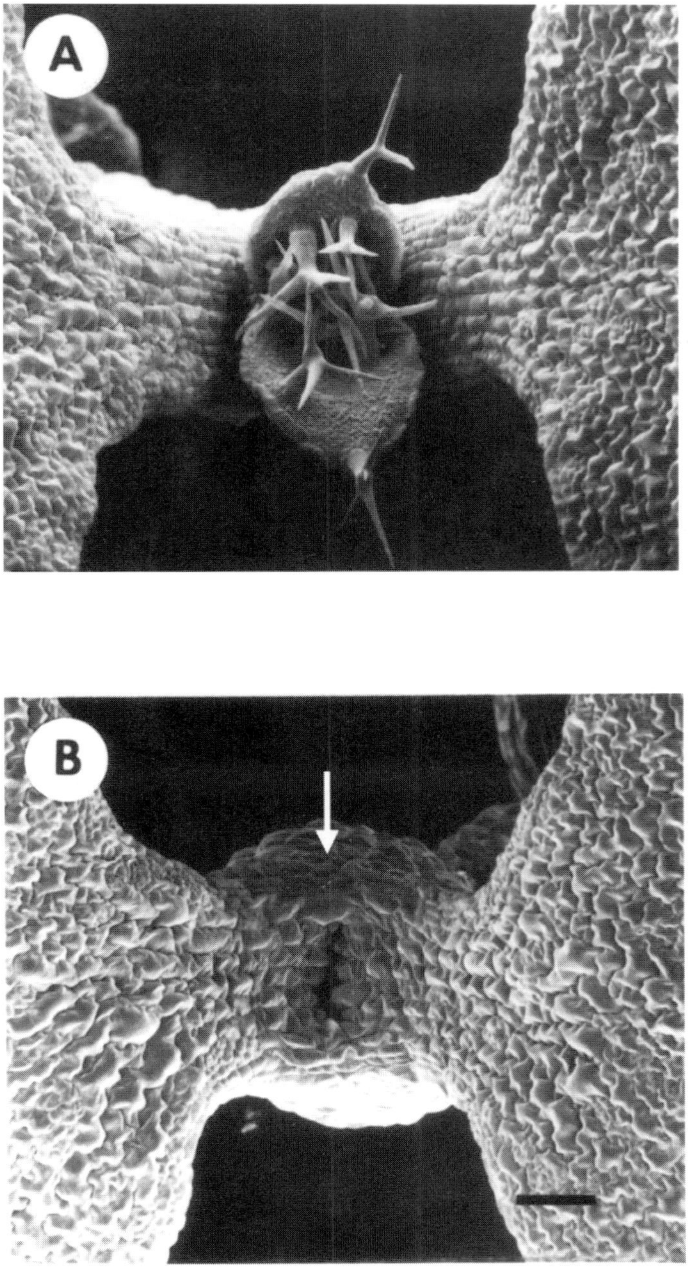

Figure 3.3 Comparison of: A) a normal vegetative shoot apex of *Arabidopsis* shortly after germination; and B) a similar staged *SHOOT MERISTEMLESS* (*stm*) plant showing the complete absence of a shoot apical meristem (SAM) between the cotyledons of the mutant (internal scale bar = 100 μm).

presumed orthologue of *CUC2*, *NO APICAL MERISTEM* (Souer *et al.,* 1996) from petunia shows analogous defects. Significantly, both of the adventitious shoots that occasionally form on these mutant plants are largely normal, suggesting that these genes function predominantly during the establishment of the SAM but are not required for its maintenance.

Three genes, *CLV1*, *CLV2* and *CLV3* (*CLAVATA*), appear to act in an interdependent way to oppose the activity of *STM,* restricting the proliferation of undetermined cells within the SAM (Clark *et al*., 1996). Loss of *CLV* expression results in dramatic increases in meristem size and concomitant changes in the pattern of organogenesis (see discussion in Section 3.3.3). The behaviour of the SAM is clearly a function of a dynamic interplay between opposing proliferation and determination processes. Several recent reviews have provided a more comprehensive discussion of these interactions and other genes that play an influential role (Evans and Barton, 1997; Clark, 1997).

3.3.2 Phyllotaxy

Phyllotaxy describes the regular spatial arrangement of leaves on a shoot (Figure 3.4). Plants in which successive leaves are positioned 180° apart from one another (e.g. maize) are said to have a 'distichous' phyllotaxy (also known as 'alternate'). Plants with a 'spiral' phyllotaxy have successive leaves

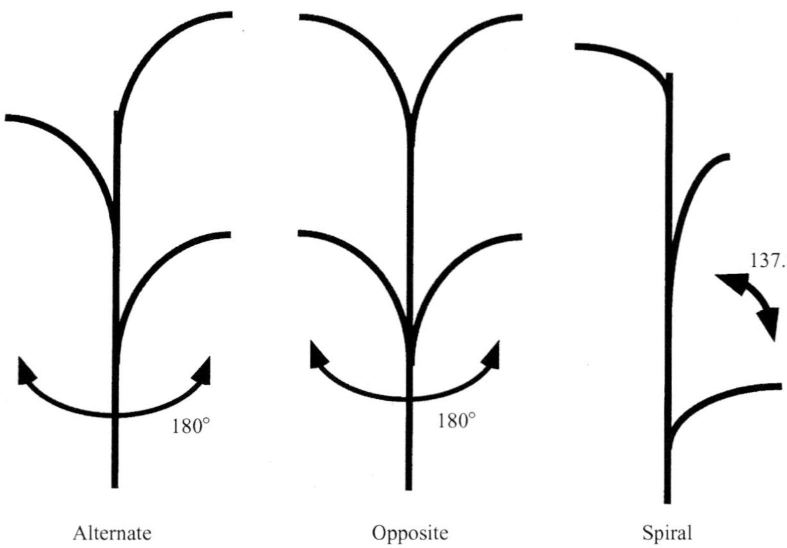

Figure 3.4 A schematic representation of three common phyllotaxies.

with a divergence angle of less than 180° (e.g. sunflower). In many cases, the angle of divergence approximates to 137.5°. A phyllotaxy in which pairs of leaves are oppositely inserted at each node is termed 'opposite' (e.g. honeysuckle); if successive pairs are inserted at right angles to each other, the phyllotaxy is termed 'decussate' (e.g. mint). Plants with more than two leaves at a single node are said to have a 'whorled phyllotaxy' (e.g. privet). While specific phyllotaxies may be characteristic of a species, some variation may occur as a function of growth conditions or the plant's age.

Regular phyllotactic patterns arise from correspondingly regular patterns of leaf initiation. For example, leaves that are positioned alternately along opposite sides of the stem are derived from primordia that form in an alternate sequence on opposite sides of the apical dome. A similar correspondence exists for other phyllotaxies. Leaf initiation is a periodic, repetitive process that begins with the formation of the first embryonic leaves and ends as the terminal meristem switches to reproductive organogenesis. For the purpose of discussing the initiation process, some terminology is useful. The term 'plastochron' describes the interval of time between successive leaf initiation events (Maksymowych and Erikson, 1960). Plastochron is also used to describe the relative age of a leaf primordium. For example, the most recent primordium initiated is described as a plastochron 1 or P_1 leaf. The next oldest leaf is termed a plastochron 2 or P_2. Leaves that have not yet emerged from the meristem are called incipient primordia and are also ordered according to their developmental progress (e.g. a P_0 is about to emerge from the meristerm and a P_{-1} is one plastochron younger than the P_0). Incipient primordia are not visible directly, but their size, position and degree of determination are inferred from experiments that will be discussed in Section 3.3.3

3.3.3 Determination of leaf initiation sites

How leaves become determined to emerge at predictable sites on the shoot apex is one of the most basic, unresolved questions in plant biology. By some means, derivatives of apical initial cells become organised into precisely positioned leaf primordia within the short interval between the apex and older, pre-existing leaves. This process can be viewed as comprising two stages. The first involves selection of a group of cells that will make up the primordium from a larger population of uncommitted cells within the shoot apex. The second involves changes in growth polarity in these cells that lead to the emergence of the primordium.

Surgical manipulations of the apical meristem have offered some insight into factors that determine the position of leaf initiation. This approach is complicated by the relative inaccessibility of the apical dome, which is typically in the order of 100 μm in diameter and which is enclosed by a series of older expanded leaves. Moreover, the contribution of untargeted wound responses to surgical manipulations is difficult to assess. Nevertheless, in a wide range of plants, including both ferns and angiosperms, radial incisions that separate predicted sites of leaf formation (e.g. P_{-2} from previously initiated primodia) lead to changes in the site of leaf emergence (Snow and Snow, 1931; Wardlaw, 1949). Typically, the leaf emerges at a site that is closer to the adjacent primordium, suggesting that the incision blocks some form of inhibition emanating from the pre-existing primordium.

The nature of this apparent lateral inhibition by older primodia is unclear. The possibility that auxin emanating from pre-existing primordia might act as a diffusible inhibitor is suggested by studies in which phyllotaxy can be perturbed by localised application of auxin transport to the apical dome (Schwabe, 1971). Although young leaves are considered a rich source of auxin, it is unclear whether inhibitors this potent growth regulator influences phyllotaxy via normal channels *vs* a more indirect path, such as altering the shape of the apex.

Alternatively, biophysical forces within the SAM could generate an ordered pattern of leaf initiation. Experiments in which the SAMs are subjected to physical stress perturb normal phyllotactic patterns, supporting this second model (reviewed by Green, 1987). In a less direct, but perhaps less disruptive approach, it has been demonstrated that small beads impregnated with the protein, expansin, a molecule that promotes the expansion of plant cell walls, can induce localised leaf formation when placed on the surface of a live SAM from tomato (Fleming *et al.*, 1997; Reinhardt *et al.*, 1998). This finding suggests that a localised change in the surface tension of the apex is sufficient to induce leaf formation, lending credence to biophysical views of meristem function.

The relationship between leaf initiation and the geometry of the shoot apex is detailed further by consideration of mutants that specifically affect both. In maize, the *abphyl* mutant (Jackson and Hake, 1999) shows a striking change from an alternate to decussate phyllotaxy, which is accompanied by a substantial increase in the size of the apical dome (Figure 3.5). Analogous changes in phyllotaxy are also seen in a number of mutants, such as *clavata,* in *Arabidopsis*, which also exhibit an increase in meristem size (Clark, 1997). Similarly, mutants that decrease meristem size may also have disturbed phyllotaxy (McConnell and Barton, 1995). Such changes are consistent with models of lateral inhibition of leaf initiation, given the altered distances over which inhibition would exert its effects.

As the site of leaf initiation becomes defined, how are the cells that make up the primordium selected? Lineage analysis in model plant systems suggests that, in angiosperms, each leaf is derived from a group of previously undetermined cells (Poethig and Sussex, 1985; Furner and Pumfrey, 1992; Irish and Sussex, 1992; Poethig and Szymkowiak, 1995; Dolan and Poethig,

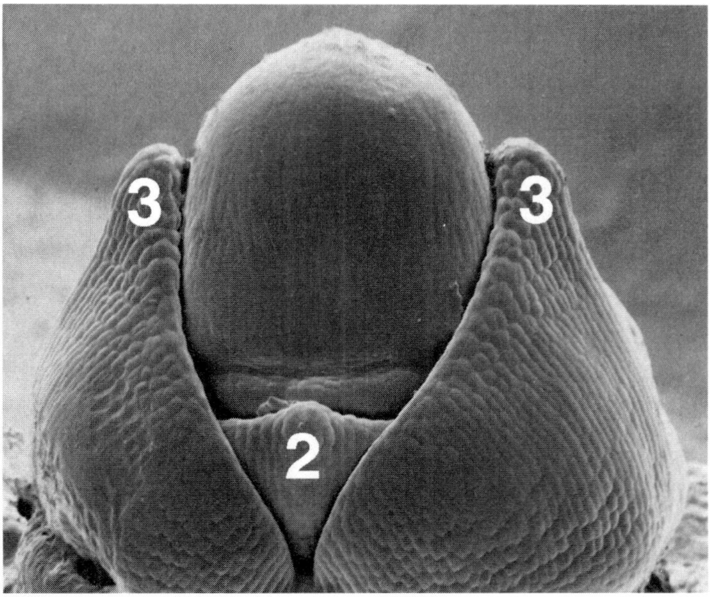

Figure 3.5 Scanning electron micrograph (SEM) view of the shoot apex of the maize *abphyl* mutant showing a decussate pattern of leaf initiation. The numbers 2 and 3 refer to successively initiated pairs of leaf primordia. (Figure kindly provided by D. Jackson and S. Hake. Printed with permission of the Company of Biologists.)

1997b). This method of analysis uses chemical or physical mutagens to induce visible mutations (e.g. albino mutations) in single cells within the shoot apex (Poethig, 1987). By assessing the size and extent of such a sector, the developmental state of the cell in which the sector was induced can be inferred. For example, a mutation in one of the apical initials would be expected to extend for a substantial length of the plant and to include many leaves. By contrast, a sector that is confined to a single leaf could be assumed to have been induced after the leaf had become determined. The fractional width of the largest sectors confined to a single leaf are inversely proportional to the number of cells in the primordium at the time of its determination. In maize, for example, the largest sectors within a single leaf are not more than 1/40 the width of the leaf, suggesting that the primordium is at most 40 cells in width at the time it becomes determined (Poethig and Szymkowiak, 1995). Similar calculations can be applied to estimate the number of cells that contribute to the other dimensions of the primordium. Estimates of the number of cells contained within the newly formed primordium are in broad agreement with cell numbers in organogenic regions of the apical dome as estimated by histological methods. The precise number of cells that contributes to the primordium varies among species, but is estimated to be typically in the order of 50–200.

3.3.4 Selection of leaf founder cell populations

Relatively little is known of the process by which this group of cells become determined, or the extent to which this recruitment is linked to the mechanisms that determine the site of leaf initiation. In maize, the earliest overt sign of leaf development is a change in the orientation of cell divisions, at a discrete point which predicts the position of the future mid-rib (Sharman, 1942). Subsequent changes in division polarity spread from this point outward to yield a leaf that is derived from superficial layers of the entire shoot circumference. Whether the determination of the cells that contribute to the leaf primordium proceeds in the same progressive fashion is unclear: *narrow sheath* mutants of maize, in which the lateral regions of the leaf appear to be missing, suggest that these domains may be secondarily determined subsequent to determination of the central mid-rib containing domain (Scanlon *et al.*, 1996; Scanlon and Freeling, 1997).

A combination of molecular and genetic analyses has provided clues to events that precede overt leaf development during which cells acquire a leaf identity. Much of the work in this area began with the analysis of a curious dominant mutant of maize, *Knotted 1* (*Kn1*), in which leaves show sporadic foci of extra growth leading to the formation of knots of tissue (Freeling and Hake, 1985) (Figure 3.6). Molecular cloning of *Kn1* by transposon tagging revealed that the gene belongs to a class known as homeobox genes (Vollbrecht *et al.*, 1991). Genes belonging to this class were first described from *Drosophila*. They were shown by genetic methods to play an important role in development by determining cell fate, and appear to function as transcription factors (Scott *et al.*, 1989; Gehring, 1987; Lawrence and Morata, 1994).

Molecular studies have revealed that, in addition to the monstrous growth associated with the dominant *Kn1* mutant, the gene appears to play an important role in the context of normal leaf development. Studies focused on where *kn1* is expressed in normal plants have revealed high levels of both the messenger ribonucleic acid (mRNA) and protein throughout the SAM, except in those cells whose position and number would suggest they comprise an incipient leaf primordium (Figure 3.7). Thus, downregulation of *kn1* appears to provide one of the earliest markers of leaf cell identity and may be an early requirement for their determination (Smith *et al.*, 1992; Jackson *et al.*, 1994). The previously discussed *SHOOT MERISTEMLESS* (*STM*), a *kn1*-like gene from *Arabidopsis*, shows a similar pattern of downregulation in the incipient leaf. Like *STM*, *kn1* is predominantly expressed in the SAM from the time of its inception during embryogenesis (Smith *et al.*, 1995; Long *et al.*, 1996).

Figure 3.6 Close-up view of the basal adaxial surface of the leaf blade of a *Knotted 1* (*Kn1*) mutant (photograph by B. Veit).

A number of genes have been described that appear capable of influencing the expression of *kn1* and *STM*. On one level are genes whose effect on the downregulation of *Kn1*-like genes appears to reflect their influence on phyllotaxy. Mutations in the *Abphyl* gene, which lead to a decussate pattern of leaf initiation, are associated with paired rather than the normal, singly occurring, downregulated domains (Jackson and Hake, 1999). Analogous changes in *STM* expression are associated with the altered phyllotaxies in *clavata* mutants with enlarged SAMs. In the *terminal ear1* mutant of maize, in which more frequent leaf initiation events give the shoot terminus an ear-like appearance, the pattern of *KN1* regulation is shifted higher on the apical dome, reflecting a shift in the site of leaf initiation (Veit *et al.*, 1998).

Despite the clear correlation between *kn1* downregulation and leaf founder cell identity, the precise role of *Kn1*-like genes in this process remains unclear. In several contexts, however, their activity can be associated with the establishment or maintenance of undetermined, meristematic populations of cells. The ectopic expression of *Kn1* in the leaves of the dominant maize mutant is associated with the proliferation of knot-like structures, from which

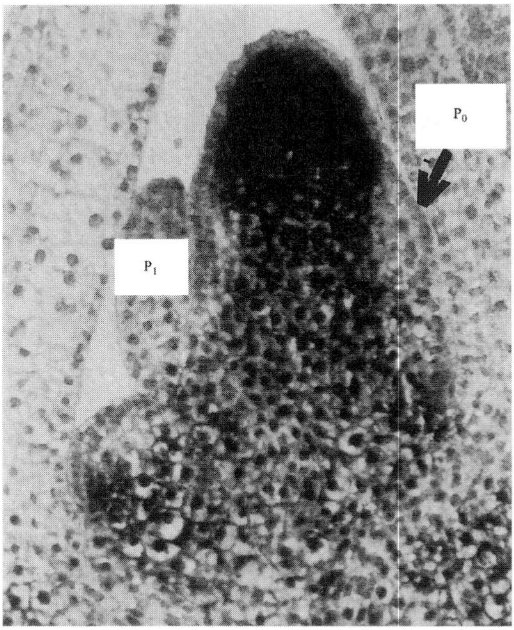

Figure 3.7 *In situ* hybridisation to show distribution of the *Knotted 1* (*KN1*) transcript (dark staining) in a median longitudinal cross-section of a normal vegetative shoot apex of maize. Note the lack of staining in regions corresponding to leaf primordia (P_0 and P_1). (Reprinted from the journal *Development*, with permission of the Company of Biologists, Ltd.)

the mutant takes its name (Smith *et al.*, 1992). Similarly, when *Kn1* is put under the control of a constitutive promoter and introduced into tobacco, *Arabidopsis*, rice or barley, its ectopic expression promotes indeterminate patterns of growth on the normally determinate leaf lamina (Sinha *et al.*, 1993; Chuck *et al.*, 1996; Matsuoka *et al.*, 1993; Williams *et al.*, 1997). In tomato, expression of *Kn1*-like genes is associated with indeterminate patterns of leaf growth (discussed in more detail in Section 3.6).

Conversely, loss of function mutations to *Kn1*-like genes limit meristematic-like growth. As previously discussed, mutations to the STM gene in *Arabidopsis* block formation of the SAM in the embryo (Long *et al.*, 1996). Analogous mutations to *Kn1* in maize perturb SAM function but do not block its formation, perhaps because of partially redundant gene activities in maize (Kerstetter *et al.*, 1997). Viewed from this perspective, the primary function of *Kn1*-like genes may be to provide a population of undetermined cells from which leaf primordia can be organised. In this context, downregulation of *Kn1* may only be a prerequisite that permits the more determinant programme of leaf development to proceed.

3.4 Establishment of pattern within the leaf

The fixation of the site at which a leaf initiates and the recruitment of cells from which it will develop is clearly only the first step in a process that will ultimately yield an organ having a complex architecture. By P_5, most non-unifacial leaves are asymmetric about the transverse (abaxial/adaxial), lateral (mid-rib/margin) and proximal/distal (tip/base) axes, although in some species, such as grasses this asymmetry is apparent much earlier (Kaplan, 1973; Steeves and Sussex, 1989; Poethig, 1997). By maturity, each region of the leaf has developed a unique pattern of venation, epidermal cell characteristics and histological organisation.

The temporal sequence with which specific facets of this architecture appear can only serve as a rough guide to understanding when these features actually become determined. For this purpose, surgical studies have offered insights into not only the initiation process but also subsequent events, including determination of dorsoventrality and the establishment of determinancy. Leaf primordia up to P_8 from ferns will develop as indeterminate structures when isolated and cultured on agar (Steeves, 1966). Similarly, horizontal cuts that isolate young leaf primordia from the shoot apex may induce them to develop as radially symmetrical or bud-like structures, which have an indeterminant pattern of growth (Wardlaw, 1949; Sussex, 1955). These experiments indicate an ongoing requirement for positional information to guide normal leaf development after the site of leaf initiation has been determined.

3.4.1 Surgical approaches to pattern formation

Surgical experiments have also suggested that the various parts of a complex leaf are determined gradually. Experiments in which leaf primordia of pea were subjected to ablation treatments at various ages indicated that the regenerative capacity of the primordium decreases with age. Over a period of approximately 1 plastochron from the time of its initiation, the primordium changes from being fully competent to regenerate excised segments to being totally incompetent, suggesting that determination of various parts occurs in this interval (Sachs, 1969).

Experiments of this type emphasise the rapidity with which basic developmental features of the leaf become determined, as well as the need for alternative approaches to this problem. Genetic analysis of the process of leaf development has provided the clearest means to break down the determination process into a readily described form. Through the careful description of mutants in which leaf development is perturbed, and through studies of how such mutants interact when combined, those aspects of development that are causally related can sometimes be inferred. Additional insight has also been afforded by the molecular cloning of the corresponding genes, providing valuable information on the mechanism of gene function. The following section describes a number of genes that influence leaf shape. No simple mechanism for the control of leaf shape emerges; instead, leaf shape appears susceptible to control by a variety of different mechanisms acting at different points during development.

3.4.2 Analysis of proximal/distal patterning in maize

Maize has provided an especially rich source of insight into the regional patterning of the leaf, especially along the proximal/distal axis, which like most grasses is divided into four morphologically distinct domains (Freeling, 1992; Freeling and Lane, 1993). These include: a proximal sheath region that wraps around the stem of the plant; a distal blade that projects away; and, between them, the auricle and ligule. These domains first become evident with the formation of the flap-like ligule, which initially appears as a ridge on the adaxial epidermis by plastochron 4–6, and spreads from the mid-rib toward the margins (Sharman, 1942; Sylvester *et al.*, 1990). Auricles develop directly distal to the ligule as triangular-shaped wedges that act as a hinge between blade and sheath.

Two loss of function mutants, *liguleless1* and *liguleless2*, specifically block the formation of the ligule and associated auricle (Sylvester *et al.*, 1990; Harper and Freeling, 1996). Analysis of genetic mosaic plants, which contain clonal sectors in which the functional gene has been deleted, indicate that the normal *LG1* gene product promotes the formation of ligule only in those cells in which it is present, while at the same time acting to propagate a 'make ligule' signal, which originates in the central domain of the leaf, towards the leaf margin (Becraft *et al.,* 1990; Becraft and Freeling, 1991). By contrast, the *LG2* gene product appears to act non-autonomously (Harper and Freeling, 1996). Molecular cloning of *LG1* and *LG2* provides some insight into the behaviour of these genes, revealing that *LG1* (Moreno *et al.*, 1997) and *LG2* (Walsh *et al.*, 1998) are transcription factor-like proteins.

Other mutations disturb the normal proximal/distal patterning of the maize leaf. The dominant morphological mutants, *Knotted1* (*Kn1*), *Rough Sheath1* (*Rs1*), *Gnarley1* (*Gn1*), *Liguleless3* (*Lg3*) and *Liguleless4* (*Lg4*), all show patches of ligule and sheath-like tissue in more distal regions of the leaf (Freeling and Hake, 1985; Sinha and Hake, 1994; Becraft and Freeling, 1994; Schneeberger *et al.*, 1995a,b; Fowler and Freeling, 1996; Foster *et al.*, 1999). Interestingly, all these mutants encode proteins belonging to the *kn1* homeobox gene family (Kerstetter *et al.*, 1994). *Knotted1* (*Kn1*), the first member of this family to be analysed (Gelinas *et al.*, 1969; Freeling and Hake, 1985), is characterised by sheath-like cells in the blade portion of the leaf (Sinha and Hake, 1994). As described previously, *kn1* is normally expressed in the meristem, but is excluded from leaf primordia from a point near the time of their inception (Vollbrecht *et al.*, 1991; Smith *et al.*, 1992). It may be concluded that the expression of genes belonging to the *Kn1* class within the primordium somehow interferes with the normal patterning process.

3.4.3 Adaxial/abaxial and lateral patterning of the leaf

There are several mutants that fail to establish asymmetry about the abaxial/adaxial axis, in which a broad lamina is abnormally narrow or even needle-shaped. The loss of function *phantastica* (*phan*) mutant of *Antirrhinum* initiates leaves normally; however, during subsequent development, the leaf develops as a radially symmetrical structure with only abaxial cell characteristics (Waites and Hudson, 1995) (Figure 3.8). Analysis of the mutant has suggested that *PHAN* may also play a role in establishing lateral domains of the leaf. In some instances, small patches of adaxial tissue develop in the mutant and, at the junction between the adaxial and abaxial tissues, a flap of tissue grows out along a new axis. These observations suggest that the juxtaposition of adaxial and abaxial surfaces may provide necessary cues for the subsequent determination and development of lateral domains. Recent cloning of *PHAN* reveals a localised pattern of expression in the developing leaf that is consistent with this model and shows that it belongs to the *MYB* class of transcription factors (Waites *et al.,* 1998).

Several mutants appear to have a specific affect on the development of either the lateral or abaxial domains of the leaf. The *lamina1* mutant of tobacco correctly establishes adaxial and abaxial surfaces but fails to initiate growth in the lateral dimension (McHale, 1993; McHale and Marcotrigiano, 1998). This defect is apparent by P_3 and indicates that certain marginal

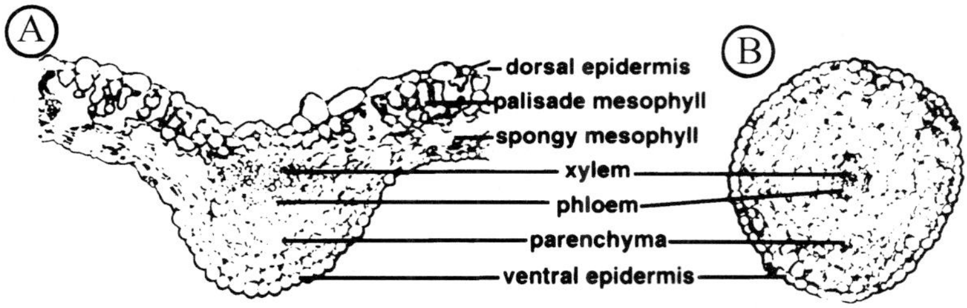

Figure 3.8 Comparison of transverse sections through the blade region of: A) a normal *vs* B) a *phan* mutant showing loss of dorsoventrality in the latter. (Reprinted from the journal *Development*, with permission of the Company of Biologists, Ltd.)

lamina founder cells fail to divide correctly. By contrast, the *fat* mutation of tobacco shows abnormal periclinal divisions of the mesophyl of the leaf during blade initiation and expansion, leading to a blade of approximately twice the normal thickness. The *narrow sheath* mutant of maize appears to be missing the lateral domains of the leaf, suggesting that the gene is normally required for the formation of these domains (Scanlon *et al.*, 1996; Scanlon and Freeling, 1997).

Recent analyses of leaf pattern defect mutants in maize suggest further intriguing links between the patterning of different axes. The *rough sheath 2* (*rs2*) mutant shows a variety of leaf defects in which certain features, such as the ligule and blade/sheath boundary, are shifted along the proximal/distal axis. These defects strongly resemble those associated with previously discussed mutations that cause ectopic expression of *kn1*-like genes. Not surprisingly, analysis of *rs2* mutants has revealed the ectopic expression of some of these same genes in developing leaf primordia, suggesting a likely cause for the aberrant development as well as the normal function of *rs2*, to suppress ectopic expression. Cloning of the *RS2* reveals that it encodes a protein highly similar to that encoded by the *PHAN* gene of *Antirrhinum* (Timmermans *et al.*, 1999; Tsiantis *et al.*, 1999).

With expression studies showing similar mRNA accumulation localised to developing leaves for *RS2* and *PHAN*, it has been proposed that these genes are homologous. This suggestion is supported by analysis of the maize mutant *leafbladeless1* (*lbl1*) mutant, which produces radially symmetrical, abaxialised leaves but which, like *rs2*, also shows ectopic expression of *kn1*-like genes (Timmermans *et al.*, 1998). Further support is offered by the observation that, in *phan* mutants, the *Antirrhinum* homologue of *KN1* is ectopically expressed in developing leaves.

The molecular mechanism by which *Kn1*-like genes are suppressed by *RS2* and *PHAN* is unclear, though the patchy nature of ectopic *KN1* expression seen in *rs2* mutants suggests that some form of clonally propagated epigenetic control may be relevant (Timmermans *et al.*, 1998). Given the ability of *Kn1*-like genes to disturb axial patterning when ectopically expressed, it is not unreasonable to suggest that they normally function to establish gradients by which basic developmental axes are defined, much as the homeobox gene, *bicoid*, acts to establish the anterior/posterior axis during *Drosophila* embryogenesis (Struhl *et al.*, 1989).

Although it would seem that downregulation of *Kn1*-like genes within the developing leaf is essential for its normal patterning, it is unclear why *phan* displays a more pronounced abaxial/adaxial pattern defect, while in *rs2* only a proximal/distal patterning defect is apparent. One explanation might be found in the fundamentally different patterns of leaf intiation between maize and *Antirrhinum*, the former involving a thin layer of cells that emerge from around the circumference of the shoot, while in the latter lateral domains form later. Alternatively, *RS2* activity may be partially redundant, as appears the case for many other maize genes. In this case, the more complete abaxialisation by the *lbl1* mutant of maize may represent a gene that acts downstream of both *RS2* and its hypothetical twin.

3.5 Leaf morphogenesis

While the observable patterns of growth during the initial phases of leaf development appear broadly conserved, subsequent growth varies significantly to yield a wide variety of leaf forms. Simple leaves display a variety of geometries that can be explained in terms of the differential growth along the proximal/distal *vs* transverse axes. Allometric analyses of a number of species indicates that this differential growth may occur at a number of points during the development of the leaf, ranging from its inception to well after its emergence (reviewed by Poethig, 1997).

Growth of the leaf primordium to produce a flattened lamina implies that the polarity of cell division is highly restricted along the lateral axis to maintain a leaf of uniform thickness. The launching point for this lateral growth has been termed the 'blastozone' (Hagemann and Gleissberg, 1966), to recognise groups of cells along the lateral margins of the primordia that have in some way been determined to proliferate laterally. In light of the previous discussion of adaxial/abaxial polarity mutants and their associated lateral defects, it might be supposed that the formation of the blastozone is a direct consequence of interactions between the abaxial and adaxial domains of the leaf. However, the once accepted view that lateral domains become established through continued divisions of cells on the margins of the leaf (e.g. Avery, 1933) appears to be largely incorrect. Lineage analysis conclusively shows that lamina growth involves intercalary cell divisions across the entire leaf width, not just in marginal regions (for discussion, see Poethig, 1987).

Cell division and expansion would also appear to be regulated along the transverse and proximal/distal axes to yield characteristic shapes. The controls that operate to maintain the patterns of these divisions are not well understood. However, clonal analyses of leaf development suggest that quite variable patterns of cell division can give rise to equivalent tissues, arguing that patterns of cell division respond to positional cues that operate on a local level to accomodate this variability (Poethig, 1989; Meyerowitz, 1996, 1997). This theme is reinforced by the relatively normal patterns of growth observed for mutations that alter the frequency (Hemerly *et al.*, 1995) or polarity (Smith *et al.*, 1996) of cell division (Figure 3.9).

Some insight into how various simple leaf shape ratios might arise is afforded by consideration of mutants that alter specific aspects of leaf shape.

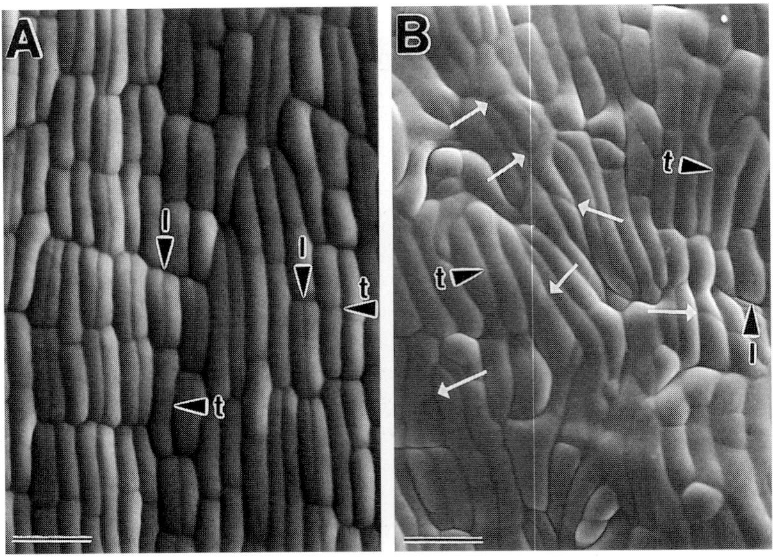

Figure 3.9 Scanning electron micrograph (SEM) photographs of the epidermis of: A) a normal *vs* B) a tangled mutant showing loss of regular, polarised cell divisions in the mutant (internal scale bar = 30 μm). Key: t, transverse division; l, longitudinal division. (Reprinted from *Development*, with permission of the Company of Biologists, Ltd.)

Two mutants of *Arabidopsis* appear capable of regulating growth along different axes independently of each other (Tsuge *et al.*, 1996). For example, loss of activity in the *ROTUNDIFOLIA3* (*ROT3*) gene, which encodes a cytochrome P-450 protein, leads specifically to decreased cell elongation along the longitudinal axis of the leaf. By contrast, loss of *ANGUSTIFOLIA* gene activity decreases expansion of cells across the width of the leaf. Additional genes appear to effect more complex patterns of growth, which yield undulating margins. Unequal growth patterns of relatively short periods across the axis of the lamina are seen in plants mutant for the *SERRATE* gene of *Arabidopsis* (Redei and Hirano, 1964). In cotton, the *Okra* mutation leads to an exaggerated pattern of the longer period undulation associated with palmate lobing (Dolan and Poethig, 1997a).

Although, in most cases, the products encoded by genes affecting leaf shape are unknown, the study of periclinal chimeras in which the contribution of gene activity from specific layers can be assessed has permitted some insight into where they may function. Such analyses have indicated that information expressed in the internal layers of the leaf plays a dominant role in specifying particular leaf shapes (Szymkowiak and Sussex, 1996).

3.6 Compound leaves

In addition to the simple forms described above, leaves may develop into complex structures in which the lamina is divided into two or more separate leaflets or pinnae (Figure 3.10). In palmately compound forms, the bases of leaflets join together at a single point at the end of the petiole. In pinnately compound forms, leaflets join the rachis individually along its length, in two files. For both forms, leaflets develop in a manner very similar to that described for a simple leaf emerging from the SAM. However, whereas leaves may be borne on a stem in a number of planes (e.g. spiral or decussate phyllotaxies), leaflets generally emerge in a single plane. This planar arrangement may be obscured in some species, where differential growth of the leaflet or its petiole yield a leaf in which leaflets project away from the rachis in a number of planes.

Certain compound leaves, such as those found in the legume family, are modified and leaflets appear to be replaced by tendril-like structures. These apparent transformations may be seen in morphological series that compare leaves initiated at different times during the plant's development (hetero-

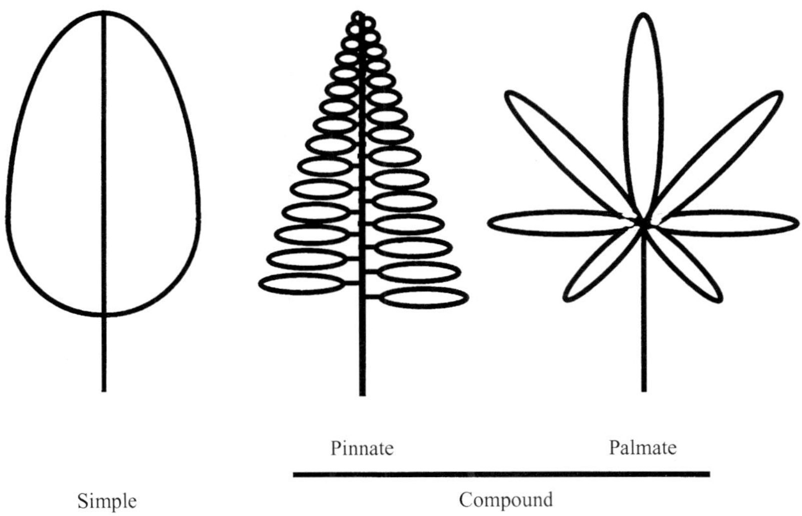

Figure 3.10 Schematic representation of simple *vs* compound leaves.

blasty, see discussion below) or different forms produced by related species. The capacity of leaves to undergo alternative patterns of development is made even more apparent by 'A Suite of Mutants that Modify Pattern Formation in Pea Leaves' (from the title by Marx, 1987), in which specific aspects of the leaf's compound organisation are perturbed. Normally, the pea plant initiates leaves which each bear from one to several pairs of oppositely arranged leaflets, as well as pairs of more terminal tendrils. In the mutant, *unifoliata* (*uni*), plants produce a single leaf blade instead of the normal combination of leaflets and tendrils, suggesting that the non-mutant form of the gene normally promotes the formation of these structures. In *tendrilless* plants, leaflets are produced in terminal positions where tendrils are normally found. In *afila* plants, the opposite transformation is shown, in which tendrils have replaced more basal leaflets.

Insights into the molecular mechanisms by which these genes might act have begun to emerge through cloning efforts. Recently, *uni* has been cloned (Hofer *et al.,* 1997) and shown to be similar to *LEAFY* of *Arabidopsis* and *FLORICAULA* (*FLO*) of *Antirrhinum*, two genes whose expression is thought to promote the transition from vegetative to floral development. This transition is thought to somehow be affected by expression of *LEAFY* or *FLO* in the SAM, leading to the activation of genes which specify a floral pattern of lateral organ development. While it is unclear how such a gene might act in pea to promote the formation of individual leaflets and tendrils, the recent characterisation of *PHANTASTICA* genes, whose expression is required for development of dorsoventrality and lateral domains of the leaf, provides some intriguing possibilities. One plausible model would suppose that the *UNI* gene product normally partially restricts the activity of *PHAN* to restricted points of the rachis, thus leading to the formation of individual leaflets rather than a single unbroken lamina. A complete repression of a *PHAN*-like gene in more terminal domains of the leaf could completely suppress dorsoventrality and concomitant lamina development, and thus provide for the formation of tendril-like structures.

The leaves of tomato afford another intriguing suite of mutants in which the compound nature of the leaf is modified. The *Lanceolate* mutation, when present as a heterozygote, leads to the production of unramified leaves that are morphologically similar to the simple leaves of *Arabidopsis*. When homozygous, plants are unable to organise a shoot meristem, though roots develop normally. By contrast, the dominant mutants, *Curl* and *Mouse-ears*, lead to an increased ramification of the normally compound leaf, with the former also forming ectopic shoots. Subsequent analyses indicate that these mutants are likely to represent altered forms of a *Kn1*-class homeobox gene, *Tkn2* (Chen *et al.*, 1997; Parnis *et al.*, 1997). Normally, the gene is expressed in the SAM, in the provascular and vascular tissue, as well as the growing points of leaf primordia. In its mutant forms, the gene is abnormally transcribed and distributed to ectopic positions in the developing leaves. These transcripts appear to promote abnormal growth of the lamina or branching at positions where they accumulate within the developing leaf.

The apparent association between expression of *Kn1*-class genes and indeterminate patterns of growth within the leaf is reinforced by the results of transgenic studies. When the *Tkn2* gene is put under the control of a constitutive promoter, increased ramificaton of the leaf results. This outcome is also observed when the maize *Kn1* gene under the control of the same promoter is introduced into tomato (Hareven *et al*., 1996) (Figure 3.11). A gene inferred to be the tomato homologue of *Kn1*, *Tkn1*, also appears to be associated with indeterminate patterns of growth, showing a similar distribution to the *Tkn2* and causing even more extreme increases in ramification when overexpressed.

Interestingly, the expression of the tomato *Kn1*-like genes in the SAM appears to differ from the basic pattern observed in maize, *Arabidopsis* and *Antirrhinum* by not showing the typical pattern of downregulation in leaf primordia at the time of their inception. Because leaves are still able to form in tomato despite this expression, it would seem that the role of *Kn1*-like genes in tomato differs from that in other model plants, perhaps in some way reflecting the more indeterminate pattern of tomato leaf development. These results suggest that, at least in some plants, an absolute downregulation is not a precondition of founder cell determination. A similar conclusion can be

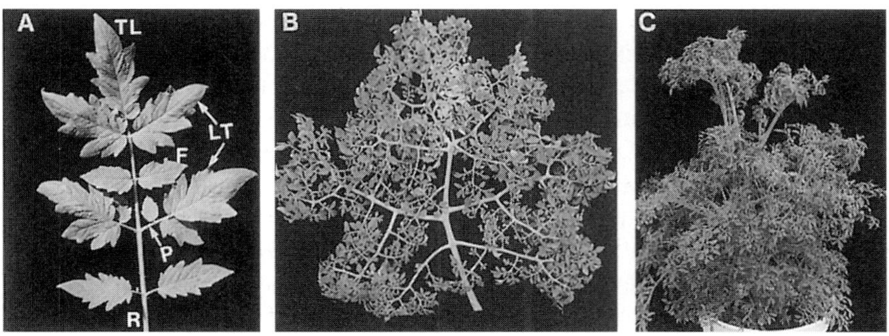

Figure 3.11 Ultracompound leaves in *kn1*-expressing tomato plants. A) Wild-type compound leaf: the prototype unit. The terminal leaflet (TL) emerges first and the pairs of lateral leaflets (LT) appear in basipetal order. Note that all leaflets are anatomically similar. Each is petiolated (P), and with serrate margins. Folioles (F) appear occasionally between leaflets along the rachis (R) or on either side of the petioles. B) A supercompound leaf of transgenic tomato plants expressing the *kn1* gene. Terminal and lateral leaflets of the prototype unit now acquire the compoundness of the wild-type leaf. C) Transgenic tomato plant expressing *kn1* under the control of the *35S* promoter. Note the bush appearance of the plant, the consequence of the lost apical dominance.

drawn from the behaviour of transgenic maize and *Arabidopsis* overexpressing *Kn1*, which in theory should be present in leaf founder cells in which *Kn1* is normally downregulated. In such plants, leaves are still initiated relatively normally, though their subsequent development is less determinate. It is possible that some downregulation of these genes still occurs in the leaf anlagen, but goes undetected by the semi-quantitative localisation methods. Alternatively, other factors may play a more important role in promoting the determination of the leaf, circumventing the need for *Kn1* downregulation, or perhaps *Kn1* activity is suppressed at a post-transcriptional level. A precedent for the latter possibility is presented by the ability of the *La* mutant to completely block the indeterminate leaf growth in tomato that would normally result from overexpression of the maize *Kn1* gene.

The two contrasting examples of compound leaf formation offered by pea and tomato, involving genes whose function was first known in the context of the SAM, raise intriguing questions concerning the evolution of plant form, including the relationship of mechanisms underlying the development of the stem and leaf (see Chapter 1). It is unclear at this point whether the involvement of *LEAFY* and *Kn1*-like genes in both contexts signifies a recruitment of versatile regulatory factors *vs* an historic association that reflects a common origin for the leaf and stem.

3.7 Vascular patterning

Very soon after its emergence from the apical dome, the leaf becomes vascularised in complex ordered patterns which are characteristic of the species (reviewed by Nelson and Dengler, 1997). Unlike the radially organised stem, the vasculature of leaves is characteristically bifacial, with xylem elements oriented to the adaxial surface while phloem is oriented towards the abaxial surface. Very little is known of how these patterns of venation become determined, though there seems to be a clear linkage with the overall shape of the leaf, since mutants which perturb leaf shape usually affect the vascular pattern as well. While changes in shape typically become apparent before associated venation, it is unclear whether the former determines the latter, or vice versa. It is possible that cells become committed to a vascular fate well before they visibly express their identity, and, during this period, they might conceivably exert an effect on the shape of the leaf. This same uncertainty applies to the issue of whether vascular organisation plays a role in establishing the site of leaf initiation. The first signs of primordia emergence precede any overt vascularisation and, in monocotyledons, once vascular tissue does begin to appear, it is discontinuous with the stem vascular system (Sharman, 1942; Esau, 1943; Russell and Evert, 1985).

Again, this apparent lag may not reflect an earlier established functional organisation in which vascular precursors exert an effect on initiation sites.

The earliest markers of vascular development have traditionally included an increased affinity to certain stains (Esau, 1965). At essentially the same time, one begins to observe a bias towards longitudinally oriented divisions that yield the long, thin cell types so characteristic of vascular tissue. These markers can be used to trace early patterns of leaf vascularisation, which typically begin with the acropetal development of the midvein. Subsequent development of lateral and higher order veins may occur in an acropetal or basipetal direction, depending on the particular species and vein order.

A number of models have been invoked that would explain how the pattern of venation might become fixed in the developing leaf, including the canalisation of signal flow hypothesis (Sachs, 1991) and variants of Turing's diffusion–reaction model (reviewed by Nelson and Dengler, 1997). The former model nominates auxin as inducing the formation of vasculature, while at the same time promoting the formation of channels that provide for its unidirectional polar transport. This model is supported by many studies, in which artificially induced auxin gradients promote the formation of vascular tissue. While this model is sufficient to explain the formation of much of the vascular network of the leaf, it falls short in providing an obvious explanation for joining the distal ends of higher order veins into a reticulate network. The alternative provided by the diffusion–reaction hypothesis is perhaps more robust in explaining this later development, but suffers for want of mechanistic detail (for a more complete review of this topic, see Nelson and Dengler, 1997).

Auxin, or possibly some other plant growth regulator, which could be invoked in signal-flow canalisation models, could also be proposed to play a central role in diffusion–reaction based models. The clearest resolution to this issue will no doubt emerge from knowledge of other components that participate in the patterning process. An analysis of several mutationally defined genes from *Arabidopsis* and tobacco, including several affecting auxin transport, may reveal the identity of such components and the mechanism by which they act.

3.8 Cellular differentiation

The previously discussed patterning and morphogenetic processes set the scene for the final stage of leaf development, when cellular identities in the various tissues of the leaf become fixed. The classical view of leaf anatomy presents just three basic tissues, vascular, epidermal and ground, which are defined both in spatial and functional terms. Studies of periclinal chimeras, which allow the clonally distinct cell lineages that make up a leaf to be traced (see previous discussion), suggest that cells remain relatively uncommitted to specific fates until relatively late in leaf development. A clear indication of this is seen where occasional periclinal cell divisions lead to the displacement of a cell into an adjacent clonally defined layer. Even when this event occurs relatively late in development, as indicated by the small size of the resulting sector, cells within the sector invariably differentiate in a manner that is consistent with their new position (Poethig, 1989).

3.8.1 Vascular differentiation

The sequence of vascular tissue development has been discussed previously as have the factors that may guide its patterning. It is unclear whether these genes may also play a role in the differentiation process itself. Given the unusual shape of vascular cells, which is manifest early in their differentiation, one may suppose that genes controlling division polarity are likely to play an important role. Certain genes that are active during early stages of tracheary development in cultured *Xinnia* cell systems, such a *TED*2 and *TED4* (Tracheary Element Differentiation), may also play key roles in the early stages of the differentiation process (Fukuda, 1997).

3.8.2 Epidermal differentiation

At maturity, the epidermis is distinguished by three or more classes of cell types that include stomata, trichomes and the pavement cells that separate them. Trichomes are hair-like structures that project away from the surface of the epidermis. They may function to modify airflow or surface properties of the leaf, or in some cases afford protection from certain classes of invertebrate herbivores. A diverse range of trichome types exist, ranging from the simple, single-celled trichomes of *Arabidopsis* to the complex multicellular forms. A large collection of genes involved in the actual differentiation of trichomes has been defined through mutant analysis (reviewed by Hülskamp *et al.*, 1999). Phenotypic and double mutant analysis has allowed the genetic pathway by which trichomes become determined and differentiated to be established.

For most plants, the distribution of trichomes differs between the adaxial and abaxial surfaces and may also be associated with specific anatomical features of the leaf (e.g. veins, margins, petioles). It is assumed that some set of regulatory factors promotes or represses the development of trichomes as a function of position within the developing leaf. A second aspect of trichome distribution is seen in their relationship to each other and their tendency not to cluster (Larkin *et al.*, 1996). Studies of clonal sector boundaries associated with trichome cells show no preferred orientation, arguing against mechanisms where a programmed set of divisions of a determined precursor cell effects separation.

A study of mutants which result in trichome clustering has suggested that a relatively uniform distribution of trichomes depends on a dual mechanism that mediates: 1) mutual inhibition between cells that would otherwise be competent to form trichomes; and, 2) lateral inhibition between potential trichome precursor cells. Several genes that appear to operate at both stages have been defined genetically, including *GLABRA1* and *TRANSPARENT TESTA GLABRA*, both of which are required for normal trichome formation (Larkin *et al.,* 1999). The products of these genes, together with those of *TRYPTYCHON*, are thought to regulate determination of trichome precursors by a mechanism that is sensitive to the stoichiometry of the *GL1* and *TTG* gene products. Presumably, this mechanism affords a means to establish and maintain lateral inhibition domains.

Stomata, like trichomes, show characteristic distribution patterns on the abaxial *vs* adaxial surfaces of the leaf. Similarly, within a given domain of the leaf, their distribution appears non-random (Larkin *et al.,* 1997), an adaptive feature which is thought to increase the efficiency of gas exchange. Two explanations have been offered to explain this non-randomness. Firstly, stomata tend not to form over veins, and this may be the consequence of a mechanism that ensures a functional positioning of stomata over intervening air chambers. The nature of this placement mechanism is unclear, though a lateral inhibition of stomatal development by vascular tissue is one obvious model. The second factor appears to be related to the programmed manner in which the cells that make up the stomatal array differentiate. Each stoma is derived from a meristemoid, a cell that undergoes a series of unequal divisions relatively late in the epidermal differentiation process to produce the stomatal complex. This ordered series not only provides the cells that will make up the functional stomatal complex but also ensures that each complex is surrounded by a ring of cells. The genetic analysis is still in its infancy, with only a handful of mutants described that specifically affect their development. Among these are those that affect the density of stomata, which possibly perturb the determination of meristemoid precursors, as well as mutants which appear to affect the idiosyncratic sequence of unequal cell divisions.

3.8.3 Differentiation of ground tissues

In most plants, the differentiation of ground tissues reflects their predominantly photosynthetic function. Analysis of periclinal chimeras of many species indicate that these tissues derive primarily from the L2 and L3 layers, though in many plants the L1 may divide periclinally to contribute to internal tissues in marginal regions of the leaf. As discussed in a recent review by Van Lijsebettens and Clark (1998), cellular differentiation within ground tissues is thought to be coordinated but not completely dependent on chloroplast development. Similarly, numerous nuclear and chloroplast encoded genes have been described in various species, including maize and *Arabidopsis*, that are required for normal chloroplast development but have relatively minor effects on other aspects of ground tissue differentiation. On the other hand, a number of mutants of genes in the so-called photomorphogenic pathway have been described that affect both types of development. These include three classes of genes that mediate photoreception, signal transduction and, finally, specific photomorphogenic responses. Among these reponses is an elongation of leaf petioles that appears to respond to cues provided via photoperception by phytochrome B. Mutations in the *PALE CRESS* gene also appear to affect both chloroplast development and cellular differentiation, not only blocking the differentiation of chloroplasts from proplastids and etioplasts, but also reducing the number of mesophyll cells and disrupting the organisation of the palisade layer.

3.9 Environmental regulation of leaf development and differentiation

Light may play an important role in regulating the process of leaf development at several stages. A detailed discussion of this topic was provided in the excellent review by Dengler (1994). However, some of the main themes discussed are summarised here.

The effect of light on the process of leaf initiation has not been intensively studied. More typically, studies have focused on leaf production (in which leaves have emerged from the bud). These studies show a general reduction in the rate of leaf production with decreased irradiance. In one study that did examine light effects on leaf initiation in *Xanthium* and *Sinapsis*, short periods of high-light conditions increased the frequency of leaf initiation relative to plants that are grown in darkness over the same interval. The significance of this observation is unclear, given the complexity of factors likely to affect the initiation process. Subsequent to initiation, light appears capable of influencing the shape and size of the leaf. In some species, reduced light conditions correlate with increased leaf size and decreased leaf thickness, while other species show the opposite response. It has been suggested that these contrasting responses may be a reflection of shade tolerance, with tolerant species showing an adaptive increase in leaf area in reponse to reduced light. For shade tolerant species, the increased thickness of so-called 'sun leaves' generally reflects an increased thickness of palisade layers as well as a thickened epidermal cuticle. As might be expected, these changes are associated with an increased photosynthetic capacity.

Changes in leaf shape have been associated with light levels. However, for those cases that have been well documented, changes in shape associated with high irradiance appear to involve an accelerated progression through an otherwise normal heteroblastic series of development.

Finally, light also appears to influence terminal differentiation of the leaf, with high irradiance being associated with increased stomatal densities and increased proliferation of chloroplasts. In C4 grasses, high irradiance levels are correlated with the development of a classic bundle/sheath Kranz anatomy (Brown, 1975). These examples, like those cited previously, appear adaptive in the sense of enhancing the photosynthetic capacity of the leaf.

3.10 Heterophylly in aquatic species

Certain aquatic plants afford some of the most dramatic examples of how environmental conditions may modify the basic pattern of leaf development and provide insight into some of the factors that mediate these changes. In *Hippuris vulgaris*, submerged leaves are long and thin, and lack stomata compared to the thicker aerial leaves that have a more typical array of stomata. This heterophyllic pattern of development can be induced relatively late in leaf development, with transfer between environments inducing a complete change as late as plastochron 10 (Goliber and Feldman, 1990). The possibility that this change may reflect altered water potentials is suggested by increased levels of abscisic acid (ABA) in aerial-grown shoots, and the observation that exogenous application of ABA to water-grown shoots induces an aerial pattern of development. A similar pattern of development is observed in the aquatic species, *Callitriche heterophylla*. Treatment of aerial leaves with gibberellins led to an increase in cell length, an effect thought to be mediated through an increase in turgor. On the other hand, treatment of submerged shoots with ABA led to less cell elongation, which appeared to be mediated by decreased extensibility rather than decreased turgor.

3.11 Heteroblastic development

Heteroblasty refers to the ontogenetic progression from one leaf type to another as the plant initiates successive leaves (Allsopp, 1967). In many cases, these differences can be correlated with changes in phyllotaxy, epidermal characters and growth habit, which are together considered to consititute the phenomenon of ‘phase change’ (see Section 3.12). As was the case for environmentally induced heterophylly, heteroblastic patterns of leaf development, may appear relatively late in development, although studies of bud scales in woody species suggest that some differences may arise close to the time of leaf initiation (Foster, 1952). Some authors have argued the significance of SAM size as an important factor in determining distinct leaf forms, as such correlations are frequently observed. Lacking experimental evidence, however, it is difficult to exclude other contributing factors. In studies of ferns, heteroblastic development would appear to be quite plastic, as indicated by the behaviour of cultured primordia whose pattern of development can be altered by changes in growth conditions. Similar conclusions have been drawn in studies of whole plants in which similar changes can be effected.

3.12 Phase change

The phenomenon of phase change offers a window from which to view the outlines of developmental programmes. During the process of phase change, specific changes to the pattern of leaf initiation, development and differentiation may be observed within a single plant as it shifts from a juvenile to adult pattern of development. As described previously, these changes may include alterations in phyllotaxy, changes in heteroblastic shape and specific changes to epidermal characters, including pigmentation, cell shape, the distribution of trichomes and stomata, and the presence or absence of cuticular waxes.

During normal development, these processes are coordinately regulated to yield a progression of juvenile to adult characteristics that may be considered in some way adaptive to distinct phases of a plant's life. Through genetic analysis, the nature of the relationship between factors which mediate specific aspects of phase change and those responsible for their coordination is beginning to emerge (Poethig, 1990). In many cases, a particular phase of development can be viewed as a superimposition of a number of independently regulated subprogrammes of development. Genetics, by providing a means to perturb the coordinate regulation of these subprogrammes, has provided a method of distinguishing the outlines of individual programmes.

An example of logic is seen in the morphological transformations effected by a class of maize mutants known as *Teopod*, whose name refers to their resemblence to primitive teosinte and pod forms of maize. Three genetically unlinked dominant mutants, *Tp1, Tp2* and *Tp3*, have been described (Poethig, 1988), all of which appear to prolong the juvenile phase of development. These mutants exhibit a diverse range of normally juvenile leaf characters, including overall shape and the presence of specific epidermal waxes and cell types, as well as a decreased apical dominance. This prolonged juvenile pattern of development does not appear to perturb the onset of reproductive development, suggesting that the two patterns of development are regulated independently. When these normally disparate programmes are made to overlap through the influence of the *Tp* mutations, the profound atavistic transformations result.

Relatively little is known of the molecular mechanisms that control phase change, though several lines of evidence implicate a role for gibberellins. Genes controlling the larger process will no doubt be cloned in the near future. Meanwhile, some insight is offered by the cloning of *GLOSSY15* (*GL15*), a gene which has been proposed to be a target of phase change regulation (Moose and Sisco, 1996). *GL15* acts autonomously to promote the development of juvenile epidermal characters in a normal context as well as the extended juvenile phase conditioned by *Teopod* mutants. Analysis of the *GL15* gene suggests that it encodes a nuclear-localised protein belonging to a class which includes *APETALA2*, a gene which influences the identity of the floral organ during reproductive development through an unknown mechanism.

Flowering could be taken as the most extreme example of phase change. During the transition to reproductive development, changes in internode elongation, leaf development and branching precede the development of the flower, in which the pattern of lateral organ initiation and development is even more radically altered. Molecular genetic analysis of flowering in model systems, such as *Arabidopsis* and *Antirrhinum*, have provided significant insights into the nature of flower development, offering support for the long-held belief that the flower represents a determinate short shoot, with floral organs representing highly modified leaves (Coen and Meyerowitz, 1991). Expression of 'floral organ' identity genes within the developing flower modifies the leaf-like ground state of lateral organs to yield the specific whorls that make up the flower. Cloning of the corresponding genes reveals that, with the exception of the previously mentioned *AP2*, the basic set of genes that mediate this specification in both *Arabidopsis* and *Antirrhinum* belong to the MADS-box class of transcription factors.

Flowering may be evoked by one or more developmentally or environmentally defined cues, reflecting the diverse growth habits of various plant species. It appears, however, that these cues are channelled into a common pathway that involves activation of genes that promote a 'floral identity' in the SAM. The first of these to be cloned, *FLORICAULA*, is expressed in the SAM in *Antirrhinum*, with loss of function resulting in partial conversion of the inflorescence into an indeterminate vegetative structure. A homologous gene product, *LEAFY* (*LFY*), appears to function in a similar manner in *Arabidopsis*, through a process that involves localisation of the protein to the nucleus, where it functions by an unknown mechanism. Recent molecular evidence, which shows that *LFY* expression is activated by gibberellins, presents an intriguing link between the juvenile/adult and vegetative/reproductive phase transitions (Blazquez *et al.*, 1999).

Like the juvenile to adult transition, the transition to flowering could be viewed as a coordinated resetting of several independent programmes of development, with some affecting phyllotaxy and with others controlling specific aspects of organ development. Through genetic approaches, significant progress has been made towards understanding the discrete nature of these programmes and the mechanisms by which they might be regulated. The analysis of such mutants has emphasised the profound transformations that can be effected by changes to a single gene and must clearly inform models how various patterns of development manifest within the plant kingdom are related. Certain characters, such as simple *vs* compound leaf architecture, appear quite plastic, being frequently lost or gained within closely related groups of plants. The plasticity of such characters is clearly compatible with what appears to be a readily malleable, hierarchical control of underlying developmental mechanisms.

3.13 Prospects

This review presents an account of recent molecular genetic studies that are beginning to yield clear insights into basic processes by which the leaf develops and differentiates. Specific themes are emerging that emphasise conserved molecular mechanisms of regulation that appear to be well adapted to specific developmental processes. Thus, MADS-box transcription factors are seen to play a major role in specifying floral organ identity states, while *Kn1*-like genes appear especially adept at providing axial patterning cues. With such a perspective, progress can be anticipated in three key areas:

- What is the molecular nature of the patterning process? Key components have been identified in both the process of leaf initiation and axial patterning. Homeodomain proteins appear to play a prominent role as they do in analogous patterning processes in animals. Will there be a similar correspondence between animals and plants for other patterning mechanisms?
- How are developmental domains that are set up by the patterning process translated into form? We find many examples in which consistent patterns of morphogenesis accommodate variable patterns of cell division. What are the mechanisms that regulate polarity of growth and cell division with such precision, enabling plants to respond to such variability?
- How are stable higher level programmes of development established and maintained? Simple autofeedback loops regulating transcription and translation have traditionally held favour. However, recent evidence implicating epigenetic modes of regulation suggests that new approaches will be required to fully understand the molecular basis of leaf development.

Acknowledgements

The authors would like to acknowledge J. Longand, A. Hudson, E. Lifschitz and L. Smith for providing copies of figures which appear in this chapter.

References

Aida, M., Ishida T. and Tasaka, M. (1999). Shoot apical meristem and cotyledon formation during *Arabidopsis* embryogenesis: interaction among the *cup-shaped cotyledon* and *shoot meristemless* genes. *Development*, **126** 1563–1570.

Allsopp, A. (1967) Heteroblastic development in vascular plants. *Advancements in Morphology*, **6** 127–171.

Avery, G.S. (1933) Structure and development of the tobacco leaf. *American Journal of Botany*, **20** 513–564.

Ball, E. (1960) Sterile culture of the shoot apex of *Lupinus albus*. *Growth*, **24** 91–110.

Becraft, P.W. and Freeling, M. (1991) Sectors of *liguleless-1* tissue interrupt an inductive signal during maize leaf development. *Plant Cell*, **3** 801–807.

Becraft, P.W. and Freeling, M. (1994) Genetic analysis of *Rough sheath1* developmental mutants of maize. *Genetics*, **136** 295–311.

Becraft, P.W., Bongard-Pierce, D.K., Sylvester, A.W., Poethig, R.S. and Freeling, M. (1990) The *liguleless-1* gene acts tissue specifically in maize leaf development. *Developmental Biology*, **141** 220–232.

Blazquez, M.A., Green, R., Nilsson, O., Sussman, M.R. and Weigel, D. (1999) Gibberellins promote flowering of *Arabidopsis* by activating the LEAFY promoter. *Plant Cell*, **10** 791–800.

Brown, W.V. (1975) Variations in anatomy, associations and origins of Kranz tissue. *American Journal of Botany*, **62** 395–402.

Chen, J., Jannssen, B., Williams, A. and Sinha, N. (1997) A gene fusion at a homeobox locus: alterations in leaf shape and implications for morphological evolution. *Plant Cell*, **9** 1289–1304.

Chuck, G., Lincoln, C. and Hake, S. (1996) *Knat1* induces lobed leaves with ectopic meristems when overexpressed in *Arabidopsis*. *Plant Cell*, **8** 1277–1289.

Clark, S.E. (1997) Organ formation at the vegetative shoot meristem. *Plant Cell*, **9** 1067–1076.

Clark, S.E., Jacobsen, S.E., Levin, J.Z. and Meyerowitz, E.M. (1996) The *CLAVATA* and *SHOOT MERISTEMLESS* loci competitively regulate meristem activity in *Arabidopsis*. *Development*, **122** 1567–1575.

Coen, E.S. and Meyerowitz, E.M. (1991) The war of the whorls: genetic interactions controlling flower development. *Nature*, **353** 31–37.

Dengler, N. (1994) In *Growth Patterns in Vascular Plants* (ed. M. Iqbal), Dioscorides Press, Portland, OR, USA.

Dolan, L. and Poethig, R.S. (1997a) Clonal analysis of leaf development in cotton. *American Journal of Botany*, **85** 315–321.

Dolan , L. and Poethig, R.S. (1997b) The *Okra* leaf shape mutation in Cotton is active in all cell layer of the leaf. *American Journal of Botany*, **85** 322–327.

Esau, K. (1943) Ontogeny of the vascular bundle in *Zea mays*. *Hilgardia*, **15** 327–368.

Esau, K. (1965) *Plant Anatomy*, 2nd edition, Wiley, New York, USA.

Evans, M.M.S. and Barton, K.M. (1997) Genetics of angiosperem shoot apical meristem development. *Annual Review of Plant Physiology and Plant Molecular Biology*, **48** 673–701.

Fleming, A., McQueen-Mason, S., Mandel, T. and Kuhlemeier, C. (1997) Induction of leaf primordia by the cell wall protein, expansin. *Science*, **276** 1415–1418.

Foster, A.S. (1952) Foliar venation in angiosperms from an ontogenetic standpoint. *American Journal of Botany*, **39** 752–766.

Foster, T., Veit, B. and Hake, S. (1999). Mosaic analysis of the dominant mutant, *gnarley 1-r*, reveals distinct lateral and transverse signalling pathways during maize leaf development. *Development*, **126** 305–313.

Fowler, J.E. and Freeling, M. (1996) Genetic analysis of mutations that alter cell fates in maize leaves: dominant liguleless mutations. *Developmental Genetics*, **18** 198–222.

Freeling, M. (1992) A conceptual framework for maize leaf development. *Developmental Biology*, **153** 44–58.

Freeling, M. and Hake, S. (1985) Developmental genetics of mutants that specify knotted leaves in maize. *Genetics*, **111** 617-634.

Freeling, M. and Lane, B. (1993) The maize leaf, in *The Maize Handbook* (eds. M. Freeling and V. Walbot), Springer-Verlag, New York, USA, pp. 17–28.

Fukuda, H. (1997) Tracheary element differentiation. *Plant Cell*, **9** 1147–1156.

Furner, I.J. and Pumfrey, J.E. (1992) Cell fate in the shoot apical meristem of *Arabidopsis thaliana*. *Development*, **115** 755–764.

Gehring, W.J. (1987) Homeoboxes in the study of development. *Science*, **236** 1245–1252.

Gelinas, D., Postlethwait, S.N. and Nelson, O.E. (1969) Characterization of development in maize through the use of mutants. II. The abnormal growth conditioned by the Knotted mutant. *American Journal of Botany*, **56** 671–678.

Goliber, T. and Feldman, L.J. (1990) Developmental analysis of leaf plasticity in the heterophyllou aquatic plant, *Hipparis vulagris*. *American Journal of Botany*, **77** 399–412.

Green, P.B. (1987) Inheritance of pattern: analysis from phenotype to gene. *American Zoologists*, **27** 657–673.

Hagemann, W. and Gleissberg, S. (1966) Organogenetic capacity of leaves: the significance of marginal blastozones in angiosperms. *Plant Systematics and Evolution*, **199** 121–152.

Hareven, D., Gutfinger, T., Parnis, A., Eshed, Y. and Lifschitz, E. (1996) The making of a compound leaf: genetic manipulation of leaf architecture in tomato. *Cell*, **84** 735–744.

Harper, L. and Freeling, M. (1996) Interactions of *liguleless1* and *liguleless2* function during ligule induction in maize. *Genetics*, **144** 1871–1882.

Hemerly, A., Engler, J.A., Bergounioux, C., Van Montagu, M., Engler, G., Inze, D. and Ferreira, P. (1995) Dominant negative mutants of the cdc2 kinase uncouple cell division from iterative plant development. *EMBO Journal*, **14** 3925–3936.

Hofer, J., Turner, L., Hellens, R., Ambrose, M., Matthews, P., Michael, A. and Ellis, N. (1997) *UNIFOLIATA* regulates leaf and flower morphogenesis in pea. *Current Biology*, **7** 581–587.

Hülskamp, M., Schnittger, A. and Folkers, U. (1999) Pattern formation and cell differentiation: trichomes in *Arabidopsis* as a genetic model system. *International Review of Cytology*, **186** 147–178.

Irish, V.F. and Sussex, I.M. (1992) A fate map of the *Arabidopsis* embryonic shoot apical meristem. *Development*, **115** 745–753.

Jackson, D. and Hake, S. (1999) Control of phyllotaxy in maize by the *abphyl1* gene *Development*, **126** 315–323.

Jackson, D., Veit, B. and Hake, S. (1994) Expression of maize *KNOTTED1*-related homeobox genes in the shoot apical meristem predicts patterns of morphogenesis in the vegetative shoot. *Development*, **120** 405–413.

Kaplan, D.R. (1973) The monocotyledons: their evolution and comparative biology. VII. The problem of leaf morphology and evolution in the monocotyledons. *Quarterly Review of Biology*, **48** 437–457.

Kerstetter, R., Vollbrecht, E., Lowe, B., Veit, B., Yamaguchi, J. and Hake, S. (1994) Sequence analysis and expression patterns divide the maize *knotted1*-like homeobox genes into two classes. *Plant Cell*, **6** 1877–1887.

Kerstetter, R.A., Laudencia-Chingcuanco, D., Smith, L.G. and Hake, S. (1997) Loss of function mutations in the maize homeobox gene, *knotted1*, are defective in shoot meristem maintenance. *Development*, **124** 3045–3054.

Larkin, J.C., Young, N., Prigge, M. and Marks, M.D. (1996) The control of trichome spacing and number in *Arabidopsis*. *Development*, **122** 997–1005.

Larkin, J.C., Marks, M.D., Nadeau, J. and Sack, F. (1997) Epidermal cell fate and patterning in leaves. *Plant Cell*, **9** 1109–1120.

Larkin, J.C., Walker, J.D., Bolognesi-Winfield, A.C., Gray, J.C. and Walker, A.R. (1999) Allele-specific interactions between *ttg* and *gl1* during trichome development in *Arabidopsis thaliana*. *Genetics*, **151** 1591–604.

Lawrence, P.A. and Morata, G. (1994) Homeobox genes: their function in *Drosophila* segmentation and pattern formation. *Cell*, **78** 181–189.

Long, A., Moan, E.I., Medford, J.I. and Barton, M.K. (1996) A member of the KNOTTED class of homeodomain proteins encoded by the *STM* gene of *Arabidopsis thaliana*. *Nature*, **379** 66–69.

Lyndon, R.F. (1998) *The Shoot Apical Meristem: Its Growth and Development*. Cambridge University Press, Cambridge, UK.

Marx, G.A. (1987) A suite of mutants that modify pattern formation in pea leaves. *Plant Molecular Biology Reporter*, **5** 311–335.

Maksymowych, R. and Erikson, R.O. (1960) Development of the lamina in *Xanthium italicum* represented by the plastochron index. *American Journal of Botany*, **47** 451–459.

Matsuoka, M., Ichikawa, H., Saito, A., Tamda, Y., Fujimura, T. and Kano-Murakami, Y. (1993) Expression of a rice homeobox gene causes altered morphology of transgenic plants. *Plant Cell*, **5** 1039–1048.

McConnell, J. and Barton, M.K. (1995) Effect of mutations in the *PINHEAD* gene of *Arabidopsis* on the formation of shoot apical meristems. *Developmental Genetics*, **16** 358–366.

McHale, N.A. (1993) *LAM-1* and *FAT* genes control development of the leaf blade in *Nicotiana sylvestris*. *The Plant Cell*, **5** 1029–1038.

McHale,N.A. and Marcotrigiano, M. (1998) *LAM1* is required for dorsoventrality and lateral growth of the leaf blade in Nicotiana. *Development*, **125** 4235–4243.

Meyerowitz, E. (1996) Plant development: local control, global patterning. *Current Opinion in Genetics and Development*, **6** 475–479.

Meyerowitz, E. (1997) Genetic control of cell division patterns in developing plants. *Cell*, **88** 299–308.

Moose, S.P. and Sisco, P.H. (1996) *Glossy15*, an *APETALA2*-like gene from maize that regulates leaf epidermal cell identity. *Genes and Development*, **10** 3018–3027.

Moreno, M.A., Harper, L.C., Krueger, R.W., Dellaporta, S.L. and Freeling, M. (1997) *liguless1* encodes a nuclear-localized protein required for induction of ligules and auricles during maize leaf organogenesis. *Genes and Development*, **11** 616–628.

Nelson, T. and Dengler, N. (1997) Leaf vascular pattern formation. *Plant Cell*, **9** 1121–1135.

Parnis, A., Cohen, O., Gutfinger, T., Hareven, D., Zamir, D. and Lifschitz, E. (1997) The dominant developmental mutants of tomato, *Mouse-ear* and *Curl*, are associated with distinct modes of abnormal transcriptional regulation of a *Knotted* gene. *Plant Cell*, **9** 2143–2158.

Pilkington, M. (1929) The regeneration of the stem apex. *New Phytologist*, **28** 37–53.

Poethig, R.S. (1987) Clonal analysis of cell lineage patterns in plant development. *American Journal of Botany*, **74** 581–594.

Poethig, R.S. (1988) Heterochronic mutations affecting shoot development in maize. *Genetics*, **119** 959–973.

Poethig, R.S. (1989) Clonal analysis of cell lineage patterns in plant development. *American Journal of Botany*, **74** 581–594.

Poethig, R.S. (1990) Phase change and the regulation of shoot morphogenesis in plants. *Science*, **250** 923–930.

Poethig, R.S. (1997) Leaf morphogenesis in flowering plants. *Plant Cell*, **9** 1077–1087.

Poethig, R.S. and Sussex, I.M. (1985) The cellular parameters of leaf development in tobacco: a clonal analysis. *Planta*, **165** 158–169.

Poethig, R.S. and Szymkowiak, E.J. (1995) Clonal analysis of leaf development in maize. *Maydica*, **40** 67–76.

Redei, G.P. and Hirano, Y. (1964) Linkage studies. *Arabidopsis Information Server*, **1** 9–10.

Reinhardt, D., Wittwer, F., Mandel, T. and Kuhlemeier, C. (1998) Localized upregulation of a new expansin gene predicts the site of leaf formation in the tomato meristem. *Plant Cell*, **10** 1427–1438.

Russell, S.H. and Evert, R.F. (1985) Leaf vasculature in *Zea mays* L. *Planta*, **164** 448–458.

Sachs, T. (1969). Regeneration experiments on the determination of the form of leaves. *Israel Journal of Botany*, **18** 21–30.

Sachs, T. (1991) *Pattern Formation in Plants*, Cambridge University Press, Cambridge, UK.

Satina, S., Blakeslee, A.F. and Avery, A.G. (1940) Demonstration of three germ layers in the shoot apex of *Datura* by means of induced polyploidy in periclinal chimeras. *American Journal of Botany*, **27** 895–905.

Scanlon, M.J. and Freeling, M. (1997) Clonal sectors reveal that a specific meristematic domain is not utilized in the maize mutant *narrow sheath. Developmental Biology*, **182** 52–66.

Scanlon, M.J., Schneeberger, R.G. and Freeling, M. (1996) The maize mutant *narrow sheath* fails to establish leaf margin identity in a meristematic domain. *Development*, **122** 1683–1691.

Schmidt, A. (1924) Histologische Studien an phanerogamen Vegetationspunkten. *Botanical Archives*, **8** 345–404.

Schneeberger, R.G., Becraft, P.W., Hake, S. and Freeling, M. (1995a) Ectopic expression of the *knox* homeobox gene, *rough sheath1*, alters cell fate in the maize leaf. *Genes and Development*, **9** 2292–2304.

Schneeberger, R.G., Becraft, P.W., Hake, S. and Freeling, M. (1995b) Correction: ectopic expression of the *knox* homeobox gene, *rough sheath 1*, alters cell fate in the maize leaf. *Genes and Development*, **9** 2870.

Schwabe, W.W. (1971) Chemical modification of phyllotaxis and its implications. *Symposia of the Society for Experimental Biology*, Company of Biologists, Cambridge, UK, **25** pp. 310–322.

Scott, M.P., Tamkun, J.W. and Hartzell, G.W. (1989) The structure and function of the homeodomian. *Biochimica et Biophysica Acta*, **989** 25–48.

Sharman, B.C. (1942) Developmental anatomy of the shoot of *Zea mays* L. *Annals of Botany*, **6** 245–282.

Sinha, N. and Hake, S. (1994) The *Knotted* leaf blade is a mosaic of blade, sheath and auricle identities. *Developmental Genetics*, **15** 401–414.

Sinha, N.R., Williams, R.E. and Hake, S. (1993) Overexpression of the maize homeobox gene, *KNOTTED-1*, causes a switch from determinate to indeterminate cell fates. *Genes and Development*, **7** 787–795.

Smith, L., Greene, B., Veit, B. and Hake, S. (1992) A dominant mutation in the maize homeobox gene, *Knotted-1*, causes its ectopic expression in leaf cells with altered fates. *Development*, **116** 21–30.

Smith, L.G., Hake, S. and Sylvester, A.W. (1996) The *tangled-1* mutation alters cell division orientations throughout maize leaf development without altering leaf shape. *Development* **122** 481–489.

Smith, L.G., Jackson, D. and Hake, S. (1995) The expression of *knotted1* marks shoot meristem formation during maize embryogenesis. *Developmental Genetics*, **16** 344–348.

Smith, R.H. and Murashige, T. (1970) *In vitro* development of the isolated shoot apical meristem of angiosperms. *American Journal of Botany*, **57** 562–568.

Snow, M. and Snow, R. (1931) Experiments on phyllotaxis. I. The effect of isolating a primordium. *Philosophical Transactions: Royal Society of London Series B*, **221** 1–43.

Souer, E., Van Houweilingen, A., Kloos, D., Mol, J. and Koes, R. (1996) The *no apical meristem* gene of petunia is required for pattern formation in embryos and flowers and is expressed at meristem and primordia boundaries. *Cell*, **85** 159–170.

Steeves, (1966) On the determination of leaf primordia in ferns, in *Trends in Plant Morphogenesis* (ed. E.G. Cutter) Longmans, London, UK, pp. 200–219.

Steeves, T.A. and Sussex, I.M. (1989) *Patterns in Plant Development*, 2nd Edition, Cambridge University Press, Cambridge, UK.

Struhl, G., Struhl, K. and Macdonald, P.M. (1989) The gradient morphogen, *bicoid*, is a concentration-dependent transcriptional activator. *Cell*, **57** 1259–1273.

Sussex, I.M. (1955) Morphogenesis in *Solanum tuberosum* L.: experimental investigation of leaf dorsiventrality and orientation in the juvenile shoot. *Phytomorphology*, **5** 286–300.

Sussex, I.M. (1964) The permanence of meristems: developmental organizers of reactors to exogenous stimuli? *Brookhaven Symp. Biol.*, **16** 1–12.

Sussex, I.M. (1989) Developmental programming of the shoot meristem. *Cell*, **56** 225–229.

Sylvester, A.W., Cande, W.Z. and Freeling, M. (1990) Division and differentiation during normal and *liguleless-1* maize leaf development. *Development*, **110** 985–1000.

Szymkowiak, E.J. and Sussex, I.M. (1996) What chimeras can tell us about plant development. *Annual Review of Plant Physiology and Plant Molecular Biology*, **47** 351–376.

Timmermans, M., Schultes, N., Jankofsky, J. and Nelson, T. (1998) *Leafbladeless1* is required for dorsoventrality of lateral organs in maize. *Development*, **125** 2813–2823.

Timmermans, M.C., Hudson, A., Becraft, P.W. and Nelson, T. (1999) ROUGH SHEATH2: a Myb protein that represses knox homeobox genes in maize lateral organ primordia. *Science*, **284** 151–153.

Tilney-Bassett, R.A.E. (1986) *Plant Chimeras*, Arnold, London, UK.

Tsiantis, M., Schneeberger, R., Golz, J.F., Freeling, M. and Langdale, J.A. (1999) The maize *rough sheath2* gene and leaf development programs in monocot and dicot plants. *Science*, **284** 154–156.

Tsuge, T., Tsukaya, H. and Uchimiya, H. (1996) Two independent and polarised processes of cell elongation regulate leaf blade expansion in *Arabidopsis thaliana* (L.). *Development*, **122** 1589–1600.

Van Lijsebettens, M. and Clark, J. (1998) Leaf development in *Arabidopsis*. *Plant Physiology and Biochemistry*, **36** 47–60.

Veit, B., Briggs, S.P., Schmidt, R.J., Yanofsky, M.F. and Hake, S. (1998) Regulation of leaf initiation by the *terminal ear1* gene of maize. *Nature*, **393** 166–168.

Vollbrecht, E., Veit, B., Sinha, N. and Hake, S. (1991) The developmental gene *Knotted-1* is a member of a maize homeobox gene family. *Nature*, **350** 241–243.

Waites, R. and Hudson, A. (1995) *phantastica:* a gene required for dorsoventrality of leaves in *Antirrhinum majus*. *Development*, **121** 2143–2154.

Waites, R., Selvadurai, H.R.N., Oliver, I.R. and Hudson, A. (1998) The *PHANTASTICA* gene encodes a *MYB* transcription factor involved in growth and dorsoventrality of lateral organs in *Antirrhinum*. *Cell*, **93** 779–789.

Walsh, J., Waters, C.A. and Freeling, M. (1998) The maize gene *liguless2* encodes a basic leucine zipper protein involved in the establishment of the leaf blade-sheath boundary. *Genes and Development*, **11** 208–218.

Wardlaw, C.W. (1949) Experiments on organogenesis in ferns. *Growth* (Suppl.), **13** 93–131.

Wetmore, R.H. (1954) The use of *in vitro* cultures in the investigation of growth and differentiation in vascular plants. *Brookhaven Symp. Biol.*, **6** 22–40.

Williams, R., Lie, Y., Hake, S. and Lemaux, P. (1997) Ectopic expression of *Knotted1* phenocopies the barley *Hooded* mutant. *Development*, **124** 3737–3745.

4 Spatio-temporal variation of leaf growth, development and function

A. Walter and U. Schurr

4.1 Introduction

4.1.1 Conceptual framework for leaf growth analysis

All plants have a principal construction plan directed by the genome but the form that this takes is dependent on interactions with the environment. The plasticity of growth during a plant's vegetative development is one of the most important characteristics, as it allows plants to alter the extent and structure of the surfaces which exchange substances with the environment. The term 'growth' depicts an irreversible increase in volume or biomass of a plant and requires the net acquisition of water, nutrients, carbon dioxide, oxygen and light energy. Substrates for growth are taken up from the environment via the root and leaf. The balanced composition of plants during the formation of new biomass is achieved through the interaction of external supplies and internal feedbacks. Thus, the regulatory processes of growth should be viewed from two perspectives: firstly, fluctuations in the availability of substrates within the environment cause changes in growth; and secondly, internal properties of the plant, under genetic control, alter during development and cause changes in the pattern and rate of growth, even when external conditions are constant. In order to identify intrinsic growth patterns, it is important to study growth when external resources are non-limiting, or when the extent of limitation is precisely controlled, as for example in controlled Ingestad nutrition situations (Hellgren and Ingestad, 1995). Otherwise, the intrinsic patterns can be confounded with and even altered by variations in environmental factors, making it impossible to unravel the underlying patterns.

Modular arrangement is a prominent feature of higher plants (Scanlon, 1998), ranging from cells constructed from organelles up to phytomers consisting of a leaf, a petiole and the adjacent node and internode. Since the entirety of a plant's vegetative leaf laminae behaves as one unit with respect to, for instance, photosynthesis, transpiration and growth distribution, this construct can be regarded as a single unit and referred to as 'total leaf area'. Total leaf area, therefore, links analysis of a single leaf to the whole plant level, and can thus contribute significantly to the bridge between small-scale analysis on individual plant parts and the whole plant performance.

As a consequence of this modular arrangement of a plant, the concept of 'scaling' is an important issue for embedding leaf growth analysis into the study of general plant growth. The plant's construction plan involves organisational mechanisms at several levels: from the whole plant level, through total leaf area to single leaf, cell(s) and finally down to molecules. A variety of regulatory mechanisms operate across these levels, the wealth of which is just beginning to be unravelled. Understanding the complexity of leaf growth in dicotyledons requires a holistic approach, including functional and structural aspects. The impact of external conditions on whole plant growth, the embedding of the individual leaf within the context of the total leaf area, the cytological status of the leaf-forming cells and the background of the leaf's genetic properties have to be considered simultaneously. The study of any one aspect in isolation is of much less value, as structural and functional aspects on different scales form the organism.

4.1.2 Technical framework for leaf growth analysis

Asking the correct questions is the first essential step in research and having the appropriate tools to answer them is the next. A measuring procedure must meet three requirements: 1) it must address a relevant property of the leaf; 2) its precision must be adequate for the aim of the investigation; and 3) it must not disturb the growth process under observation.

In this case, the relevant properties refer to aspects of growth, development and function. Growth can be observed by any one of several plant properties. Some, e.g. dry weight, can only be quantified by destructive measurements. The high variability between individual plants makes destructive measurements unsuitable for the study of small-scale patterns, e.g. diurnal growth patterns or gradients of growth along a leaf axis; and, by definition, once measured, the growth process has been radically disturbed! In dicotyledonous leaves, area and fresh weight increase in parallel to each other from leaf emergence to maturity (Figure 4.1), i.e. their specific leaf area on a fresh weight basis is conservative over time. Thus, the non-destructive measurement of leaf area can be surrogate for monitoring changes in individual leaf biomass over time and, in certain circumstances, can even be used to estimate whole-plant growth processes (Walter and Schurr, 1999).

Growth processes are both spatially and temporally organised. Classical methods allow either a high temporal resolution, e.g. measurements of midrib length with linearly variable displacement transducers, (LVDTs) or, to a lesser degree, spatial resolution, e.g. physically marking locations on a leaf and studying their subsequent movements of growth. Both approaches have significant limitations and, for technical reasons, cannot be combined (Schurr,

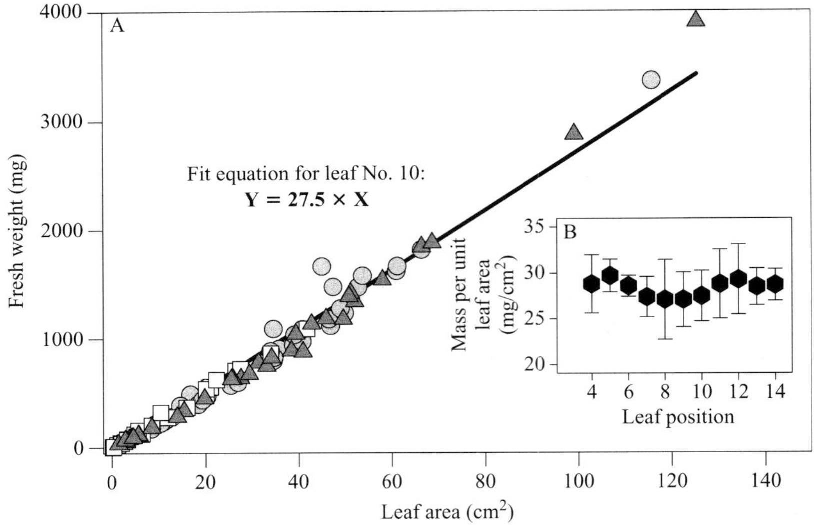

Figure 4.1 A) Relationship between fresh weight and leaf area of leaf number 10 of *Nicotiana tabacum* plants. B) Variation of mass per unit leaf area with leaf position along the stem (position 1 and 2: cotyledons). The oscillating tendency is due to: 1) variation in the relative mass of veins to interveinal areas; and, 2) the slight increase in leaf thickness with leaf maturation.

1997). Therefore, new approaches are needed that yield spatio-temporal data sets on growth processes.

The precision with which aerial growth rates can be measured is determined by the precision of the length detection; time resolution is usually no limitation. The time interval between successive measurements has to be increased to maintain the required precision when growth rates are too small to yield sufficient length differences.

Disturbance of growth by the measurement procedure is often neglected. Leaves are sensitive to mechanical stimulation, e.g. thigmomorphogenesis (Jaffe *et al.*, 1985); even touching evokes strong signalling events (e.g. Knight *et al.*, 1991). Tensile forces, for instance, as applied when using LVDTs, can significantly alter leaf growth (Walter and Schurr, unpublished). Furthermore, diurnal nyctinastic movements are common (Bünning, 1948; Herbert, 1992), and at least partly due to differential growth of the upper and lower surface of the leaf (Else and Jackson, 1998). How fixation and, thus, hindrance of nyctinastic movements affects growth patterns is a hitherto unanswered question.

New, non-invasive approaches that can provide both spatial and temporal data simultaneously are required. Modern optical techniques meet these requirements in part: precision and remote sensing (Schmundt *et al.*, 1998;

Jaffe *et al.*, 1985; Granier and Tardieu, 1998; see also Section 4.4). However, fixation of the leaf is still needed and additional control measurements are necessary to check for effects on leaf performance.

4.2 Growth of the total leaf area

The expansion of the total leaf area can be used as a reference point for the analysis of lower level organisation (leaves, internodes, etc.) and as an estimator of total plant growth. The spatial distribution of leaf expansion rates within total leaf area can be followed with ease using the simple biometric relationships between length and width of each leaf lamina and leaf area (Schurr, 1997). Total leaf area, the sum of all individual leaf areas, approaches its maximum value asymptotically (Figure 4.2). There is an initial phase of constant relative growth rate (RGR; exponential growth), followed by a second phase in which RGR steadily declines. In an otherwise constant environment, plant nutrition affects the duration and intensity of the exponential and subsequent growth phases and, of course, the underlying leaf area distribution, i.e. the plant's architecture (Roggatz *et al.*, 1999; Walter and Schurr, 1999). The exponential phase is characterised by constant patterns of biomass allocation: for instance, in tobacco, 50% of the plant's increase in total fresh weight is allocated to leaf area at any time during this phase. Exponential growth of total leaf area on a single stem is achieved by an increasing number of individual leaves simultaneously expanding, each

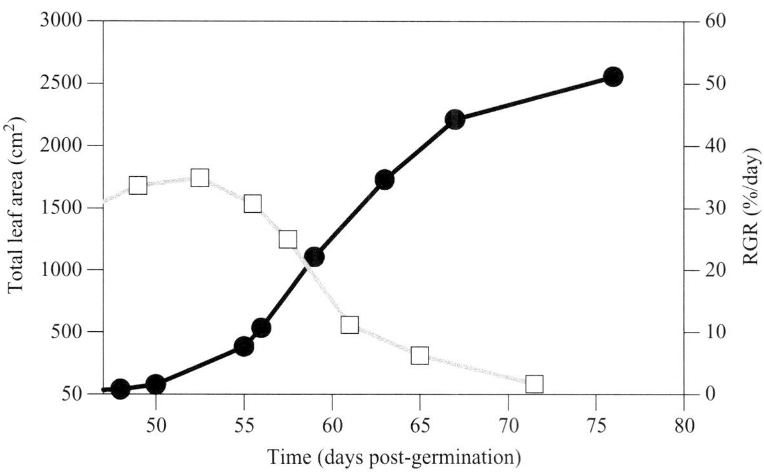

Figure 4.2 The change in total leaf area (●—●) and relative growth rate (RGR; □—□) with time after germination in *Nicotiana tabacum*.

individual leaf following a sigmoidal expansion of area, and each successive leaf initiated reaching a different (first increasing; and then, towards flowering, decreasing) final size.

The distribution of leaf area up the shoot axis can be described by so-called normalised distribution functions of leaf area (Figure 4.3): the percentage contribution of an individual leaf to the total leaf area is plotted against its position on the shoot axis. By analogy, the increases in area of each leaf over successive days can be related to the increments in total leaf area during the same time intervals. The relative contribution of individual leaves to the total leaf area follows a slightly asymmetrical binomial distribution along the stem during the entire experiment. With increase in plant size, the distribution broadens as more leaves contribute to total leaf area. The maximum of this distribution declines almost linearly, and is shifted to higher positions on the stem with time and, therefore, increasing plant size. The contribution of a particular leaf to the daily gain in leaf area also follows a binomial distribution, but is narrower and shifted by one to two leaf positions up the stem compared to the distribution of absolute leaf areas. Again, the distribution broadens with time and the maximum declines. At steady-state nutrition with the amount of available nutrient rising proportionally to plant biomass (Hellgren and Ingestad, 1995), both distributions are independent of the actual value of growth rate and show intrinsic patterns of growth in total leaf

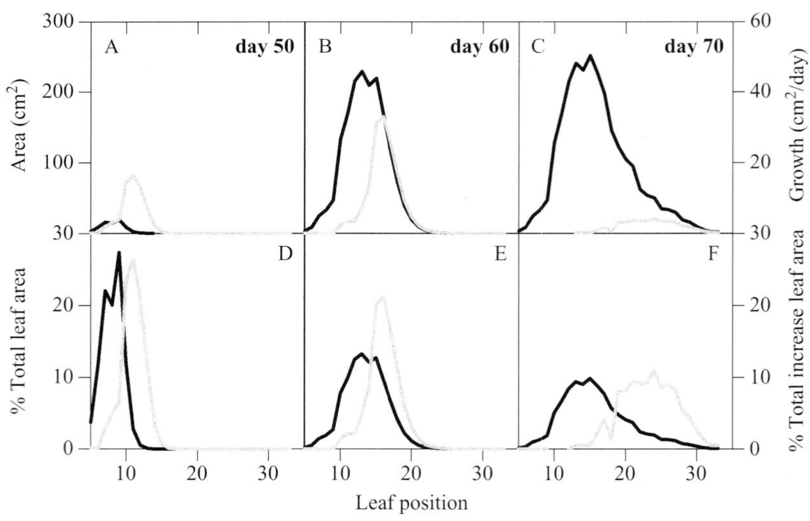

Figure 4.3 The distribution of leaf area (black line) and growth rate of leaf area (grey line) up the stem of *Nicotiana tabacum* plants at 50 (A,D), 60 (B,E) and 70 (C,F) days post-germination. Results are expressed as absolute areas and rates (A–C), and as percentages of the total leaf area and total increase in leaf area (D–F).

area. Different overall growth rates are associated with differences in final leaf size and different times to reach a specific plant size but the normalised distribution functions are unchanged (Walter and Schurr, 1999). Therefore, higher growth rates are obtained by acceleration or deceleration of a unified growth pattern.

Growth distribution along the stem of a dicotyledon can thus be described as an acropetally moving wave, which flattens and broadens with increasing plant size, leaving behind a binomially shaped distribution of leaf area along a monopodial shoot axis. This has consequences for the distribution of morphological and functional properties along the stem (Walter and Schurr, 1999).

Leaves at lower positions are, of course, initiated earlier than leaves at higher nodes. Thus, leaves at different positions show a temporal shift in their phasing of development and growth relative to each other, and are growing with slightly different intensities at a given time. This is in contrast to monocotyledons, in which only a few leaves contribute to the increase in total leaf area at any one time (Groeneveld *et al.*, 1998). Leaf initiation plays a crucial role in modifying the growth rate by determining the speed of initiation of new modules, while duration of cell division determines the final size of the leaves (Roggatz *et al.*, 1999).

4.3 Growth of the single leaf

Within a single leaf, growth and differentiation proceed as a basipetally moving wave (Maksymowych, 1973; Van Volkenburgh, 1987; Granier and Tardieu, 1998). A similar acropetally moving wave of growth of total leaf area occurs up the main shoot axis. Apical parts of the leaf are released from the shoot apical meristem some days earlier than the basal parts of the leaf (Poethig, 1997). Whereas the initiation of the modules comprising a single leaf is a continuous process, albeit with a temporal delay between the parts forming the tip and those forming the base of the leaf, the initial modules contributing to total leaf area are initiated discontinuously, i.e. internodes interrupt the initiation of successive leaves.

The time delay of initiation and differentiation processes within the leaf blade (Avery, 1933; Poethig and Sussex, 1985; Walter and Schurr, 1999) causes gradients of growth rate along the midrib (Figure 4.4): the basal segments of the tobacco leaf midrib grow faster than the apical segments. Comparison of the growth rates of leaf tip and leaf base show a developmental delay between base and tip of approximately 4 days. This gradient flattens with time, mirroring the decline in the overall growth rate of the leaf. In contrast, the side veins do not show significant gradients of growth rate from the midrib to the leaf margin (Walter and Schurr, 1999).

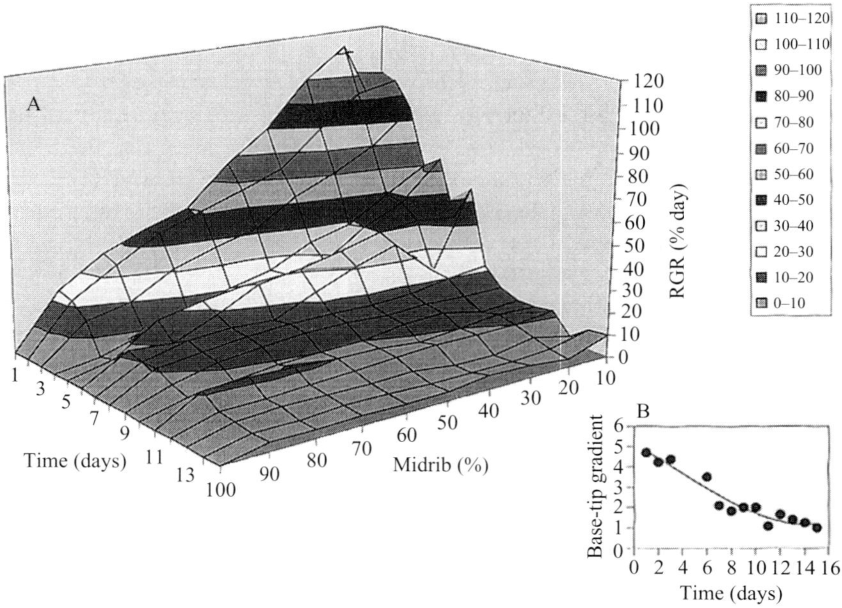

Figure 4.4 A) Variation of relative growth rate along the midrib of a *Nicotiana tabacum* leaf over a 2 week period (day 0: midrib length 2 cm). Relative growth rate (RGR) was analysed for individual midrib segments by means of a caliper. B) Development of the ratio between relative growth rate at the leaf base (10% of midrib length) and the leaf tip (90% of midrib length) over the same period.

Observation clearly shows that growth within the leaf is linked to the development of the veinal system. A rectangular coordinate system, as often used for analysis of the distribution of growth within dicotyledonous leaves, is inappropriate (Avery, 1933; Poethig and Sussex, 1985; Maksymowych, 1973; Granier and Tardieu, 1998). The veinal system itself provides a more appropriate, natural coordinate system for the analysis of leaf growth. Hejnowicz and co-workers (Hejnowicz and Hejnowicz, 1991; Hejnowicz and Karczewski, 1993) have already proved that the natural coordinate system of the cell files is superior to a rectangular basis for the quantitative analysis of growth in root tips. Clonal analysis of the origin of cells within tobacco leaves indicates that cell files that originate from a single mother cell lie parallel to class II veins (Poethig and Sussex, 1985). Within this veinal coordinate system growth processes are isotropic. Recent developments in image analysis techniques have the potential to replace the original, rectangular approaches with a system based on these natural 'landmarks' within the leaf (Schmundt *et al.*, 1998). Further evidence for the veins to be the natural coordinate system is provided by the phenomenon of interveinal buckling. The veinal system is the major transport route for growth substrates as well as providing the mechanical skeleton for the planar expansion of leaf area. When leaf expansion is not balanced by concomitant development of the veins, then buckling of interveinal areas occurs, rather than increasing the projected leaf area. This effect is most prominent in small leaves emerging from the bud.

Temporal variation in the rate of leaf growth is a prominent feature in many species (Table 4.1). The timing of maxima and minima in growth rates also differs between species and growing conditions. Tobacco and *Ricinus* show a distinct diurnal variation of growth activity (Figures 4.5 and 4.6). In both cases, strong nyctinastic movements are observed, causing significant three-dimensional deformation of the leaf blade. In such species, bending makes direct measurements of leaf extension (e.g. by LVDTs) problematical. Quantitative analysis can be preformed by using the ‘thread measurement method’ (Figure 4.7). A fine thread is fixed with Parafilm to the apical end of the leaf petiole and laid along the midrib. At the beginning of the measurement, the thread is cut a few millimeters proximal to the leaf tip and a small piece of tape is stuck to it, perpendicularly, to give an exact mark for the length measurement. The variable distance between the edge of the tape and the leaf tip can then be measured repeatedly with a slide caliper (precision 50 μm) and added to the constant length of the thread. This technique reduces the risk of interaction with basic growth processes, as no permanent tensile stress is applied as in LVDT measurements. The thread follows the shape of the midvein and thus allows measurement of the actual length variation, a feature of great importance in all leaves that show extensive bending.

Table 4.1 Temporal variation of growth rates in leaves of monocotyledons and dicotyledons

Species	RGR-maximum		Measuring method	Reference
	Day	Night		
Monocotyledons				
Zea	X		LVDT	Watts (1974)
	X		Ruler	Acevedo *et al.* (1979)
Triticum	X		LVDT	Christ (1978)
Sorghum	X		Ruler	Acevedo *et al.* (1979)
Festuca		X	LVDT	Durand *et al.* (1995)
Oryza	X		LVDT	Seneweera *et al.* (1995)
Dicotyledons				
Helianthus		X	Ruler	Boyer (1968)
Glycine		X	Ruler	Bunce (1977)
Phaseolus	X		LVDT	Davies and van Volkenburgh (1983)
Vitis	X		LVDT	Shackel *et al.* (1987)
Ricinus	X		As described	Schmundt *et al.* (1998)
Nicotiana	X		As described	herein

Abbreviations: RGR, relative growth rate; LVDT, linearly variable displacement transducers.

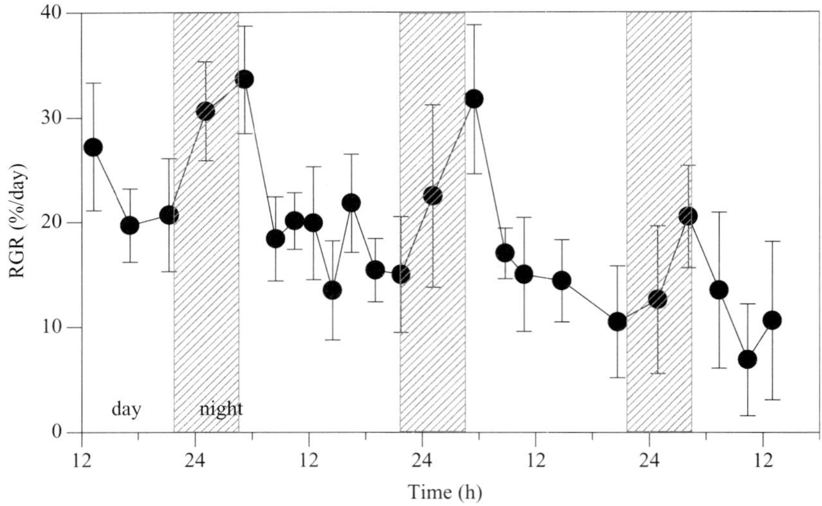

Figure 4.5 Diurnal variation of relative growth rate (RGR) of a *Nicotiana tabacum* leaf over three days, measured using the 'thread measurement method' (see Figure 4.7). Means and standard deviations of five leaves at position number 10. Cross-hatched bars indicate the periods of darkness.

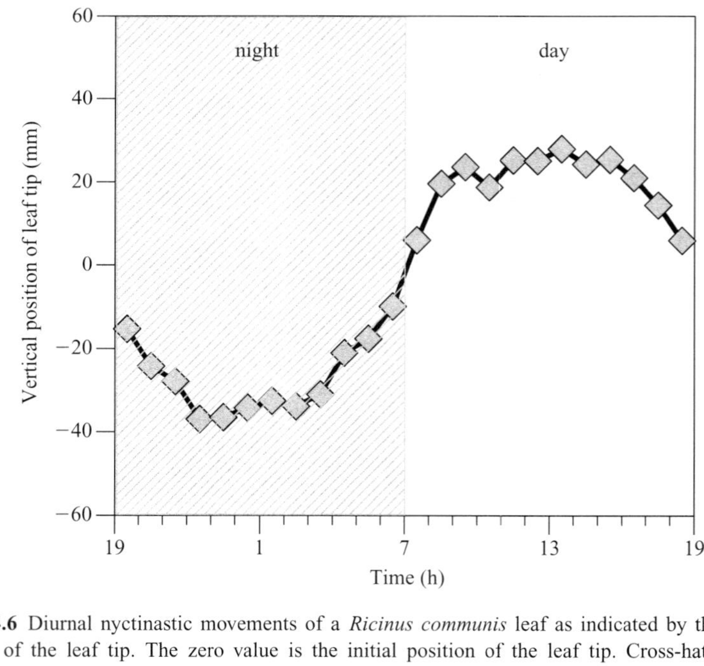

Figure 4.6 Diurnal nyctinastic movements of a *Ricinus communis* leaf as indicated by the vertical position of the leaf tip. The zero value is the initial position of the leaf tip. Cross-hatched area indicates the period of darkness.

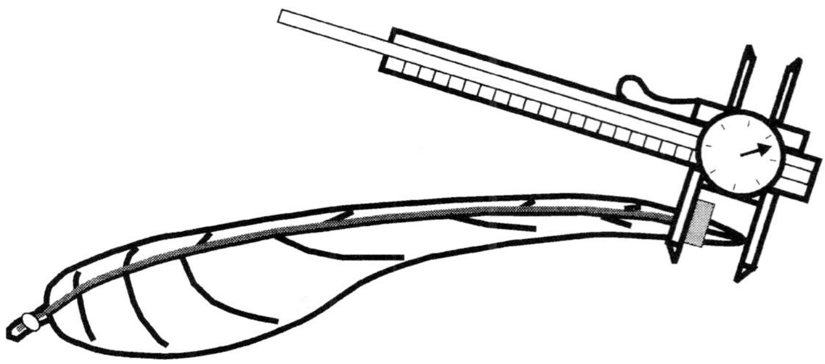

Figure 4.7 Scheme of the 'thread measurement method'. A thread is positioned along the midrib of a leaf and the distance between the end of the thread and the leaf tip is measured by means of a caliper. This method allows for bending of the leaf lamina and accurate analysis of leaf expansion.

4.4 Spatio-temporal analysis using digital image processing

Recent developments in methods based on image sequence analysis now make it possible to study spatial distributions of growth patterns within the leaf lamina and changes in these patterns over time, simultaneously.

4.4.1 Image sequence analysis

A leaf is fixed in a wire-frame, the proportions of which are customised to the particular leaf being measured (Figure 4.8); this is the one significant requirement of the present method, i.e. the need to constrain the leaf surface to a single planar surface. A charge-coupled device (CCD)-camera views the leaf lamina perpendicularly, illuminated by an array of infrared-diodes ($\lambda = 950$ nm) attached to the camera. The camera is sensitive to near-infrared radiation and a filter cuts off radiation of shorter wavelengths. The camera is connected via a frame grabber card to a standard personal computer (PC). Acquisition of digital video images is controlled by a program provided in the image processing software package (Heurisko, Aeon Verlag, Hanau). Images are recorded at regular intervals (typically 1–5 min). Each individual image recorded is the result of summing 80 images taken at the video capture rate of the camera (30 images per second). This improves the signal to noise ratio considerably. The frame size of the camera system used is 640×480 pixels. Depending on the magnification of the optical system used, one pixel typically corresponds to lengths of between 25 μm (if the object size is 16×12 mm) and 250 μm (16×12 cm) in the present applications. The use of light in the infrared allows images to be recorded in complete darkness, avoiding any disturbance of the underlying physiological processes and sensing mechanisms by the measurement procedure.

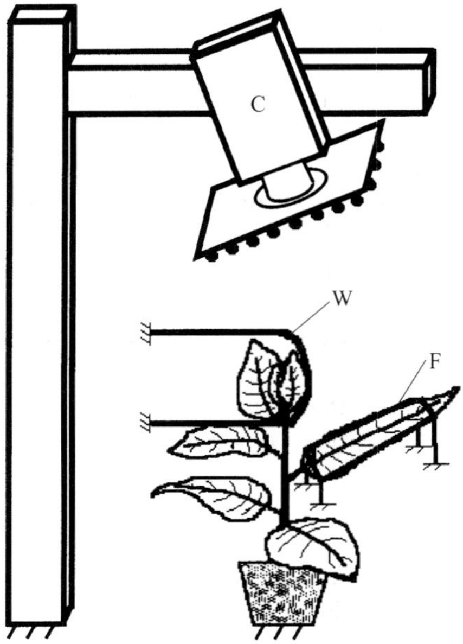

Figure 4.8 Scheme of the image acquisition system used to obtain image sequences for the optical analysis of leaf growth. A charge-coupled device (CCD) camera (C) is mounted on a strong stand, which prevents movement of the camera relative to the imaged leaf. A wire frame (F) keeps the imaged leaf in a fixed position perpendicular to the optical axis of the camera. Later-formed leaves are kept away from the field of view by wires (W).

4.4.2 Determination of growth rate from digital image sequences

Calculation of the spatio-temporal distribution of leaf growth rate consists of three major steps. In the first step, 'motion analysis', the displacement vector fields are calculated: a velocity of displacement is attributed to every pixel in each image of the sequence. Calculation of displacement vector fields is not limited to tracking individual landmarks through the entire image sequence; rather, any structure that results in a suitable grey value contrast is used to contribute to the computation of motion fields (Schmundt and Schurr, 1999; Haußecker and Spies, 1999). In the second step, 'divergence fields' are calculated from the corresponding displacement vector fields produced in the first stage. The magnitude of the divergence is equivalent to the local growth rate. Again, this is carried out for each pixel in each of the images in the time sequence. In the final step, pixel-wise calculated growth rates are smoothed and transformed to a natural leaf coordinate system, e.g. the veinal network.

Motion analysis is the most abstract step of the procedure. Images of the leaf surface viewed with near-infrared radiation highlight morphological features similar to those seen in visible light (Figure 4.9). Veins form dark lines, with the regions between looking brighter and somewhat patchy. This patchiness is caused by the presence of smaller veins of higher order. The brightness of the interveinal regions is due to the abundance of intercellular spaces, especially in the spongy parenchyma (Schmundt and Schurr, unpublished). A significant proportion of infrared radiation is reflected at interfaces between cells and air. Depending on the magnification, trichomes (epidermal hairs) and similarly sized structural elements can be resolved. These structures move with the surface; distances between them change due to local growth and can, thus, be used as reference points to estimate local growth rates. A structure, in terms of digital image processing, is a distinctive edge in the matrix of grey values that form an image. When the leaf grows, these edges move within the imaged plane. Stacking successive pictures produces a three-dimensional object with two spatial and one temporal axis (Figure 4.10). Persistent structures form characteristic patterns, indicating both the direction and intensity of motion. For example, a bar-shaped structure that moves with time yields a plane inclined relative to the time-axis (Figure 4.10), while a bar at rest produces a plane that is parallel to the time axis. A point-like structure results in a rod in the stacked images. The two inclination angles of a plane or line relative to the time axis in the two spatial planes define the x- and y-components of the velocity of that structure.

Figure 4.9 Near-infrared image of a *Nicotiana tabacum* leaf (7 cm long; original image size 48 × 36 mm) fixed in a rectangular wire frame. The leaf base is at the left hand border of the picture.

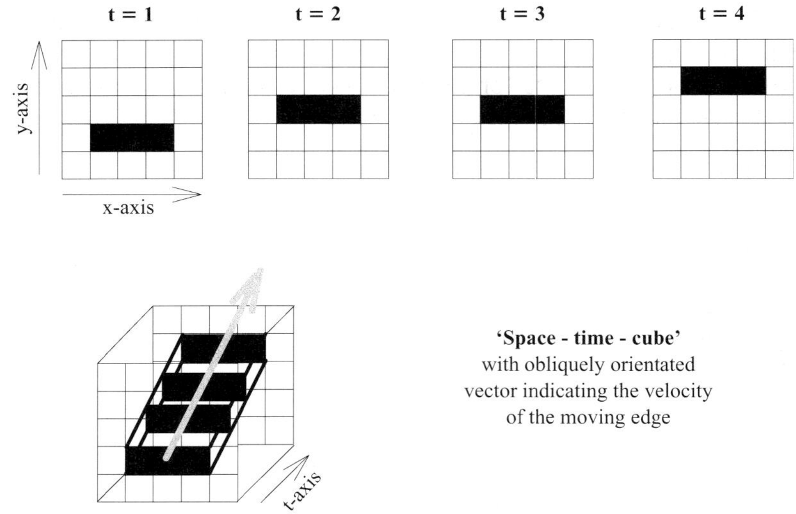

Figure 4.10 Scheme of the motion analysis in the time-space domain as performed by the structure tensor. In this example, a black bar (3 pixels wide) moves vertically by 1 pixel in time step 1–2 and 3–4, and stays constant at time step 2–3. Movement is indicated by the slope of the plane formed by the bar in the time-space cube.

In the implementation used in the following, the overall image stack is subdivided into cubes of $7 \times 7 \times 7$ pixels (x-, y- and time-direction) for analysis of motion therein. Since natural structures never produce rectangular grey value edges but smeared distributions of finite slope, motion is detectable in subpixel resolution of down to 0.01 pixel/frame (Haußecker and Spies, 1999). The grey value distribution inside this cube of 343 pixels is analysed by means of a 'structure tensor' (Bigün and Granlund, 1987). All grey value gradients are detected and a vector perpendicular to all the gradient vectors is fitted. If fitting leads to a clearly preferred orientation of the change in the grey values with time, the components of the calculated vector are assigned to the central pixel of the spatio-temporal cube, and thus one element of the displacement vector field is calculated. The calculation is repeated for the neighbouring pixels of the original central pixel by shifting the cube to the next adjacent pixel. Depending on the richness and distinctness of the structures present, displacement vectors are assigned to 20–50% of all pixels by the algorithm. This results in several thousand points within each image that contribute to the final velocity information (Figure 4.11A). Missing velocity information is then filled in by interpolation (Schmundt and Schurr, 1999).

In the second step of the procedure, the divergence of the displacement vector field is calculated pixel-wise for each image. The divergence is the sum of the spatial derivatives of velocity in x- and y-directions, and is thus linked quantitatively to the local growth rate at this position (Figure 4.11B). Time series of growth rate for specified locations within the image can be plotted, on demand, by the software (Schmundt *et al.*, 1998).

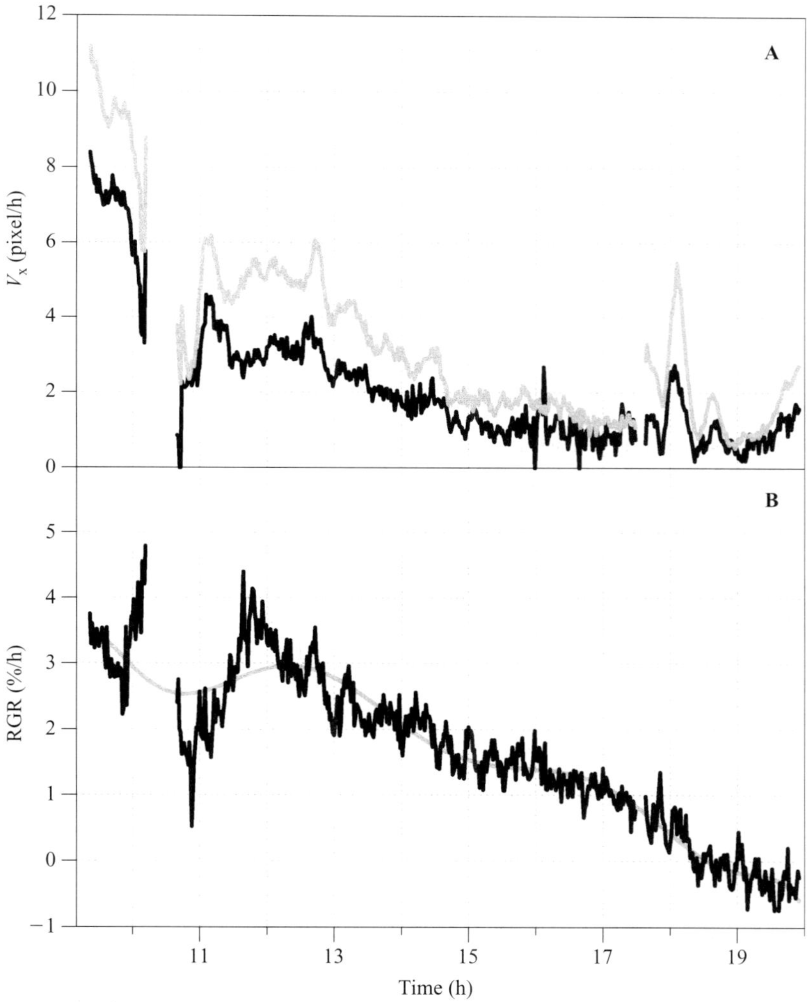

Figure 4.11 A) Time series of velocity detected in x-direction V_x, along the leaf midrib for two image points of the sequence from which Figure 4.10 is extracted. The faster movement near the leaf tip compared to that at the point near the leaf base results from growth of the lamina. The non-zero value of V_x for the leaf base results from growth of the petiole. Key: grey line for tip; black line for base. B) Time series of the relative growth rate (RGR) of an intermediate image point in the centre of the midrib. This value is calculated by taking the difference in V_x and V_y of neighbouring image points. Key: —, calculated divergence; —, polynomial fit (8th order).

In the third and final step, the calculated growth rates defined in Cartesian coordinates of the camera are then projected onto the veinal, the natural, coordinate system of the leaf. The image analysis software delivers maps of growth within the reference frame of the camera system. With growth, the object (leaf) is moving through the image; the growth rate indicated at a certain pixel position corresponds to different regions of the leaf during the sequence. Fragments of the automatically generated time series at different image positions are merged to follow the development of growth rate linked to a physical leaf region. Small scale fluctuations are smoothed with standard curve-fitting procedures and mean values of growth rate can be calculated for (e.g. hourly) intervals of segments along the leaf. Up to now, time series of development of growth rate have been studied for positions along the midrib of leaves only. Reference points are given in quantiles of the total actual midrib length (10% distance from base to tip; 20%, ...) (Figure 4.12). The next step will be to investigate time series along side veins of leaves.

4.4.3 *Diurnal variation of the spatial pattern of growth rate*

Using this method, small-scale variations in both spatial and temporal distributions of growth rate can be detected along the leaf midrib (Figure 4.12). In tobacco, growth is maximal several hours before the end of the night, and a minimum is reached at the end of the daylight period. A distinct, linear gradient of growth rate from base to tip is consistently present.

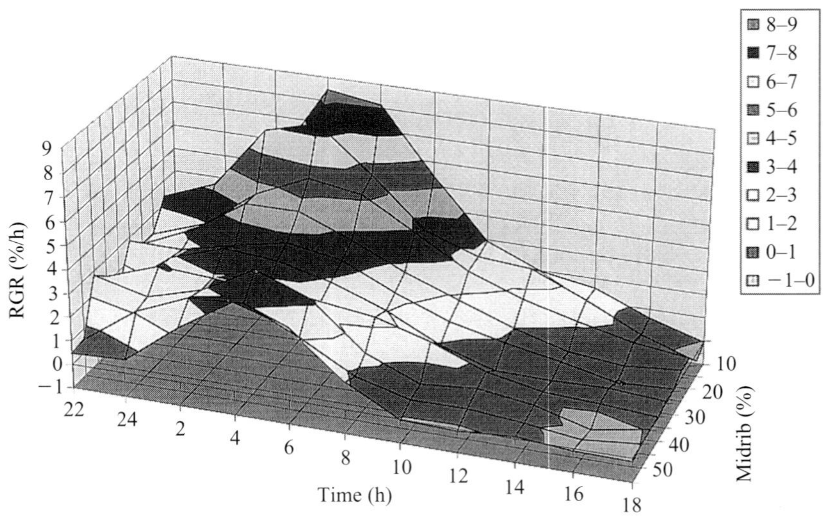

Figure 4.12 Diurnal variation of relative growth rates (RGR) along the midrib of a *Nicotiana tabacum* leaf determined by image sequence analysis. Midrib length is expressed as percentage of the entire length of the midrib. During this experiment, the period of darkness lasted from 8 pm to 8 am.

Changes in growth activity throughout the day between different regions of the leaf are temporally synchronised. An analogous pattern is found in *Ricinus*, with maximal growth rate somewhat closer to the end of the dark phase than in tobacco (Schmundt *et al.*, 1998).

The diurnal variation in expansion growth in *Ricinus* and tobacco is linked to a concomitant net influx of water into the leaf cells. Cell division has almost ceased in leaves of the size used in the studies presented above (Heckenberger *et al.*, 1998; Roggatz *et al.*, 1999; Walter and Schurr, unpublished). This implies that cell expansion must be regulated in a diurnal manner. As cell expansion is controlled by the interaction of turgor, *P*, and cell wall properties, then one or both of these features must also vary diurnally.

Water potential and turgor often increase during the night because of the reduced transpiration of the leaves (Bunce, 1977; Boyer, 1968). Changes in turgor pressure cause concomitant alterations in growth rate (for overview, see Schurr, 1997). However, turgor per se does not seem to control local cell expansion, as it does not show the spatial and temporal differentiation into growth zones that could explain the spatio-temporal patterns of growth rate observed (e.g. Spollen and Sharp, 1991; Tomos and Pritchard, 1994; Schurr, 1997; Schmundt *et al.*, 1998). It appears that spatial and temporal changes in cell wall properties are responsible for the organisation of these patterns *in vivo* (Cosgrove, 1986; Dale, 1988). Diurnal variations of cell wall properties have not yet been reported in the literature—these properties need to be considered in future studies.

The constant turgor in a rapidly growing tissue implies significant osmotic flux in the growing regions. In contrast to mature tissues, a constancy of concentration or osmotic potential in the growing region of the plant implies significant deposition of substances, to compensate for the increase in volume or biomass at the expanding site. Quantification of local deposition rates requires the combination of local analysis of growth rates and distribution of concentrations by means of kinematic growth analysis, as determined by Silk and co-workers (Silk and Erickson, 1979; Silk, 1984). Up to now, because of the lack of suitable methods to resolve aerial growth rates spatially, kinematic growth analyses could only be performed on linearly growing organs, such as monocotyledonous leaves (Hu and Schmidhalter, 1998) or roots (Silk *et al.*, 1986). The new methods enable the deposition rates and mass fluxes within dicotyledonous leaves to be calculated, and quantitative studies of, for example, sink-source transition processes are now possible.

4.5 Cytological development

The bases of macroscopic growth are the cellular growth processes, namely cell division, cell expansion and—often neglected—formation of intercellular spaces (Dale, 1988). Cellular growth processes also show dynamic spatio-temporal gradients during leaf development (Heckenberger *et al.*, 1998; Roggatz *et al.*, 1999).

Cell density is the result of the interactions of all three growth processes and illustrates the gradients of the cytological status of a tissue. Cell density—even in the different tissue layers—can be easily detected by counting stained cell nuclei per unit area in cleared leaf samples (Roggatz *et al.*, 1999).

After an initial constant phase, cell density declines with leaf maturation (Figure 4.13). This implies that cell division ceases prior to cell elongation—cell elongation without simultaneous cell division results in declining cell density. Decline in cell density at the leaf tip precedes the decline at the leaf base. In tobacco, the temporal delay observed is about 4 days; the same time interval as the delay in development of macroscopic growth between the leaf tip and base. This indicates that growth processes in tip and base follow the same cytological programme. The shift in the 'timetable' of the programme in the leaf base compared to the leaf tip is presumably a reflection of the delay in the initiation between the two leaf regions in the shoot apical meristem.

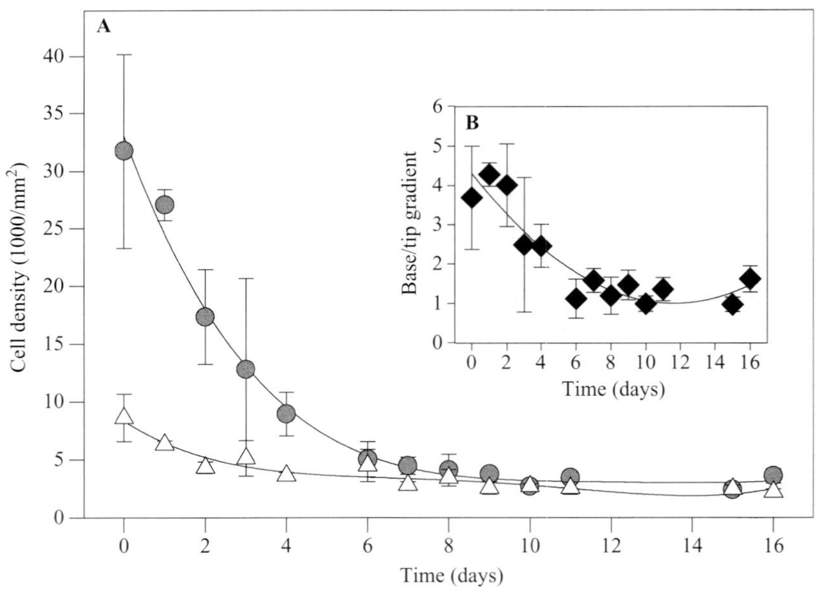

Figure 4.13 A) Development of cell density at the leaf base (●) and the leaf tip (△) with time during the development of leaf number 10 of *Nicotiana tabacum*. B) Ratio of the relative growth rates at the leaf base to that at the leaf tip during the same period.

4.6 Parallel development of morphology and function

Functional development takes place in parallel with growth processes in plants. Perhaps the best known are sink-source transitions in carbon relations. However, many other metabolic processes are altered during maturation of the leaf tissue (e.g. nitrogen metabolism and secondary metabolism). Development of gas exchange properties as a leaf expands and matures is a good example. Transpiration rates per unit leaf area increase with leaf size (Figure 4.14A) and with formation of stomata (Heckenberger *et al.*, 1998; Roggatz *et al.*, 1999). However, tip-to-base gradients are not very pronounced. In contrast, tip-to-base gradients are very prominent in assimilation rate (Figure 4.14B). While the tip already has positive assimilation rates in very small leaves (*Ricinus* 4.5 cm), the base respires up to a significantly bigger leaf size (*Ricinus*: 10 cm).

Classical cuvette techniques—as used in the above study—are limited in their spatial resolution and can have significant impact on the boundary layers of the leaf, and thus on the exchange properties themselves. New developments—again based on image sequence analysis—such as differential thermography (Kümmerlen *et al.*, 1999) or chlorophyll fluorescence imaging (Siebke and Weis, 1995), in combination with growth analysis, offer new opportunities for gaining insight into the spatio-temporal organisation of structure–function relationships in growing leaves.

4.7 Spatio-temporal analysis: prospects and future tasks

In contrast to general impression, plants are highly dynamic organisms, as everyone who has ever seen time-lapse videos of plants is aware. This dynamic power is not only due to the fact that plants cannot escape their fluctuating environment but is closely linked to the basic construction plan of the plant. The modular organisation requires a continuing development of new organs, e.g. to match new environmental conditions. Therefore, spatio-temporal analysis is a prerequisite to understanding the plant's basic organisation as well as its response to the changing environment. However, this requires quantitative measurements and suitable concepts for analysis. Technical problems have limited analysis to a few systems, such as monocotyledonous leaves and roots, which are simpler but only at first view: they also grow in three dimensions, undergo rapid and complete alteration of structure and function in their growth zones, and are not even readily accessible in their natural environment. Dicotyledonous leaves are directly observable during most of their development without disturbance, as described here. As a first step, the diurnal variation of expansional growth distribution has been analysed. However, this is just a starting point, from which completely new

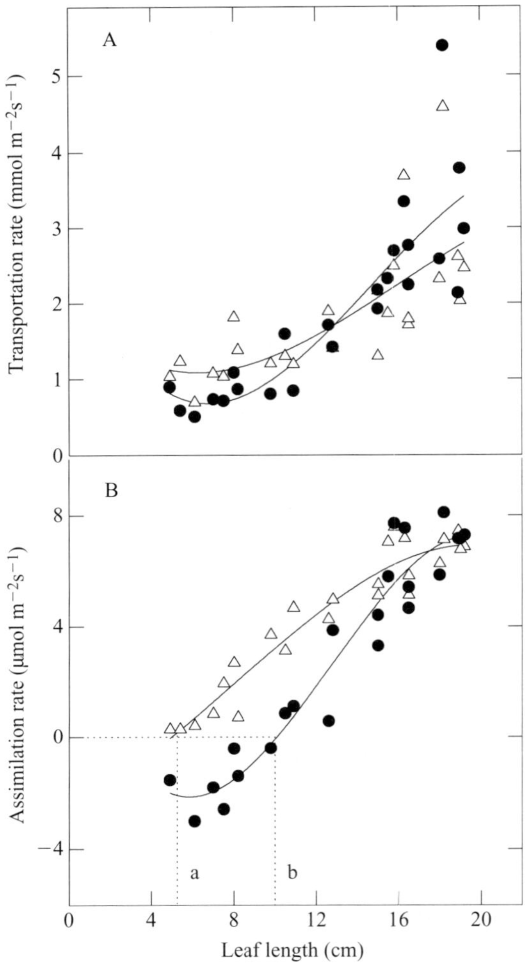

Figure 4.14 The changes in: A) transpiration rate, and B) assimilation rate at the leaf base (●) and the leaf tip (△) of *Ricinus communis* as the leaf expands. Gas exchange was measured using a small cuvette system.

insights can emerge into both basic knowledge of plant organisation and interaction of the plant with the environment. Growth is more than expansion and the destructive analysis of cell density indicates that there are important spatio-temporal patterns of cell division and of intercellular spaces.

Spatial and temporal resolution is intrinsic to image sequences. Imaging techniques for additional processes are still to come and will provide a more complete picture of the structural aspects of growing leaves. With increasing availability of modern imaging equipment, quantitative image analysis will provide the basis for a more complete understanding of growth processes. A good example of this is the use of modern molecular biological techniques, such as reporter gene imaging and other non-destructive methods (Fricker and Oparka 1999). Molecular techniques of this kind also provide new insights into spatio-temporal distribution of functional development. However, these techniques must be further improved, as the present techniques only allow the imaging of strongly expressed genes. The present reporter gene products are stable, which makes them less suitable for dynamic imaging. Moreover, functional understanding cannot be based on reporter gene analysis alone but requires direct support for physiological relevance. Chlorophyll fluorescence imaging and differential thermography are initial achievements in this direction. With the incorporation of an increasing number of physical and physicochemical approaches into botanical research, the methods available will spread further and will help to build an integrated view rather than a mosaic of independent observations. In this context, the development of new methods supports the holistic approaches needed to understand systems as a whole rather than as a collection of independent pieces and processes.

Acknowledgements

The authors thank Uwe Roggatz, Dominik Schmundt, Uwe Heckenberger and Regina Feil for help during and before preparation of this chapter and the Deutsche Forschungsgemeinschaft (DFG) for kind provision of funding in the context of SFB 199 and the research initiative 'Image Sequence Analysis for the Study of Dynamic Processes'. The ongoing interest and support of Mark Stitt is gratefully acknowledged.

References

Acevedo, E., Fereres, E., Hsiao, T.C. and Henderson, D.W. (1979) Diurnal growth trends, water potential and osmotic adjustment of maize and sorghum leaves in the field. *Plant Physiology*, **64** 476–480.

Avery, G.S. (1933) Structure and development of the tobacco leaf. *American Journal of Botany*, **20** 565–592.

Bigün, J. and Granlund, G.H. (1987) Optimal Orientation Detection of Linear Symmetry. *Proceedings of the First International Conference on Computer Vision*, ICCV, London, UK, June 8–11, 1987.

Boyer, J.S. (1968) Relationship of water potential to growth of leaves. *Plant Physiology*, **43** 1056–1062.

Bunce, J.A. (1977) Leaf elongation in relation to leaf water potential in soybean. *Journal of Experimental Botany*, **28** 156–161.

Bünning, E. (1948) *Entwicklungs-und Bewegungsphysiologie der Pflanze*, Springer, Berlin, Germany.
Christ, R.A. (1978) The elongation rate of wheat leaves. *Journal of Experimental Botany*, **29** 603–610.
Cosgrove, D.J. (1986) Biophysical control of plant cell growth. *Annual Review of Plant Physiology*, **37** 377–405.
Dale, J.E. (1988) The control of leaf expansion. *Annual Review of Plant Physiology and Plant Molecular Biology*, **39** 267–295.
Davies, W.J. and van Volkenburgh, E. (1983) The influence of water deficit on the factors controlling the daily pattern of growth of *Phaseolus* trifoliates. *Journal of Experimental Botany*, **34** 987–999.
Durand, J.L., Onillon, B., Schnyder, H. and Rademacher, I. (1995) Drought effects on cellular and spatial parameters of leaf growth in tall fescue. *Journal of Experimental Botany*, **46** 1147–1155.
Else, M.A. and Jackson, M.B. (1998) Transport of 1-aminocyclopropane-1-carboxylic acid (ACC) in the transpiration stream of tomato (*Lycopersicon esculentum*) in relation to foliar ethylene production and petiole epinasty. *Australian Journal of Plant Physiology*, **25** 453–458.
Fricker, M.D. and Oparka, K.J. (1999) Imaging techniques in plant transport. *Journal of Experimental Botany*, **50** 1089–1100.
Granier, C. and Tardieu, F. (1998) Spatial and temporal analysis of expansion and cell cycle in sunflower leaves. *Plant Physiology*, **116** 991–1001.
Groeneveld, H.W., Bergkotte, M. and Lambers, H. (1998) Leaf growth in the fast-growing *Holcus lanatus* and the slow-growing *Deschampsia flexuosa*: tissue maturation. *Journal of Experimental Botany*, **49** 1509–1518.
Haußecker, H. and Spies, H. (1999) Motion, in *Handbook of Computer Vision and Applications* (eds. B. Jähne, H. Haußecker and P. Geißler), Academic Press, San Diego, CA, USA, pp. 310–369.
Heckenberger, U., Roggatz, U. and Schurr, U. (1998) Effect of drought stress on cytological status of *Ricinus communis*. *Journal of Experimental Botany*, **49**, 181-189.
Hejnowicz, Z. and Hejnowicz, K. (1991) Modelling the formation of root apices. *Planta*, **184** 1–7.
Hejnowicz, Z. and Karczewski, J. (1993) Modelling of the meristematic growth of root apices in a natural coordinate system. *American Journal of Botany*, **80** 309–315.
Herbert, T.J. (1992) Geometry of heliotropic and nyctinastic leaf movements. *American Journal of Botany*, **79** 547–550.
Hellgren, O. and Ingestad, T. (1995) A comparison between methods used to control nutrient supply. *Journal of Experimental Botany*, **47** 118–122.
Hu, Y. and Schmidhalter, U. (1998) Spatial distributions and net deposition rates of mineral elements in the elongating wheat leaf under saline soil conditions. *Planta*, **204** 212–219.
Jaffe, M.J., Wakefield, A.H., Telewski, F., Gulley, E. and Biro, R. (1985) Computer-assisted image analysis of plant growth, thigmomorphogenesis and gravitropism. *Plant Physiology*, **77** 722–730.
Knight, M.R., Campbell, A.K., Smith, S.M. and Trewavas, A.J. (1991) Transgenic plant aequorin reports the effects of touch, cold-shock and elicitors on cytoplasmic calcium. *Nature*, **352** 524–526.
Kümmerlen, B., Dauwe, S., Schmundt, D. and Schurr, U. (1999) Thermography to measure water relations of plant leaves, in *Handbook of Computer Vision and Applications* (eds. B. Jähne, H. Haußecker and P. Geißler), Academic Press, San Diego, CA, USA, pp. 763–781.
Maksymowych, R. (1973) *Analysis of Leaf Development*. Cambridge University Press, Cambridge, UK.
Poethig, R.S. (1997) Leaf morphogenesis in flowering plants. *Plant Cell*, **9** 1077–1087.
Poethig, R.S. and Sussex, I. M. (1985) The cellular parameters of leaf development in tobacco: a clonal analysis. *Planta*, **165** 170–184.
Roggatz, U., McDonald, A.J.S., Stadenberg, I. and Schurr, U. (1999) Effect of nitrogen deprivation on cell division and expansion of *Ricinus communis* (L). *Plant, Cell and Environment*, **22** 81–90.
Scanlon, M.J. (1998) Force fields and phyllotaxy: an old model comes to age. *Trends in Plant Science*, **3** 413–414.

Schmundt, D. and Schurr, U. (1999) Plant-leaf growth, in *Handbook of Computer Vision and Applications* (eds. B. Jähne, H. Haußecker and P. Geißler), Academic Press, San Diego, CA, USA, pp. 719–735.

Schmundt, D., Stitt, M., Jähne, B and Schurr, U. (1998) Quantitative analysis of the local rates of growth of dicot leaves at a high temporal and spatial resolution, using image sequence analysis. *The Plant Journal*, **16** 505–514.

Schurr, U. (1997) Growth physiology and measurement of growth, in *Progress in Botany* (eds. H.D. Behnke, U. Lüttge, K. Esser, J.W. Kadereit and M. Runge), Springer Verlag, Berlin, Germany, **59** pp. 355–373.

Seneweera, S.P., Basra, A.S., Barlow, E.W. and Conroy, J.P. (1995) Diurnal regulation of leaf blade elongation in rice by CO_2. *Plant Physiology*, **108** 1471–1477.

Shackel, K.A., Matthews, M.A. and Morrison, J.C. (1987) Dynamic relation between expansion and cellular turgor in growing grape (*Vitis vinifera* L.) leaves. *Plant Physiology*, **84** 1166–1171.

Siebke, K. and Weis, E. (1995) Assimilation images of leaves of *Gleochoma hederaceae*: analysis of non-synchronous stomata-related oscillations. *Planta*, **196** 148–165.

Silk, W.K. (1984) Quantitative descriptions of development. *Annual Review of Plant Physiology*, **35** 479–518.

Silk, W.K. and Erickson, R.O. (1979) Kinematics of plant growth. *Journal of Theoretical Biology*, **76** 481–501.

Silk, W.K., Hsiao, T.C., Diedenhofen, U. and Matson, C. (1986) Spatial distributions of potassium solutes and their deposition rates in the growth zone of the primary corn root. *Plant Physiology*, **82** 853–858.

Spollen, W.G. and Sharp, R.E. (1991) Spatial distribution of turgor and root growth at low water potentials. *Plant Physiology*, **96** 438–443.

Tomos, D. and Pritchard, J. (1994) Biophysical and biochemical control of cell expansion in roots and leaves. *Journal of Experimental Botany*, **45** 1721–1731.

Van Volkenburgh, E. (1987) Regulation of dicotyledonous leaf growth, in *Physiology of Cell Expansion During Plant Growth* (eds. D.J. Cosgrove and D.P. Knievel), The American Society of Plant Physiologists, Rockville, USA, pp. 193–201.

Walter, A. and Schurr, U. (1999) The modular character of growth in *Nicotiana tabacum* plants under steady-state nutrition. *Journal of Experimental Botany*, **50** 1169–1177.

Watts, W.R. (1974) Leaf extension in *Zea mays*. *Journal of Experimental Botany*, **25** 1085–1096.

5 Plant architecture and light signals

Harry Smith

5.1 Introduction

The architecture of the individual plant, and thereby that of the canopy in a population of plants, is determined largely by the radiation environment. Light energy is required for photosynthesis and the strategy of most plants is to optimise the acquisition of light energy by placing the photosynthetic structures, mainly the leaves, into those parts of the spatial environment where radiation is plentiful. In essence, the plant operates a space-filling technique that results in leaves being disposed optimally for radiation interception. In the complex radiation environments posed by plant communities, individual plants compete to fill the space available, and it is quite obvious, even to the casual observer, that this phenomenon must involve interaction between competing plants. The objective of this chapter is to review the mechanisms through which plants detect the proximity of neighbours and operate processes to avoid mutual shading.

5.2 Shade avoidance

Plants have evolved two contrasting strategies that enhance survival under conditions of competition for light. The most ancient strategy is that of shade tolerance, a combination of mechanisms that improves the biochemical efficiency of light capture whilst reducing the expenditure of energy through respiratory activity. Shade tolerance is a 'passive' approach to coping with limited light and is exemplified by many of the vascular cryptogams, whose energy requirements for growth and development are relatively low and whose demands on photosynthetic energy acquisition are correspondingly conservative. Their efficiency of light capture and utilisation often allows such plants to survive in deeply shaded environments. Some higher plants assume this strategy but, for most, coping with shade is a matter of avoidance rather than tolerance. As plants became more complex throughout evolution, energy requirements correspondingly rose and, concomitantly, mechanisms that actively improve light energy capture acquired important adaptive value. Shade avoidance is a syndrome of developmental responses, which collectively allows plants to maintain photosynthetic structures in light conditions that are not limiting for growth. Shade-avoiding plants have high resource capture requirements and are generally profligate with the utilisation of resources. In other words, shade avoiders are aggressive, competitive plants, whose existence depends on the ability to dominate in the battle for the resource of light. This may sound a melodramatic description of plant life, usually considered in somewhat more pastoral terms, but competition for resources between plants is essentially completely ruthless. Successful shade avoiders dominate the community by monopolisation of resources, particularly the resource of light, thereby shading out their competitors. The hypothesis of 'asymmetric competition'—more simply described as 'first-come, first-served'—assumes that individuals that anticipate and respond to the earliest signals of competition are able to appropriate resources and gain a competitive advantage that keeps them ahead of the pack (Harper, 1976; Weiner, 1990). Shade avoidance is a classic example of the dictum 'get ahead and stay ahead'. By early detection of the presence of neighbours, a successful shade avoider ensures the supply of radiant energy; and thus it grows rapidly, extends its resource capture capability and ultimately dominates—much like human competition.

Shade avoidance is one of the most important competitive strategies. The fact that plants grow long and slender when grown in dense stands has probably been known for centuries, and for that period it has generally been assumed that such plants were 'growing towards the light'. In fact, it is known that plants grow away from the shade rather than towards the light (Smith, 1982). Shade avoidance involves the perception of specific light signals reflected from neighbours. The photoreceptors responsible are the phytochromes, and they have been supremely designed through evolution to detect the reflection of radiation from nearby vegetation. The phytochromes are unique amongst biological photoreceptors, as they respond to the relative amounts of two radiation signals rather than to the absolute amount of a single signal. The phytochromes are chromoproteins with linear tetrapyrrole chromophores existing in two isomeric forms. The absorption maxima of the two forms are different; one (Pr) absorbs red photons preferentially, whilst the other (Pfr) absorbs far-red photons. Absorption of a photon by either form causes conversion to the other form, according to the following scheme:

$$\text{Pr} \underset{\text{Far-red}}{\overset{\text{Red}}{\leftrightarrows}} \text{PfR} \rightarrow \text{biological action}$$

These photochemical properties endow the phytochromes with the capacity to measure the relative amounts of red (R) and far-red (FR) radiation incident upon the plant. Perception of the R:FR comparative signal provides the plant with unambiguous information on the proximity of neighbours.

5.3 Information in the radiation environment

The radiation environment to which plants are exposed is very complex. Variation in amount, spectral distribution, direction, timing and polarisation occurs both randomly and predictably during daily and annual cycles (Smith, 1982). Some of these components also vary with latitude, altitude or aspect. The light environment underwater is equally complex, and solar radiation even penetrates to significant depths in some types of soil. There is, therefore, no shortage of signals in the light environment, but it is important to determine which signals provide information of potential ecological significance. Random signals, such as those associated with unpredictable weather conditions (i.e. clouds, rainstorms, etc.) logically provide little in the way of meaningful information. Predictable signals, such as those associated with daily or seasonal cycles, obviously have the potential to be important indicators of environmental conditions. This article is concerned primarily with the responses of plants to the presence of competing neighbours and, therefore, the first priority is to identify those signals that provide reliable information on the proximity of other vegetation. Clearly, signals of proximity will depend upon the interaction of light with vegetation, so first it is important to understand the general nature of solar radiation, and how the properties change upon interaction with vegetation.

5.3.1 Solar radiation

Radiation is emitted from the sun in accordance with the Planck radiation distribution formula, which describes the spectral distribution of emitted energy (or photons) as a function of the surface temperature of the sun. The sun approximates to a 'black body', with an effective surface temperature of 5800°K; this means that the wavelength of peak photon emission is c. 600 nm.

The radiation incident upon a particular point is composed of both direct solar radiation and indirect radiation scattered from the sky. The proportions of direct and scattered radiation vary with time of day, time of year, altitude and latitude, as all of these factors affect the path length of solar radiation through the atmosphere. Scattering by atmospheric molecules (Rayleigh scattering) increases greatly as an inverse function of wavelength ($R \propto 1/\lambda^4$; where, R = scattering, and λ = wavelength). Shorter wavelengths are, therefore, preferentially scattered, accounting for the blue colour of the sky. As the sun's elevation decreases below c. 10° above the horizon, the shorter wavelengths are increasingly attenuated from the direct beam, but scattering is increased towards the earth.

5.3.2 Daylight spectra

The radiation incident upon the earth's surface is modified as a result of absorption and scattering by constituents in the atmosphere (Figure 5.1). Important absorbing molecules include water vapour, oxygen and carbon dioxide. At shorter wavelengths, ozone is an important absorber of UV-radiation. In spite of all these modifications of solar radiation, the spectral distribution of global radiation during daytime (i.e. the period when the sun is more than c. 10° above the horizon) is relatively constant (Holmes and Smith, 1977a). Even overcast clouds or heavy rain, which may attenuate the total radiation by up to 99%, have little effect on the spectrum detected by a horizontal, cosine-corrected sensor. This is important, as will be emphasised below, because the constant daylight spectrum provides a norm against which ecologically important light signals may be compared.

5.3.3 Interaction with vegetation

When radiation interacts with vegetation, selective reflection and absorption occur. The photosynthetic and other pigments within vegetative structures absorb radiation with characteristic spectral dependencies, whilst the particu-

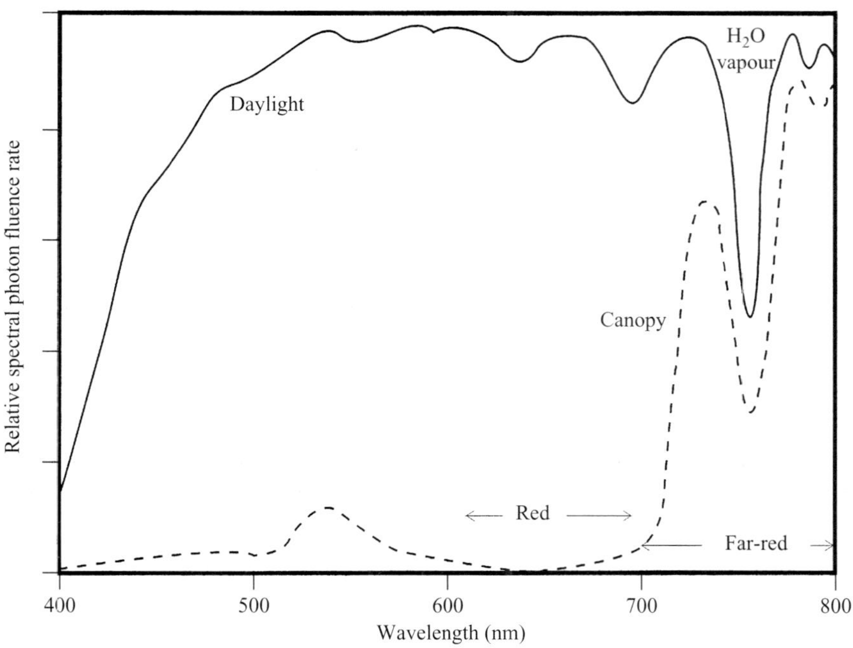

Figure 5.1 The spectral distribution of radiation in the open and within a vegetation canopy.

lar physical structure of plant organs results in specific reflection properties. Wavelengths in the blue and red are strongly absorbed by vegetation, whilst those in the green and in particular those beyond the red are strongly reflected or scattered (Holmes and Smith, 1977b). The human eye is optimally sensitive to radiation at about 550 nm and is insensitive beyond c. 700–730 nm. Correspondingly, leaves appear green, rather than far-red—if we could see wavelengths beyond the red we would have a completely different perception of the natural world!

5.3.4 Reflection signals

Figure 5.1 shows plots of global radiation outside and within a plant canopy. Thus, the light climate within vegetation canopies is dominated by radiation that is scattered through or from the surface of leaves. It is sensible, therefore, to think in terms of 'reflection signals' (Smith *et al.*, 1990), since the specific spectral changes associated with reflection from vegetation potentially provide important information on the presence, proximity and magnitude of neighbouring plants.

5.4 Phytochromes and light signal perception

The phytochromes are a family of chromoprotein photoreceptors (Clack *et al.*, 1994) that, as far as is known, all have similar chromophores. The signal perception capacities of the phytochromes are determined by their absorption characteristics and by the kinetic relationships between the two stable forms of the molecule, i.e. Pr and Pfr.

5.4.1 Absorption spectra, photoconversions and photoequilibria

Figure 5.2 shows the absorption spectra of phytochrome purified from etiolated oat seedlings taken after far-red radiation (i.e. Pr) and after red irradiation (i.e. predominantly Pfr). Pr absorbs strongly in the red with an absorption maximum (λ_{max}) at c. 665 nm, and Pfr absorbs maximally in the far-red with a broader λ_{max} at c. 730 nm. Although conversions between Pr and Pfr begin with very rapid photochemical reactions to form a primary photoconversion intermediate, each is followed by slower, thermochemical reactions that result in the formation of the final stable state (i.e. Pfr or Pr). The actual kinetics of photoconversion are, in practice, surprisingly slow even at high photon fluence rates. For example, it can take upwards of 60 s to reach photoequilibrium when etiolated seedlings are irradiated with a helium-neon laser beam at 633 nm providing 125 $\mu mol\ m^{-2}\ s^{-1}$, and even at 1250 $\mu mol\ m^{-2}\ s^{-1}$ equilibrium takes 15 s or so (Smith and Fork, 1992). Photoperception by the phytochromes, therefore, is substantially slower than

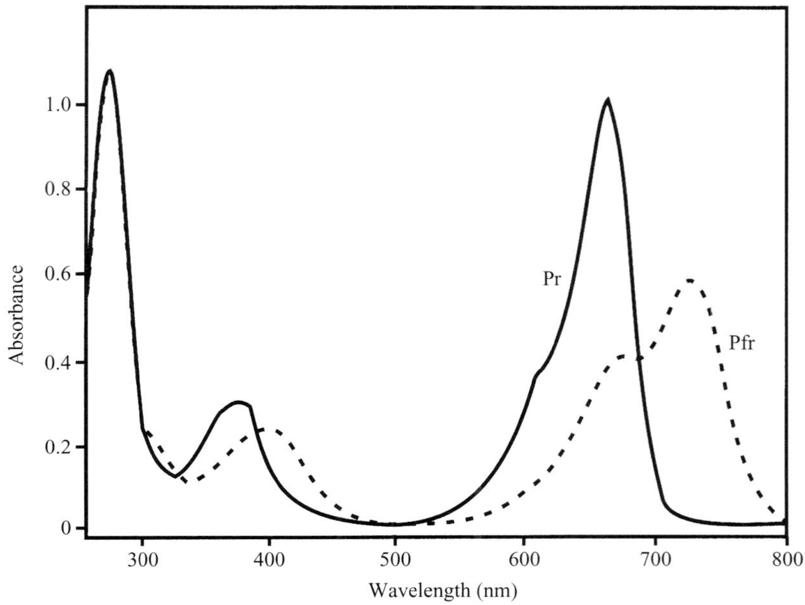

Figure 5.2 The absorption spectra of the phytochromes. Abbreviations: Pr, phytochrome that absorbes red photons preferentially; Pfr, phytochrome that absorbs far-red photons.

photoperception by the human visual system. Furthermore, the thermochemical reactions between the intermediates in the two photoconversions can substantially retard the overall rates at which photoequilibrium is reached, and can cause strange results. At irradiances similar to midday sunlight (i.e. 1500–2000 $\mu mol\, m^{-2}\, s^{-1}$), a large proportion (c. 50% in some cases) of the total phytochrome can be held in the form of photoconversion intermediates (Smith and Fork, 1992). Thus, whereas at low irradiance, red light ultimately establishes a photoequilibrium (i.e. Pfr/P where P = total phytochrome) of c. 0.87, at sunlight irradiances only c. 45% of the total phytochrome may be in the Pfr form. Under these circumstances, the relative proportions of Pr and Pfr are similar at both high and low irradiance, the rest of the phytochrome being present in the form of short-lived intermediates.

5.5 Phytochrome response modes

5.5.1 Responses to light pulses

The classical notion of phytochrome is that the photoconversion of Pr to Pfr initiates the biological action; in other words, red light is a signal perceived by phytochrome and transduced to a biological message that induces developmental responses. This idea of phytochrome as a switch is well established in the literature but is a misleading oversimplification. There are some situations in which one can think in terms of a switch, but these are generally unimportant in the real world, where plants are exposed not to single pulses of radiation but to continuous light. Many early investigations showed that the relationship between Pfr/P (established by a pulse of light) and the response evoked by that pulse, described a curve of exponentially decreasing sensitivity. Later research showed that, in many cases, the relationship was more accurately described by two exponential relationships, one providing an exceptionally sensitive response at low fluences and the other a less-sensitive response at higher fluences (Mandoli and Briggs, 1981; Cone *et al.*, 1985). These two response modes are now conventionally known as the 'very low fluence response' (VLFR) and the 'low fluence response' (LFR). They can be observed for the germination of many seeds in response to light and for the inhibition of hypocotyl elongation in seedlings grown in the dark and exposed to light. In light-grown plants, the LFR can be seen to operate if pulses of red or far-red are given after the end of a daily light period, the so-called 'end-of-day' (EOD) response. The VLFR is saturated at extremely low total fluence and, realistically, can only be important in nature in very restricted circumstances, such as the germination of seed that is buried under the soil but exposed to brief light, as might happen when an animal disturbs the surface (Scopel *et al.*, 1991).

The best-known criterion that phytochrome is responsible for a particular light-mediated response is R/FR reversibility. Unfortunately, R/FR reversibility in the classical sense can only be used for responses that fall into the LFR category. Because Pr absorbs radiation at least up to 730 nm, even FR radiation establishes a small proportion of Pfr, and this is sufficient to saturate responses in the VLFR range.

5.5.2 Responses to continuous irradiation

The VLFR and the LFR are responses induced by pulses of radiation. When plants are exposed to continuous radiation, the biological response becomes a complex function of the steady-state concentration of Pfr and of the irradiance. In etiolated tissues, the so-called 'High irradiance response' (HIR) (Hartmann, 1966) is manifested, as a response in which maximum activity is observed at wavelengths that establish low concentrations of Pfr. High irradiance responses are so-called because they exhibit increasing activity with increasing irradiance, although simultaneously there is a relationship with wavelength. In the classical HIR, the wavelength of maximum activity is in the far-red, between the absorption maxima of Pr and Pfr; at this wavelength, there is a log-linear relationship between photon irradiance and biological response. The HIR is known to regulate processes such as seed germination, hypocotyl elongation, cotyledon expansion and the synthesis of anthocyanins.

In light-grown plants, the phytochromes mediate responses to continuous light that bear little relationship to those observed in etiolated seedlings, although it is possible to account for the responses on the basis of phenomena studied in seedlings. In the natural environment, plants are exposed not to radiation of restricted wavelength range but to broadband radiation, spanning the whole 400–800 nm range. At wavelengths shorter than c. 730 nm, the absorption spectra of Pr and Pfr overlap (Figure 5.2), which means that broadband radiation establishes a photostationary equilibrium between Pr and Pfr. In principle, the photoequilibrium between Pr and Pfr can be calculated from the absorption cross-sections of Pr and Pfr, and the spectral photon flux distribution of the actinic radiation. In practice, it is simpler to relate Pfr/P to the relative amounts of red and far-red (i.e. the R:FR ratio) in the actinic radiation (Smith and Holmes, 1977). Conventionally, R:FR ratio is measured as the ratio of photon flux in 10 nm wide bands centred on 660 nm and 730 nm, and these data can be readily obtained from a spectroradiometric scan. By demonstrating that a particular biological response is a function of R:FR ratio, and therefore of Pfr/P, it is possible to show that the response is mediated by one or more of the phytochromes. Such R:FR ratio responses provide the experimental basis for concluding that shade avoidance responses are mediated by the phytochromes (Smith, 1982, 1995).

5.6 Phytochromes in the natural environment

5.6.1 Pfr/P vs *R:FR*

The relationship between the R:FR ratio of natural or artificial broadband radiation and Pfr/P established by that radiation is both experimentally and theoretically a rectangular hyperbola (Figure 5.3). The R:FR ratio of unfiltered solar radiation (with a solar angle of $> 10°$) approximates to 1.12, and is largely unaffected by random fluctuations in the total amount of radiation caused by weather conditions (Smith and Holmes, 1977). This value of R:FR ratio establishes a photoequilibrium of c. 0.55, which is on the shoulder of the hyperbola in Figure 5.3, and represents a norm against which the plant may compare light signals of potentially important environmental conditions. The R:FR ratio is potentially a signal of a number of important environmental conditions, including depth underwater, the day–night or night–day transition, and the proximity of neighbouring vegetation. For good reasons, however, it is likely that only the latter signal is of real significance.

Whenever light interacts with vegetation, there is a large change in spectral

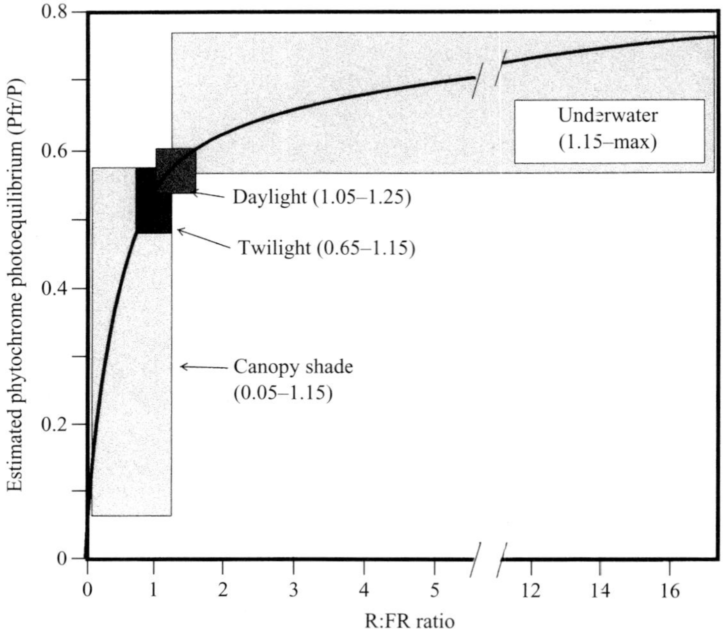

Figure 5.3 The relationship between R:FR ratio and the phytochrome photoequilibrium. Abbreviations: R, red; FR, far-red; P, total phytochrome; Pfr, phytochrome that absorbs far-red.

distribution, with transmitted or scattered radiation being deficient in red and blue wavelengths and relatively (sometimes absolutely) enhanced in the far-red. These changes in R:FR occur on the steep slope of the rectangular hyperbola and, therefore, could potentially act as highly sensitive signals of shade, or impending shade. A small reduction in R:FR from that of the daylight norm causes a relatively large change in Pfr/P, the exact requirement of a sensitive detection system. Demonstration that the phytochromes do operate as a function of R:FR ratio, however, demands experimental evidence that distinguishes responses to spectral changes from those potentially caused by changes in the total amount of light.

5.6.2 Phytochromes as proximity sensors

Within natural vegetation stands, both the amount of radiation and the spectral distribution of that radiation vary as a function of the proximity and mass of vegetation with which the radiation interacts (Gilbert *et al.*, 1995). In nature, therefore, it is impossible to determine whether the developmental responses that comprise shade avoidance are caused by changes in irradiance or in spectrum. In controlled conditions, on the other hand, it is possible to reduce the R:FR ratio not by diminishing the radiation in the red wavelengths but by elevating the radiation in the far-red spectral region. As far-red has little effect on photosynthetic activity, if shade avoidance is caused by phytochrome-mediated perception of reduced R:FR, then shade avoidance responses should be evoked under such artificial conditions. In other words, by increasing the total amount of radiation through far-red supplementation, shade avoidance responses should be observed.

Figure 5.4 shows the first published example of shade avoidance responses observed in plants growing in FR-supplemented white light. A uniform white light background was maintained in all the cabinets, whereas the far-red supplementation increased from left to right. The plants are *Chenopodium*

Figure 5.4 Simulating shade avoidance by adding FR rather than removing R. These *Chenopodium album* plants were grown in specialised growth cabinets in which the photosynthetically active radiation (i.e. PAR, 400–700 nm) was held uniform in each cabinet, but in which the R:FR ratio was progressively reduced by adding FR, giving the estimated Pfr/P values shown. For abbreviations see legend to Figure 5.3.

album, an aggressive weed species, and the most striking response is the increase in stem elongation seen at low R:FR ratios. Since these early observations, many species have been tested and all show qualitatively similar responses, indicating that R:FR ratio detection is universal amongst plant species. When the growth rates were plotted against the Pfr/P calculated to be established by the light regimes, the relationships with Pfr/P were found, in all cases, to be linear (Figure 5.5) (Morgan and Smith, 1976). The slope of the relationship between growth rate and Pfr/P differs between species, of course, but the variability is systematic. Species that are genetically adapted to shaded habitats, and whose normal strategy is shade tolerance, show shallow slopes, whereas aggressive, competitive plants show steep slopes (Figure 5.6) (Morgan and Smith, 1979).

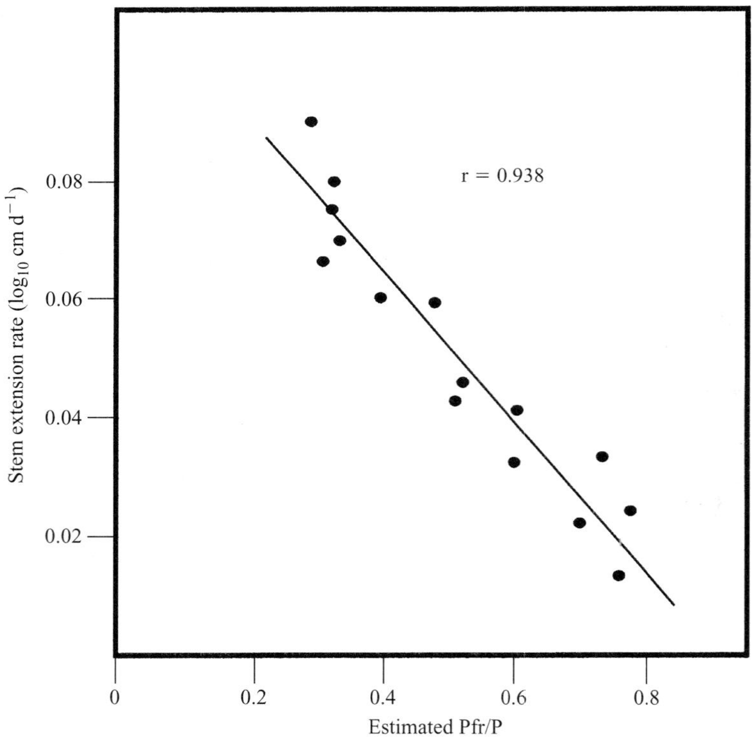

Figure 5.5 Decreasing the phytochrome photoequilibrium concentration causes linear increases in extension growth. The data for the *Chenopodium* experiment shown in Figure 5.4 are plotted here as stem extension rate *vs* estimated phytochrome photoequilibrium (Pfr/P). For abbreviations see legend to Figure 5.3.

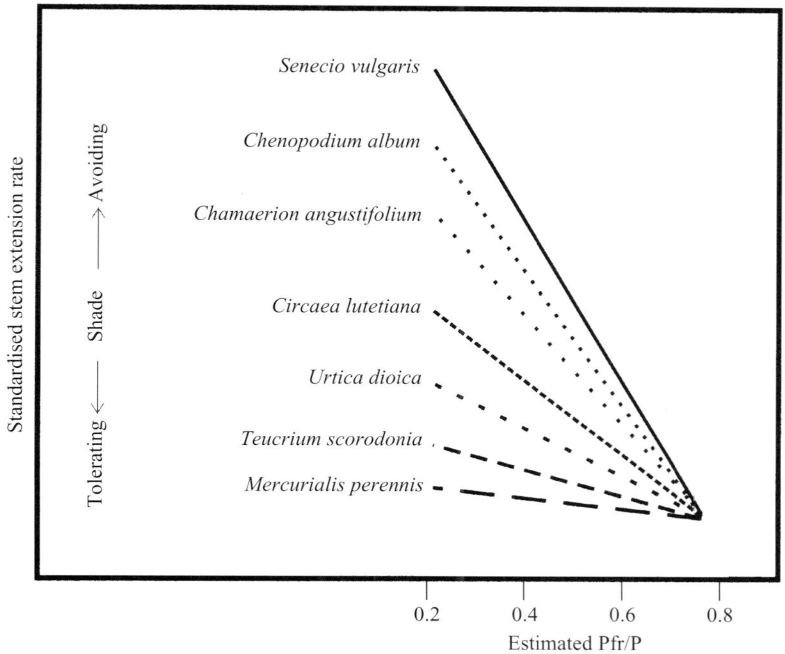

Figure 5.6 The strength of shade avoidance varies systematically with ecological habitat. All plants that have been tested using the simulation approach of Figure 5.4 show linear relationships similar to that for *Chenopodium* in Figure 5.5. The strength of the response to low R:FR ratio, however, depends on the genetic adaptation of the species (or ecotype) to shaded or open habitats. For abbreviations see legend to Figure 5.3.

5.7 Nature of shade avoidance

Shade avoidance is described here, and elsewhere, as a syndrome of responses. It is, perhaps, most simple to think of a syndrome of responses as being mediated by a single sensor from which branching transduction pathways lead to a range of potential responses. On this basis, whether or not a specific response is observed will depend upon the particular condition of the target cells, tissues or organs. A somewhat less comfortable hypothesis is that syndromic responses are causally unrelated, each having a specific sensor and a specific transduction pathway. As yet, the jury is out on this question regarding shade avoidance (see below). At the very least, though, it is certain that the responses are all induced by the same environmental signal. Shade avoidance responses are, by definition, those responses that occur when plants are grown in close proximity. They include: increased stem and petiole elongation; reduced leaf development and accelerated senescence; elevated leaf angle; enhanced apical dominance and reduced branching; accelerated flowering and fruit production; reduced seed set; and, overall, a redistribution of assimilates from leaf and storage organ development towards extension growth (Table 5.1). All of these phenomena can be observed under natural shade avoidance conditions, and can be simulated by growing plants in cabinets in which R:FR ratio is reduced by supplementation with far-red.

Shade avoidance represents a systematic modification of developmental pattern, in response to a simple environmental signal. In this respect, shade avoidance exemplifies the plasticity of plant development (Schmitt and Wulf, 1993). Although, obviously, shade avoidance is genetically determined, the capacity of the plant to modify its developmental pattern in response to environmental signals is nowhere more spectacular. It is plasticity that allows individual plants to react to environmental stimuli in such a way that they can maximise their chances of survival. Plant physiologists tend to think in terms of averages, but there is no such thing as an average plant. Individuality is crucially important for survival, not only of the individual but ultimately of

Table 5.1 The shade avoidance syndrome

Physiological process	Response to shade (i.e. reduced R:FR ratio)
Germination	**Retarded**
Extension growth	**Accelerated**
Internode extension	Rapidly increased (lag c. 5 min)
Petiole extension	Rapidly increased
Leaf extension	Increased in cereals
Leaf development	**Retarded**
Leaf area growth	Marginally reduced
Leaf thickness	Reduced
Chloroplast development	**Retarded**
Chlorophyll synthesis	Reduced
Chlorophyll *a:b* ratio	Balance changed
Apical dominance	**Strengthened**
Branching	Inhibited
Tillering (in cereals and grasses)	Inhibited
Flowering	**Accelerated**
Rate of flowering	Markedly increased
Seed set	Severe reduction
Fruit development	Truncated
Senescence	**Accelerated**
Leaf senescence	Advanced
Leaf abscission	Advanced
Assimilate distribution	**Marked change**
Storage organ deposition	Severe reduction

the species. Such considerations are of major significance when it comes to trying to establish the cellular and molecular mechanisms through which plastic processes, such as shade avoidance, operate.

It has to be accepted that the basic developmental plan of a plant carries within it the capacity for profound modification in response to the perception of environmental signals. Light signals are by no means the only such environmental signals evoking massive modifications in developmental programmes, but they are exemplars of the mechanisms for expressing the inherent plasticity of plants. How such plasticity has evolved and how it has been maintained through evolution remain intriguing questions outside the scope of this article (see Smith 1990; Callahan *et al.*, 1997; Pigliucci and Schlichting, 1997).

5.7.1 Shade avoidance happens quickly

Shade avoidance is a very rapid phenomenon. Using position sensitive transducers, it has been shown that many plants respond to reductions in R:FR ratio by increases in stem extension rate within a few minutes of the stimulus being applied (Morgan and Smith 1978; Morgan *et al.*, 1980; Child and Smith, 1987). Mustard (*Sinapis alba*) seedlings were grown in constant white light but exposed to brief pulses of far-red from a fibreoptic probe directed at the uppermost growing internode. Growth rate was repeatedly and reversibly responsive to the far-red stimulus, and the period of elevated growth rate was related to the period of exposure to the stimulus. An interesting point is that, with very short stimuli, the resulting brief peak of growth rate occurred after the stimulus had been removed, and therefore in conditions in which the R:FR ratio should have determined a reduced growth rate. These observations provide some limits to the mechanisms involved. It is clear that responses occur within a finite time interval after the onset of the stimulus; in other words, although the responses are rapid (for a plant), the transduction chains require several minutes to convert the environmental stimulus into a modification of growth rate. The overall picture is that the phytochromes are continuously modulating the output from a mechanism that controls extension growth rate. An appropriate analogy might be a rheostat, rather than the switch so favoured by those who write about the phytochromes in textbooks.

5.8 Roles of individual phytochromes in shade avoidance

As stated above, the phytochromes are a family of photoreceptors, and it is important to determine which of the phytochromes is responsible for the perception of R:FR ratio signals and the induction of shade avoidance responses. The most compelling evidence comes from studies of photomorphogenic mutants but studies of transgenic plants have also been useful.

5.8.1 Evidence from mutants

There are now many mutants that are deficient in specific aspects of photomorphogenesis. Some of these have been known for many years, whereas others have been selected recently on the basis of increasingly sophisticated screening strategies. Photomorphogenic mutants are defined as mutants lacking specific light-mediated responses, and can be classified into three main groups: photoreceptor mutants; chromophore synthesis mutants; and transduction chain mutants. The most useful for the purpose of this article are the photoreceptor mutants, in which the lesion is within the gene encoding the photoreceptor apoprotein. Mutants in this category are available in several plants but *Arabidopsis* provides the most complete range. *Arabidopsis* mutants are now available that have lesions in the genes encoding phytochromes (phy) A, B, D and E. Some of these mutants are completely devoid of the specific members of the family in which the mutation is located, and therefore provide excellent material for assessing the roles of individual phytochromes in specific responses. The allocation of function to phytochrome family members is an ongoing project at the time of writing, but a number of conclusions can be reached from studies of the *Arabidopsis* mutants (Whitelam and Harberd, 1994; Smith, 1995).

Thus, mutant evidence shows that phyA is responsible for the VLFR and the FR-HIR in germinating seeds and de-etiolating seedlings; no other member of the phytochrome family can mediate these two specialist functions (Botto *et al.*, 1996; Casal *et al.*, 1997). Phytochrome A appears to have limited function in the light-grown plant, as shown by the almost complete lack of phenotype in mature plants of the *phyA* mutant (Whitelam *et al.*, 1993). On the other hand, phyA does function in the perception of extended photoperiods in long-day plants (Bagnall *et al.*, 1995), and it is possible to demonstrate, using double *phyA: phyB* mutants, that phyA does have a minor control over elongation growth in de-etiolated seedlings.

Phytochrome B has many functions in photomorphogenesis, including the LFR regulating seed germination (Botto *et al.*, 1995) and seedling de-etiolation (Reed *et al.*, 1994). Phytochrome B has a major function in the regulation of extension growth in mature plants, and has been closely linked to the shade avoidance syndrome. It is certain that phyB is involved in shade avoidance, but it is also clear that phyB is insufficient for shade avoidance (Smith and Whitelam, 1997). The latest evidence implicates both phyD and phyE as important collaborators with phyB in mediating shade avoidance. Phytochrome D has a high sequence similarity to phyB (Matthews and Sharrock, 1997), and may indeed be described as phyB-2, as has been suggested for other species, such as tomato and tobacco (Pratt *et al.*, 1995). Mutants at the *PHYD* and *PHYE* loci have shown that both of these phytochromes have contributory functions in the perception of the light signals that evoke shade avoidance (Devlin *et al.*, 1998, 1999).

Thus, the evidence from mutants is that the shade avoidance syndrome is probably mediated by multiple phytochromes (Smith and Whitelam, 1997). It may be found that individual components of the shade avoidance syndrome can be mediated by individual phytochromes, but at present it is more prudent to think in terms of overlapping functions. Such overlap of function within the members of a gene family has been proposed as the basis of functional plasticity (Smith, 1990).

5.9 Ecology of shade avoidance

The ecological significance of shade avoidance lies in its importance as a competitive strategy for plants growing in crowded environments (Schmitt and Wulf, 1993). Using reflection signals from neighbours, plants can obtain information on the proximity of neighbours, and therefore on the competitive threat posed. The operation of shade avoidance in the field has been neatly demonstrated by experiments carried out in Argentina, where seedlings of *Sinapis alba*, *Datura ferox* and *Chenopodium album* were grown in the open, close to reflective screens of vegetative material, so-called ‘green fences’ (Ballaré *et al.*, 1987). The presence of a green fence caused increased internode extension in the exposed plants, compared to those grown close to screens of identical material but which had been bleached by the action of a herbicide. Following-up this ingenious experiment, the same group grew plants in the open with transparent collars placed around the growing internodes (Ballaré *et al.*, 1990). The collars could be filled with liquid; if the liquid was copper sulphate, which is a strong absorber of FR radiation, the presence of a nearby ‘green fence’ had no effect on internode elongation. These two field experiments demonstrate that plants respond to the presence of neighbours by increasing internode elongation rate, and that the signal is reflected FR radiation. Similar inferences can be drawn from experiments on agricultural crops, which are commonly grown at high density in regular arrays. Early work by Kasperbauer and co-workers (see Kasperbauer, 1987) demonstrated the importance of transmitted and reflected FR radiation in determining the elongation rate, and the allocation of resources, in crop plants growing in ordered rows. Clearly, then, plants are highly plastic in their responses to the natural light environment, and this plasticity seems to be related to ecological pressure, such as competition at high densities. However, demonstration of the occurrence of a phenomenon, and its correlation with an environmental signal, does not necessarily prove causality.

Phenotypic plasticity can only be adaptive if the phenotype induced by each relevant environment has a high fitness in that environment. Thus, although it is tempting to conclude that the plasticity that is represented by shade avoidance responses has significant adaptive value, it is necessary to show that shade avoidance enhances fitness in crowded environments. To test this proposal is no simple matter, as a proper test requires examination of the relative fitness of individuals, or populations, in which the plasticity is either present or absent. Natural populations do not provide such experimental material, but individuals in which specific aspects of plasticity have been modified, either by mutation or by transgenic means, can, in principle, be used as powerful discriminators. Such a test has been performed for the adaptive significance of shade avoidance.

Using mutant *Brassica rapa* deficient in phytochrome B (the *ein* mutant), and transgenic tobacco plants overexpressing a *PHYA* complementary deoxyribonucleic acid (cDNA), Schmitt and co-workers (1995) showed that phytochrome-mediated elongation responses to foliage shade can have important impact on growth and relative fitness. Suppression of the elongation response in the transgenic *PHYA* overexpressers resulted in decreased relative performance in crowded conditions, whilst on the other hand the constitutive elongation of the *ein* mutant was disadvantageous at low density. These results were as predicted by the shade avoidance hypothesis. Whereas the *PHYA* and *PHYB* genes may not directly underlie genetic variation in plastic shade avoidance responses, transgenic overexpression or mutational lesion of these genes disables the normal plastic responses to crowding, permitting the comparison of inappropriate phenotypes that would not be possible with natural populations. Even so, there is variation in shade avoidance responses within closely related plant populations, as has been observed by Dudley and Schmitt (1996) in studies of *Impatiens capensis*. In these experiments, it was also found that selections with strong elongation responses to crowding were, indeed, fitter in crowded conditions but were at a disadvantage at low density, in comparison to other selections with low elongation capacities. The overall conclusion is that phytochrome-mediated shade avoidance responses allow plants to react sensitively to signals from neighbours with a morphological response appropriate for the level of competition they are experiencing. There is, however, clearly a cost, as individuals with a high capacity to respond competitively to crowded conditions may be at a disadvantage under conditions in which neighbour competition is unlikely.

5.10 Shade avoidance and canopy architecture in the field

Although the most dramatically obvious manifestation of shade avoidance is the regulation of internode elongation, the correlative nature of plant development results in major changes in the whole pattern of development. When shade avoidance is invoked, leaf expansion and, in particular, growth in leaf thickness tend to be retarded. The normally clear separation of leaves into mesophyll cell layers may break down, particularly under natural situations where low R:FR signals are accompanied by low total irradiances. As long ago as 1978, McLaren and Smith demonstrated an interaction between R:FR and irradiance in *Rumex obtusifolius*, showing that both phytochrome-mediated effects and effects presumably dependent on photosynthesis regulated leaf development. When low R:FR was combined with low total irradiance, the normal structure of the leaf was indistinguishable, and there were no clear divisions between palisade and spongy mesophyll layers. On the other hand, such environmental conditions appear to have little direct effect on photosynthetic characteristics. Smith *et al.*, (1993) grew wild-type and *aurea* tomato mutants in combinations of low R:FR and low irradiance and observed no regulation of photosynthetic capacity by the phytochromes—the *aurea* mutant is a chromophore mutant that is deficient in all phytochrome family members.

Shade avoidance also affects leaf orientation (Whitelam and Johnson, 1982). The angle subtended by petioles to the main axis is reduced, causing the leaves to be orientated more upright than when the plant is grown in the open. In many species, leaf angle is under circadian control, but the phytochromes—particularly phytochrome B—exert an overriding regulation, which seems to be independent of responses to low R:FR. Hudson and Smith (1998) measured circadian leaf movements in *Nicotiana plumbaginifolia* and in a mutant (*hlg-1*) that lacked one of the two B phytochromes present in the wild-type. Leaves of the wild-type plants displayed a small oscillation in their angle, but those of the *hlg-1* mutant displayed a strong rhythmic oscillation in angle, with a period of between 24–26 h. Under low R:FR conditions (simulating shade) leaf angle was elevated both in mutant and wild-type plants, but the circadian oscillations were lost. The orientation of leaves is presumably regulated at the cellular level via a mechanism similar to that operating in specialised pulvini, and a phytochrome-dependent potassium channel activity has been described in protoplasts isolated from pulvini (Cote, 1995).

Alert observers of plant communities will have noticed, particularly with trees, that plants tend to grow away from each other; in other words, the orientation of stem growth is often such that the main trunk of a tree on the outside of a stand will be directed away from the canopy and towards the open aspect. Similar responses at a different magnitude are observable with herbaceous plants (e.g. Novoplansky *et al.*, 1990). At first sight, this appears to be an example of phototropism, classically defined as the reorientation of the direction of stem growth towards (or sometimes away from) the direction of unilateral incident radiation. Phototropism in etiolated seedlings is mediated by a blue-absorbing flavoprotein photoreceptor; in recent years, a breakthrough has become possible using mutants of Arabidopsis resulting in the gene encoding this photoreceptor being cloned and characterised (Christie *et al.*, 1998). However, research with herbaceous model plants indicates that the control of growth orientation of plants in canopy situations is yet another component of the phytochrome-mediated shade avoidance syndrome. Ballaré and co-workers (1992) grew wild-type and phyB-lacking mutants of cucumber in the proximity of other vegetation and demonstrated that a large proportion of the reorientation was absent in the mutant. Previous research (Iino, 1990; Iino *et al.*, 1984; Parker *et al.*, 1989) had demonstrated that establishing a gradient of Pfr across an internode or coleoptile can cause a tropic curvature similar to that caused by unilateral blue radiation. Thus, the assertion made earlier that ‘plants grow away from the shade rather than towards the light’ is borne out.

When plants are grown in close proximity but isolated from root competition, it can be shown that the dynamics of canopy development can be explained entirely on the basis of phytochrome-mediated shade avoidance responses (Gilbert *et al.*, 1995). Stands of a rapidly growing poplar cultivar (*Populus trichocarpa* × *deltoides* ‘Beaupré’) were grown at different planting densities in large independent pots, and growth parameters were related to parameters of the radiation environment within the stands. Elongation growth rate was found to be directly related to the R:FR of radiation propagated horizontally within the tree stands, as detected by vertical, cylindrical radiation sensors. In a recent extension of this work, similar relationships have been shown for birch, willow and sycamore stands (I. Gilbert, P.G. Jarvis, H. Smith, unpublished). In mixed stands of the four species (including poplar), even though the basic growth rates of the individual species differed markedly, each demonstrated an identical response to the R:FR signal, such that the overall dynamics of the development of the mixed canopy could be explained in terms of phytochrome-mediated responses.

The significance of phytochrome-mediated shade avoidance for the development of plant canopies is clear. Moreover, phytochrome-mediated shade avoidance allows for continual monitoring of the radiation environment, so that each individual plant, and perhaps each individual leaf, can be placed in the most appropriate position available.

5.11 Application of the shade avoidance syndrome

5.11.1 Genetic engineering of shade avoidance

The isolation of *PHY* genes, both genomic and cDNA, has made possible an approach towards the modification of plant architecture based upon transgenic expression. Both sense and anti-sense expression of *PHY* sequences have been reported, for several source genes and host plants (Robson and Smith, 1997). The first *PHY* cDNAs to be isolated were those encoding phyA in oats and rice, and transgenic expression of the oat *PHYA* cDNA, driven by the 'constitutive' CaMV35S promoter in both tobacco and tomato, yielded extremely dwarf, dark-green plants (Keller *et al.*, 1989; Boylan and Quail, 1989). Later, *PHYB* cDNA expression also yielded dwarf plants both in tobacco and *Arabidopsis*, although the extent of the phenotype varied depending on the source of the cDNA (i.e. *Arabidopsis* or rice) (Wagner *et al.*, 1991; McCormac *et al.*, 1993). Anti-sense expression of *PHY* cDNAs has not been extensively carried out, although Jackson and co-workers (1996) expressed an anti-sense potato *PHYB* fragment in *Solanum andigena*, a potato species that requires short days for tuberisation. In the anti-sense plants, tuberisation occurred under all daylength conditions; the phenotype also included other characteristics similar to those observed in *phyB* mutants.

Overexpression of *PHYA* sequences, however, has so far been the principal source of potential improvements to crop plant performance. Observations of transgenic tobacco expressing an oat *PHYA* cDNA to relatively high levels revealed an entirely unexpected result; these plants exhibited negative shade avoidance (McCormac *et al.*, 1991; McCormac *et al.*, 1992a). In other words, whereas the wild-type plants elongated much more rapidly in white light with added far-red (i.e. W + FR) than they did in white light alone, the transgenic plants were dwarfed under W + FR conditions. The explanation for this initially bizarre result requires an elaboration of what is known about the properties of phytochrome A.

Phytochrome A is very different from the other members of the phytochrome family. It is synthesised in relatively large quantities in dark-grown seedlings, and in etiolated oats, for example, reaches levels that allow for ready extraction and purification by standard protein fractionation techniques. It is for this reason that almost everything known about the chemical properties of the phytochromes comes from analyses of phyA, mainly from oats. The other phytochromes are present in much lower concentrations than phyA, and generally resist extraction and purification because of their very low abundance. Although phyA accumulates in etiolated seedlings, once the plant is transferred to the light, it is degraded rapidly and, furthermore, its synthesis is strongly downregulated. The half-life of the loss of phyA after transfer to the light is very short, in some cases as short as 15–20 min. This means that, after 12–24 h, there is very little phyA left in the cells, and indeed the concentration of phyA at steady-state in a light-grown plant is much less than that of phyB. Phytochrome B and, by extension, the other phytochromes are not subject to such degradation in the light, and are therefore known as light-stable phytochromes.

Phytochrome A is commonly described as being 'light-labile' but, like much other scientific jargon, this description is not strictly correct. What happens is that the Pfr form of phyA (i.e. Pfr_A) is very unstable; thus, after Pfr_A is formed from Pr_A by the action of light, it is rapidly degraded, whether or not the light treatment is maintained. If, however, the light treatment is extended, the continual photoconversion of Pr_A to Pfr_A results in the loss of total phyA down to very low levels, at which the very slow synthesis is matched by degradation at steady-state. The downregulation of phyA synthesis is caused by a direct repression by Pfr itself. The overall dynamics are shown by the following scheme:

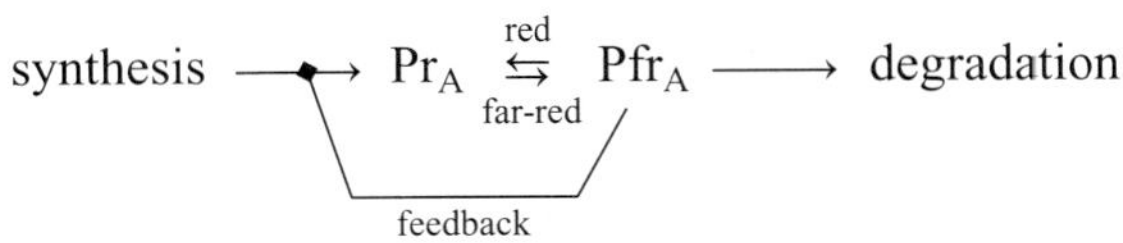

One of the implications of this is that the actions of phyA on seedling development during de-etiolation and establishment as a photoautotrophic organism are short-lived. Thus, the high irradiance response (HIR) can manifest itself as very strong inhibition of hypocotyl elongation under continuous far-red light for seedlings that have been grown in continuous darkness, but this FR-HIR is lost once the seedlings have been exposed to light that establishes a high proportion of Pfr_A for some hours. The loss of action of phyA during de-etiolation appears to be essential, as otherwise the normal shade avoidance responses, which are mediated by phytochromes other than phyA, would not be possible. In this respect, the actions of phyA and the other phytochromes are antagonistic, such that under shade conditions, with enhanced relative far-red radiation, phyA would inhibit elongation growth, whereas the phytochromes that mediate shade avoidance would enhance elongation growth (Smith *et al.*, 1997). Therefore, the regulation of the relative concentrations of phyA and the other phytochromes can be envisaged as being vital components of the proper establishment of the seedling and the development of the ability of avoid shade. Thus, as the phyA-mediated FR-HIR is lost upon de-etiolation, the shade avoidance responses become possible (McCormac *et al.*, 1992b). It has been proposed that phyA is essential to allow for satisfactory de-etiolation even under shade conditions; if shade avoidance was allowed to operate as seedlings de-etiolate, then reserves would quickly run out and the seedling would be in danger (Yanovsky *et al.*, 1995).

In seedlings overexpressing a transgenic *PHYA* cDNA, the dynamics of phyA concentration are expected to be very different. On the one hand, expression driven by the CaMV35S promoter will be independent of the feedback regulation that downregulates the activity of the endogenous phyA promoter. Furthermore, the mechanism of phyA degradation, known to be via the ubiquitin pathway, may be sufficiently specific that the heterologous phyA may be spared degradation. Both of these predictions are fulfilled, and it is known in transgenic tobacco expressing an oat *PHYA* cDNA, for example, that the half-life of the heterologous phyA is close to 4 h, whereas that of the native phyA is less than 1 h (McCormac *et al.*, 1992a). This results in the persistence of the heterologous phyA through de-etiolation; and indeed in the transgenic tobacco mentioned above, the existence of a strong phenotype in adult plants demonstrates that the phyA remains at sufficient concentration to be active throughout the life of the plant. Overexpressed phyA, therefore, is able to maintain a FR-HIR after the plants have de-etiolated, resulting in the suppression of normal shade avoidance responses to supplementary far-red, and indeed to the negative shade avoidance response described above. This being so, the opportunity exists to exploit the transgenic expression of phyA to suppress shade avoidance in the field.

5.11.2 Transgenic suppression of shade avoidance in the field

Suppression of shade avoidance can be expected to provide a number of advantages for crop plants grown at high densities (Smith, 1992). Virtually all known crops are strong shade avoiders and plant breeding through the centuries has apparently not bred out elongation responses to crowding. Even the dwarf cereals that have led to marked improvements in agricultural productivity via the so-called 'Green Revolution' still show very strong shade avoidance responses. As most crops are grown at high density, then shade avoidance results in elongated stems and petioles, reduced leaf development, accelerated flowering and reduced seed set, and overall the allocation of resources to parts of the plant that are not harvested. Shade avoiding crops grow long and have weak stems, making them susceptible to lodging.

Field experiments using tobacco as a model crop plant have confirmed that transgenic expression of a *PHYA* cDNA suppresses shade avoidance in the field (Robson *et al.*, 1996). For these experiments to be successful, it was necessary to select transgenic lines that expressed the heterologous cDNA to intermediate levels, since high-level expression caused unacceptable dwarfing. It was found that expression causing increases in the concentration of phyA in etiolated plants of about two- to three-fold that of the native phyA did not cause dwarfing in ordinary daylight. However, when the plants were grown at high densities, then the negative shade avoidance phenomenon described above caused the plants to be significantly dwarfed compared to the wild-types. Strong negative shade avoidance is seen in the two transgenic lines. Furthermore, the allocation of resources to stem growth was markedly reduced in the transgenic lines at high density, in comparison with the wild-types. It is not yet known whether such impressive improvements in performance will be obtained with more conventional crops, but at the time of writing a major effort is being undertaken worldwide to transfer this approach to other crops.

5.12 The future

The past 30 years have seen an enormous increase in our understanding of the mechanisms through which plants sense the potential competitive threat from neighbours and respond appropriately. The account of this research here has concentrated on the enlightenment that has come from the application of physiological reductionism to an otherwise extremely complex set of ecological questions. Reductionism divides complex, apparently unanswerable questions into sets of simpler partial questions that are capable of solution. The challenge, of course, is to synthesise those partial answers into a comprehensive, and comprehensible, whole. Molecular genetics has given us powerful approaches that have yielded otherwise unobtainable information on the photoreceptors that sense proximity signals and on the genes that encode those photoreceptors, and has provided means of using that knowledge in biotechnological manipulation. Intelligent application of mutant and transgenic plant material in real or simulated natural environments has led to new ways of thinking about plant interactions, and will certainly continue so to do. On the other hand, much information is still required on the intervening processes between signal perception and the myriad adaptive responses evoked at the plant architecture level. Molecular genetics, through the selection of new mutants via sophisticated screens, will undoubtedly contribute to the elucidation of these pathways, but the cloning of yet more genes, although hugely important, will only provide additional, individual steps. For the overall picture, an understanding is needed of how the many genes and components encoded by those genes interact in a quantitative manner to provide the plasticity so manifest in shade avoidance. To this end, a very old approach, recently brought to prominence through marriage to molecular genetics, namely the analysis of quantitative trait loci, will be of paramount significance (Koornneef *et al.*, 1997). Such analysis will generate new genotypes that can be used in real ecological situations, and this will at last make it possible to piece together the individual strands of incomplete knowledge that presently exist.

References

Bagnall, D., King, R.W., Whitelam, G.C., Boylan, M.T., Wagner, D. and Quail, P.H. (1995) Flowering responses to altered expression of phytochrome in mutants and transgenic lines of *Arabidopsis. Plant Physiology*, **108** 1495–1503.

Ballaré, C.L., Scopel, A.L., Sanchez, R.A., Casal, J.J. and Ghersa, C.M. (1987) Early detection of neighbour plants by phytochrome perception of spectral changes in reflected sunlight. *Plant, Cell and Environment*, **10** 551–557.

Ballaré, C.L., Scopel, A.L. and Sanchez, R.A. (1990) Far-red radiation reflected from adjacent leaves: an early signal of competition in plant canopies. *Science*, **247** 329–332.

Botto, J.F., Sanchez, R.A. and Casal, J.J. (1995) Role of phytochrome B in the induction of seed germination by light in *Arabidopsis thaliana. Journal of Plant Physiology*, **146** 307–312.

Botto, J.F., Sanchez, R.A. and Casal, J.J. (1996) Phytochrome A mediates the promotion of seed germination by very low fluences of light and canopy shade light in Arabidopsis. *Plant Physiology*, **110** 439–444.

Boylan, M.T. and Quail, P.H. (1989) Oat phytochrome is biologically active in transgenic tomatoes. *Plant Cell*, **1** 765–773.

Callahan, H.S., Pigliucci, M. and Schlichting, C.D. (1997) Developmental phenotypic plasticity: where ecology and evolution meet molecular biology. *Bioessays*, **6** 519–525.

Casal, J.J., Sanchez, R.A. and Yanovsky, M.J. (1997) The function of phytochrome A. *Plant, Cell and Environment*, **20** 813–819.

Child, R. and Smith, H. (1987) Phytochrome action in light-grown mustard: kinetics, fluence-rate compensation and ecological significance. *Planta*, **172** 219–229.

Christie, J.M., Reymond, P., Powell, G.K., Bernasconi, P., Raibekas, A.A., Liscum, E. and Briggs, W.R. (1998) *Arabidopsis* NPH1: a flavoprotein with the properties of a photoreceptor for phototropism. *Science*, **282** 1698–1701.

Clack, T., Mathews, S. and Sharrock, R.A. (1994) The phytochrome apoprotein family in *Arabidopsis* is encoded by five genes: the sequences and expression of *PHYD* and *PHYE. Plant Molecular Biology*, **25** 413–427.

Cone, J.W., Jaspers, P.A.P.M. and Kendrick, R.E. (1985) Biphasic fluence-response curves for light-induced germination *of Arabidopsis thaliana* seeds. *Plant, Cell and Environment*, **8** 605–612.

Cote, G.G. (1995) Signal transduction in leaf movement. *Plant Physiology*, **109** 729–734.

Devlin, P.F., Patel, S.R. and Whitelam, G.C. (1998) Phytochrome E influences internode elongation and flowering time in *Arabidopsis. Plant Cell*, **10** 1479–1487.

Devlin, P.F., Robson, P.R.H., Patel, S.R., Goosey, L., Sharrock, R.A. and Whitelam, G.C. (1999) Phytochrome D acts in the shade-avoidance syndrome in *Arabidopsis* by controlling elongation growth and flowering time. *Plant Physiology*, **119** 909–915.

Dudley, S.A. and Schmitt, J. (1996) Testing the adaptive plasticity hypothesis: density-dependent selection on manipulated stem length in *Impatiens capensis. American Naturalist*, **147** 445–465.

Gilbert, I.R., Seavers, G.P., Jarvis, P.G. and Smith, H. (1995) Photomorphogenesis and canopy dynamics: phytochrome-mediated proximity perception accounts for the growth dynamics of canopies of *Populus trichocarpa* × *deltoides* 'Beaupré. *Plant, Cell and Environment*, **18** 475–497.

Harper, J.L. (1976) *Population Biology of Plants*. Academic Press, London, UK.

Hartmann, K.M. (1966) A general hypothesis to interpret 'high energy phenomena' of photomorphogenesis on the basis of phytochrome. *Photochemistry and Photobiology*, **5** 349—366.

Holmes, M.G. and Smith, H. (1977a) The function of phytochrome in the natural environment. I. Characterisation of daylight for studies in photomorphogenesis and photoperiodism. *Photochemistry and Photobiology*, **25** 533–538.

Holmes, M.G. and Smith, H. (1977b) The function of phytochrome in the natural environment. II. The influence of vegetation canopies on the spectral energy distribution of natural daylight. *Photochemistry and Photobiology*, **25** 539–545.

Hudson, M. and Smith, H. (1998) The phytochrome B encoded by the *HLG* locus of *Nicotiana plumbaginifolia* is required for detection of photoperiod: *hlg* mutants show altered regulation of flowering and circadian movement. *Plant Journal*, **15** 281–287.

Iino, M. (1990) Phototropism—mechanisms and ecological implications. *Plant, Cell and Environment*, **13** 633–650.

Iino, M., Schäfer, E. and Briggs, W.R. (1984) Photoperception sites for photochrome-mediated phototropism of maize mesocyotyls. *Planta*, **162** 477–479.

Jackson, S.D., Heyer, A., Dietze, J. and Prat, S. (1996) Phytochrome B mediates the photoperiodic control of tuber formation in potato. *Plant Journal*, **9** 159–166.

Kasperbauer, M. (1987) Far-red light reflection from green leaves and effects on phytochrome-mediated assimilate partitioning under field conditions. *Plant Physiology*, **85** 350–354.

Keller, J.M., Shanklin, J., Vierstra, R.D. and Hershey, H.P. (1989) Expression of a functional monocotyledonous phytochrome in transgenic tobacco. *EMBO Journal*, **8** 1005–1012.

Koornneef, M., Alonso-Blanco, C. and Peeters, A.J.M. (1997) Genetic approaches in plant physiology. *New Phytologist*, **137** 1–8.

Mandoli, D.F. and Briggs, W.R. (1981) Phytochrome control of two low-irradiance responses in etiolated oat seedlings. *Plant Physiology*, **67** 733–739.

Matthews, S. and Sharrock, R.A. (1997) Phytochrome gene diversity. *Plant, Cell and Environment*, **20** 666–671.

McCormac, A.C., Cherry, J.R., Hershey, H.P., Vierstra, R.D. and Smith, H. (1991) Photoresponses of transgenic tobacco plants expressing an oat phytochrome gene. *Planta*, **185** 162–170.

McCormac, A.C., Whitelam, G.C. and Smith, H. (1992a) Light-grown plants of transgenic tobacco expressing an introduced oat phytochrome A gene under the control of a constitutive viral promoter exhibit persistent growth inhibition by far-red light, *Planta*, **188** 173–181.

McCormac, A.C., Whitelam, G.C., Boylan, M.T., Quail, P.H. and Smith, H. (1992b) Contrasting responses of etiolated and light-adapted seedlings to red far-red ratio: a comparison of wild-type mutant and transgenic plants has revealed differential functions of members of the phytochrome family. *J Plant Physiol*, **140** 707–714.

McCormac, A.C., Wagner, D., Boylan, M.T., Quail, P.H., Smith, H. and Whitelam, G.C. (1993) Photresponses of transgenic *Arabidopsis* seedlings expressing introduced phytochrome B-encoding cDNAs: evidence that phytochrome A and phytochrome B have distinct photoregulatory functions. *Plant Journal*, **4** 19–27.

McLaren, J.S. and Smith, H. (1978) The function of phytochrome in the natural environment. VI. Phytochrome control of the growth and development of *Rumex obtusifolius* under simulated canopy light environments. *Plant Cell and Environment*, **1** 61–67.

Morgan, D.C. and Smith, H. (1976) Linear relationship between phytochrome photoequilibrium and growth in plants under simulated natural radiation. *Nature*, **262** 210–212.

Morgan, D.C. and Smith, H. (1978) Simulated sunflecks have large, rapid effects on plant stem extension. *Nature*, **273** 534–536.

Morgan, D.C. and Smith, H. (1979) A systematic relationship between phytochrome-controlled development and species habitat for plants grown in simulated natural radiation. *Planta*, **145** 253–259.

Morgan, D.C., O'Brien, T and Smith, H (1980) Rapid photomodulation of stem extension in light-grown *Sinapis alba* (L): studies on kinetics, site of perception and photoreceptor. *Planta*, **150** 95–101.

Novoplansky A., Cohen, D. and Sachs, T. (1990) How *Portulaca* seedlings avoid their neighbors. *Oecologia*, **82** 490–493.

Parker, K., Baskin, T.I. and Briggs, W.R. (1989) Evidence for a phytochrome-mediated phototropism in etiolated pea-seedlings. *Plant Physiology*, **89** 493–497.

Pigliucci, M. and Schlichting, C.D. (1997) On the limits of quantitative genetics for the study of phenotypic evolution. *Acta Biotheoretica*, **45** 143–160.

Pratt, L.H., Cordonnier-Pratt, M.-M., Hauser, B. and Caboche, M. (1995) Tomato contains two differentially expressed genes encoding B-type phytochromes, neither of which can be considered an ortholog of *Arabidopsis* phytochrome B. *Planta*, **197** 203–206.

Reed, J.W., Nagatani, A., Elich, T.D., Fagan, M. and Chory, J. (1994) Phytochrome A and phytochrome B have overlapping but distinct functions in *Arabidopsis* development. *Plant Physiology*, **104** 1139–1149.

Robson, P.R.H. and Smith, H. (1997) Fundamental and biotechnological applications of phytochrome transgenes. *Plant, Cell and Environment*, **20** 831–839.

Robson, P.R.H., McCormac, A.C., Irvine, A.S. and Smith, H. (1996) Genetic engineering of harvest index in tobacco through overexpression of a phytochrome gene. *Nature Biotechnology*, **14** 995–998.

Schmitt, J. and Wulf, R.D. (1993) Light spectral quality, phytochrome, and plant competition. *Trends in Ecology and Evolution*, **8** 47–51.

Schmitt, J., McCormac, A.C. and Smith, H. (1995) A test of the adaptive plasticity hypothesis using transgenic and mutant plants disabled in phytochrome-mediated elongation responses to neighbors. *American Naturalist*, **146** 937–953.

Scopel, A.L., Ballaré, C.L. and Sânchez, R.A. (1991) Induction of extreme light sensitivity in buried weed seeds and its role in the perception of soil cultivations. *Plant, Cell and Environment*, **14** 501–508.

Smith, H. (1982) Light quality, photoperception and plant strategy. *Annual Review of Plant Physiology*, **33** 481–518.

Smith, H. (1990) Signal perception, differential expression within multigene families and the molecular basis of phenotypic plasticity *Plant, Cell and Environment*, **13** 585–594.

Smith, H. (1992) Ecology of photomorphogenesis: clues to a transgenic programme of crop plant improvement. *Photochemistry and Photobiology*, **56** 815–822.

Smith, H. (1995) Physiological and ecological function within the phytochrome family. *Annual Review of Plant Physiology and Plant Molecular Biology*, **46** 289–315.

Smith, H. and Fork, D.C. (1992) Direct measurement of phytochrome photoconversion intermediates at high photon fluence rates. *Photochemistry and Photobiology*, **56** 599–606.

Smith, H. and Holmes, M.G. (1977) The function of phytochrome in the natural environment. III. Measurement and calculation of phytochrome photoequilibrium. *Photochem. Photobiol*, **25** 547–550.

Smith, H. and Whitelam, G.C. (1997) The shade avoidance syndrome: multiple responses mediated by multiple photoreceptors. *Plant, Cell and Environment*, **20** 840–844.

Smith, H., Casal, J.J. and Jackson, G.M. (1990) Reflection signals and the perception by phytochrome of the proximity of neighbouring vegetation. *Plant, Cell and Environment*, **13** 73–78.

Smith, H., Samson, G. and Fork, D.C. (1993) Photosynthetic acclimation to shade: probing the role of phytochromes using photomorphogenic mutants of tomato *Plant, Cell and Environment*, **16** 929–937.

Smith, H., Xu, Y. and Quail, P.H. (1997) Antagonistic but complementary actions of phytochromes A and B allow optimum seedling de-etiolation. *Plant Physiology*, **114** 637–641.

Wagner, D., Tepperman, J.M., Quail, P.H. (1991) Overexpression of phytochrome-B induces a short hypocotyl phenotype in transgenic *Arabidopsis*. *Plant Cell*, **3** 1275–1288.

Weiner, J. (1990) Asymmetric competition in plant populations. *Trends in Ecology and Evolution*, **5** 360–364.

Whitelam, G.C. and Harberd, N.P. (1994) Action and function of phytochrome family members revealed through the study of mutant and transgenic plants. *Plant Cell and Environment*, **17** 615–625.

Whitelam, G.C. and Johnson, C.B. (1982) Photomorphogenesis in *Impatiens parviflora* and other plant species under simulated natural canopy radiation. *New Phytologist*, **90** 611–618.

Whitelam, G.C., Johnson, E., Peng, J., Carol, P., Anderson, M.L., Cowl, J.S. and Harberd, N.P. (1993)

Phytochrome A null mutants of *Arabidopsis* display a wild-type phenotype in white light. *Plant Cell*, **5** 757–768.

Yanovsky, M.J., Casal, J.J. and Whitelam, G.C. (1995) Phytochrome A, phytochrome B and *HY4* are involved in hypocotyl growth responses to natural radiation in *Arabidopsis*: weak de-etiolation of the *phyA* mutant under dense canopies. *Plant, Cell and Environment*, **18** 788–794.

6 Development of the photosynthetic apparatus

Andrew N. Webber

6.1 Introduction

The primary photosynthetic events of light energy capture, electron transport and reduction of carbon dioxide occur in a specialized plastid organelle called the chloroplast. The chloroplast is delimited by a double membrane system, the outer and inner chloroplast envelope. A third membrane system within the chloroplast is named the thylakoid membrane. Chloroplasts arise from a progenitor organelle, called a proplastid, present in meristematic cells. Proplastids contain only a rudimentary internal membrane system and lack most components of the photosynthetic apparatus. Meristematic cells contain 10–20 proplastids. As leaf cells develop and expand, chloroplast development proceeds with the rapid synthesis of chlorophyll and thylakoid membranes, and the number of chloroplasts increases to approximately 100 chloroplasts per mesophyll cell. Light is critical for chlorophyll biosynthesis and chloroplast development in angiosperms. When seedlings are grown in darkness, the proplastid develops into an etioplast that contains the prolamellar body predominantly composed of the reduced form of nicotinamide adenine dinucleotide phosphate (NADPH): protochlorophyllide oxidoreductase complexed with NADPH and protochlorophyllide.

The accumulation and development of a photochemically competent chloroplast population has an enormous effect on the photosynthetic competence of leaves. The photosynthetic productivity of many crops and of natural vegetation can often be determined by environmental effects on the rate of chloroplast development. Although the chloroplast has been a widely studied biological system, and a detailed understanding of its components, structure and function now exist, knowledge of the factors determining the rate of chloroplast development, and the expression and assembly of the photosynthetic apparatus, are still relatively poor. Chloroplast development is a highly complex process involving coordination and regulation of organelle replication and the synthesis and assembly of photosynthetic membranes and protein complexes. This chapter reviews recent advances in our understanding of chloroplast development, focusing on the synthesis and assembly of chloroplast protein complexes and the influence of the light environment on this process.

6.2 Overview of the photosynthetic apparatus

The thylakoid membrane is the site of light energy transduction into chemical energy. The thylakoid membrane contains four major protein complexes named photosystem II (PSII), photosystem I (PSI), the cytochrome b_6f complex and the adenosine triphosphate (ATP) synthase (Figure 6.1). There is also evidence that thylakoid membranes contain an NADP oxidoreductase complex and an alternative oxidase, similar to that found in mitochondria, although the physiological role of either complex is unclear.

PSII, PSI, cytochrome b_6f and ATP synthase are multimeric protein complexes containing from 6 polypeptides in the cytochrome b_6f complex to over 26 polypeptides in PSII. The genes encoding these proteins are located both in the nucleus and the chloroplast (Figure 6.1). PSII uses light energy to oxidize water and reduce plastoquinone. The PSII complex contains several different co-factors that are involved in mediating the transfer of electrons from water to plastoquinone (reviewed by Diner and Babcock, 1996; Ruffle and Sayre, 1998). These include manganese, a tyrosine, the reaction center chlorophyll *a* (P_{680}), pheophytin *a* (chlorophyll *a* lacking the central Mg) and quinone. Additional co-factors in PSII are antenna chlorophyll *a* and cytochrome b_{559}. Electron transfer is initiated when excitation energy reaches P_{680}. The excited state reaction center chlorophyll, P^*_{680}, transfers an electron to pheophytin to form the charge separated state, $P^+_{680}Pheo^-$. Subsequently, the electron is rapidly transferred across the PSII complex to the quinone acceptor, Q_A. This charge separation across the membrane serves to minimize

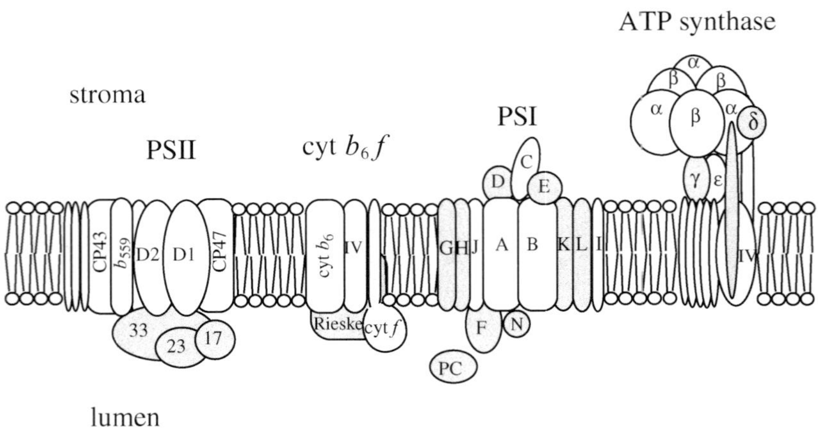

Figure 6.1 Photosynthetic complexes of the thylakoid membrane. The shaded subunits are nucleus-encoded and the unshaded subunits are chloroplast encoded. Abbreviations: PSI, photosystem I; PSII, photosystem II; cyt, cytochrome; ATP, adenosine triphosphate.

the rate of wasteful charge recombination, increasing the quantum yield of electron transfer to close to unity. The electron is then transferred from Q_A to a second quinone, Q_B, that serves as a two electron acceptor and is only loosely associated with PSII at the Q_B site. Each electron transfer to Q_B is also associated with a protonation of the quinone. The doubly reduced Q_B molecule dissociates from PSII and migrates randomly through the thylakoid membrane until it reaches a cytochrome b_6f complex. Approximately nine unbound Q_B molecules (both reduced and oxidized) per PSII complex form the plastoquinone pool.

The PSII complex contains numerous (>26) individual polypeptide components that coordinate co-factors and/or serve a structural role. Several key proteins in PSII make up what is known as the reaction center core. In particular, two related proteins, D1 and D2 (encoded by the chloroplast *psbA* and *psbD* genes, respectively), coordinate the co-factors involved in electron transfer, i.e. manganese, P_{680}, pheophytin, Q_A and Q_B. The D1 and D2 polypeptides are also closely associated with the heme containing cytochrome b_{559} complex. While cytochrome b_{559} is redox active, its role in PSII is unknown, although some evidence suggests a role in photoprotection of the reaction center. Additional chlorophyll *a* molecules are coordinated by proteins named CP43 and CP47 (encoded by the chloroplast *psbC* and *psbB* genes). CP43 and CP47 function as antenna proteins funneling excitation energy to the reaction center. On the lumenal side of PSII are several proteins that form the oxygen evolving complex (OEC) and are required for water oxidation; these are called OEC33, OEC23, OEC16, etc.

Water oxidation by PSII requires the removal of four electrons from two molecules of water. The oxidation-reduction potential of water is $+0.82$ mV, so it is an extremely difficult compound to oxidize. In PSII, the water oxidation is driven by P_{680}^+, which has a midpoint potential estimated at $+1.2$ mV. P_{680}^+ oxidizes a tyrosine amino acid, Y_Z, on the D1 protein. The Y_Z radical cation is then able to oxidize the Mn ions of the OEC by an as yet unknown mechanism. Each electron of the four is removed sequentially from components of the OEC, with oxygen evolution occurring only after the fourth electron is removed.

The doubly reduced Q_B dissociates from PSII and becomes a part of the plastoquinone pool. The PQH_2 diffuses to the cytochrome b_6f complex, where it is oxidized, concomitantly releasing protons that are deposited in the thylakoid lumen. The cytochrome b_6f complex contains cytochrome b_6 (encoded by *petB*), cytochrome *f* (encoded by *petA*), Reiske protein (encoded by *PetC*), subunit IV (encoded by *petD*) and several additional polypeptides. Electrons are next transferred from the cytochrome b_6f complex to PSI by plastocyanin, a small soluble Cu-containing protein located in the thylakoid lumen.

The PSI complex catalyzes the transfer of an electron from plastocyanin to ferredoxin. PSI consists of approximately 13 individual proteins and contains close to 100 chlorophyll *a* (reviewed by Webber and Bingham, 1998). The structure of the PSI complex has been solved at 4 Å resolution, providing useful information on the organization of the different co-factors (Krauss *et al.*, 1996). In PSI, the antenna chlorophyll *a*, and chlorophyll *a* components of the electron transfer chain (6 chlorophyll *a*) are associated with two proteins, PsaA and PsaB, that form the reaction center heterodimer. Light energy is transferred from the antenna chlorophyll *a* to the reaction center chlorophyll *a* dimer (P_{700}). Excitation of P_{700} results in electron transfer to the first electron acceptor, a chlorophyll *a* molecule, named A_0. An additional chlorophyll *a* molecule, termed A, is positioned between P_{700} and A_0, but has not yet been detected as an electron transfer intermediate. Electrons are then transferred from A_0 to ferredoxin through several electron carriers, called A_1 (a phylloquinone), F_X, F_A and F_B. F_X, F_A and F_B are iron-sulfur clusters. The detailed structure of PSI indicates that there are two potential branches for electron transfer between P_{700} and F_X. The structure shows that P_{700} is a dimer and that there are four chlorophyll *a* molecules between P_{700} and F_X. These chlorophylls reside on either side of a symmetry axis running through P_{700} and F_X. The two chlorophylls furthest from P_{700} represent A_0. While two quinone molecules have been identified in the PSI crystal structure, only a single quinone radical (per P_{700}) is detected spectrophotometrically, raising the intriguing possibility that only one of the electron transfer branches is operative.

Each thylakoid protein complex contains a precise stoichiometric number of polypeptide subunits. Therefore, there must be precise coordination of the expression of the nuclear and plastid genes. In addition, thylakoid membrane protein complexes contain a large number of non-proteinaceous co-factors, such as chlorophylls, pheophytins, quinones, hemes, manganese, iron and iron-sulfur clusters. These co-factors are also bound in precise stoichiometric amounts; the synthesis and assembly of these co-factors must also be controlled, and may even be involved in coordinating the synthesis and assembly of the apoprotein subunits. A further level of control must also exist to maintain the overall stoichiometry of the four major complexes in the thylakoid membrane, although this stoichiometry may vary, depending on environmental variables, primarily the quality and quantity of light.

6.3 Chloroplast gene expression

6.3.1 Messenger ribonucleic acid (mRNA) synthesis and stability

Associated with chloroplast development are large changes in the accumulation of both nuclear and chloroplast encoded mRNAs for thylakoid proteins

(Link, 1988). The steady-state level of mRNAs from different chloroplast genes may be determined by the relative transcription rate and/or message stability (reviewed by Mullet, 1988, 1993; Gruissem, 1989; Gruissem and Tonkyn, 1993). The accumulation of mRNA from different chloroplast genes, at the same developmental stage, is known to vary by as much as 900-fold in some cases (Rapp *et al.*, 1992).

Chloroplast transcription is mediated by two different types of RNA polymerase (reviewed by Maliga, 1998). One RNA polymerase resembles the bacteriophage T7 and SP6 single subunit enzyme and is nucleus-encoded (Maliga, 1998). A second multisubunit RNA polymerase is chloroplast encoded. The chloroplast *rpoA*, *rpoB*, *rpoC1* and *rpoC2* genes encode putative homologues of the *Escherichia coli* core RNA polymerase subunits α, β, β' and β''. In addition, the *E. coli* RNA polymerase also contains a sigma factor that confers the holoenzyme with the ability to recognize the -10 and -35 promoter motifs, and thus provide specificity to transcription initiation. Genes that might encode such sigma factors have not be identified in the chloroplast genome, and they are thought to be encoded by nuclear genes. Indeed, nuclear genes encoding sigma factors have been isolated (Tan and Troxler, 1999). Presumably, differential expression of the sigma factors is important for the light regulation and developmental expression of some chloroplast genes (Tiller and Link, 1993; Tan and Troxler, 1999).

The mechanism by which light regulates leaf and chloroplast development through the action of red-light (phytochromes) and blue-light (cryptochrome) receptors is becoming well understood. However, factors that control light regulation of chloroplast transcription have not been characterized in as much detail. Some chloroplast genes can be modulated during chloroplast development and in response to the quality and quantity of light (Deng *et al.*, 1989; Haley and Bogorad, 1990; Klein and Mullet, 1990; Christopher *et al.*, 1992). One of the more interesting features to emerge from studies of light-induced chloroplast mRNAs is that the light induction often involves the use of alternative chloroplast promoters (Haley and Bogorad, 1990; Sexton *et al.*, 1990; Christopher *et al.*, 1992). The best characterized of the light-regulated promoters is that for *psbD*, a gene encoding the D2 protein of the PSII reaction center. In 4.5 day-old, dark-grown barley, co-transcription of the *psbK-psbI-psbD-psbC* gene loci gave rise to at least six mRNAs, when isolated mRNA was analyzed by northern hybridization with DNA probes specific for *psbD* or *psbC* (Gamble *et al.*, 1988). Upon irradiation for 108 h, the total mRNA from this transcript declined and was composed of only two mRNA species of 4.0 and 3.2 kb (Gamble *et al.*, 1988).

Thus, only a subset of the transcripts from the *psbD-psbC* operon are maintained in the light. Biochemical analysis indicates that the transcripts maintained in the light originate from a second promoter, located within the *psbD-psbC* cluster. This second promoter is virtually inactive in the dark, and is induced by blue-light (Gamble and Mullet, 1989; Sexton *et al.*, 1990; Christopher and Mullet, 1994). Transcription from the *psbD* blue-light responsive promoter requires a prokaryotic −10 promoter element but not the −35 element. Instead, a *trans*-acting activating complex binds to an AAG DNA motif immediately upstream of −35 (Kim and Mullet, 1995). A second sequence specific *trans*-acting factor binds further upstream between −71 and −100. The binding of the *trans*-acting complex was shown to be inhibited by ADP-dependent phosphorylation of the complex (Kim *et al.*, 1999).

Run-on transcription assays show that differential mRNA stability also plays an important role in determining the steady-state level of chloroplast mRNAs (Deng and Gruissem, 1987; Mullet and Klein, 1987). The stability of several chloroplast mRNAs has been observed to change *in vivo* with leaf age (Klaff and Gruissem, 1991) and chloroplast development (Kim M.Y. *et al.*, 1993). It is important to determine how selective stabilization of chloroplast mRNAs is regulated during chloroplast development *in vivo*. Biochemical and genetic evidence indicates that both *cis*-acting elements and *trans*-acting factors are involved in mRNA stabilization. The 5′ and 3′ untranslated regions of chloroplast transcripts can serve as RNA stabilizing elements (Stern and Gruissem, 1987; Sakamoto *et al.*, 1993; Salvador *et al.*, 1993; Nickelsen *et al.*, 1994; Lee *et al.*, 1996; Bruick and Mayfield, 1998). At the 5′ and 3′ untranslated region, inverted repeat sequences, which can fold into a stable stem-loop structure, help protect the mRNA from degradation, or are targets for specific binding of RNA-binding proteins. Deletion analysis of the 5′ and 3′ regions has demonstrated that these untranslated regions are required for mRNA stability *in vivo* (Stern and Kindle, 1993; Lee *et al.*, 1996; Bruick and Mayfield, 1998). It has recently been shown that sequences within the 3′ untranslated region can serve as sites for polyadenylation (Kudla *et al.*, 1996; Lisitsky *et al.*, 1996). Polyadenylation of the chloroplast transcripts was shown to make them more susceptible to degradation *in vitro*.

Genes that may encode RNA-binding proteins have not been identified in any of the chloroplast genomes that have been entirely sequenced. It is most probable that control of mRNA stability within chloroplasts will be controlled by the expression of nuclear genes. In line with this, photosynthesis-deficient nuclear mutants of *Chlamydomonas reinhardtii*, which are unable to accumulate transcripts from the *psbB* (Sieburth *et al.*, 1991; Monod *et al.*, 1992), *psbC* (Sieburth *et al.*, 1991), *psbD* (Kuchka *et al.*, 1989) and *atpB* (Drapier *et al.*, 1992) genes, have been isolated. Each mutant affects accumulation of only one specific transcript, suggesting that each chloroplast message requires at least one nuclear-encoded factor for stabilization. One *Chlamydomonas reinhardtii* nuclear mutant, *nac2-26*, has provided evidence that 5′ mRNA sequences of *petD* may interact with the nuclear mRNA stabilizing factor (Kuchka *et al.*, 1989; Rochaix, 1992). These nuclear-encoded factors are presumably related to the RNA-binding proteins identified in chloroplast extracts.

6.3.2 *Translation*

Chloroplast mRNAs encoding many thylakoid proteins are present in the dark, although the polypeptides, particularly those involved in chlorophyll binding, do not accumulate (Klein and Mullet, 1986, 1987; Kreuz *et al.*, 1986; Hird *et al.*, 1991). In barley, mRNA from *psaA-psaB* and *psbA* are found in dark-grown leaves but the apoproteins are absent (Klein and Mullet, 1987). In wheat, mRNA from the co-transcribed *psbB* and *psbH* genes, encoding the 47 kDa chlorophyll *a*-binding protein and the 10 kDa phosphoprotein of PSII, are present in etiolated leaves (Hird *et al.*, 1991). While the 10 kDa phosphoprotein is present in etioplasts, the 47 kDa chlorophyll *a*-binding protein only accumulates after exposure to light (Hird *et al.*, 1991), demonstrating that there is not an overall block on chloroplast mRNA translation but, rather, specific regulation of certain proteins, such that they are either not translated in the dark or are rapidly degraded.

Chlorophyll is expected to play a key role in mediating chlorophyll-binding protein accumulation. Availability of chlorophyll could stabilize newly synthesized proteins, or control translation initiation or elongation. Pulse labeling of proteins in barley etioplasts with ^{35}S-methionine was unable to detect the synthesis of chlorophyll apoproteins, suggesting a block in translation or very rapid degradation of the newly synthesized protein (Klein and Mullet, 1986). The mRNAs encoding chlorophyll *a*-binding apoproteins were associated with polysomes in dark-grown leaves (Klein *et al.*, 1987). In the light, there was no shift in the polysome distribution of mRNAs, indicating that elongation of the polypeptide chain was inhibited, rather than a block in translation initiation. The importance of chlorophyll synthesis in apoprotein accumulation has been directly tested in etioplasts, where it was demonstrated that addition of chlorophyll biosynthesis precursors, chlorophyllide-*a* and phytylpyrophosphate, which induce chlorophyll synthesis in the dark, results in accumulation of chlorophyll-binding protein (Eichacker *et al.*, 1990). It is possible that binding of chlorophyll *a* to the apoprotein facilitates translation and/or release of the protein from the ribosome complex. Kim *et al.* (1991) detected D1 protein translation intermediates associated with polysomes in pulse-labeled chloroplasts, suggesting that a lack of chlorophyll causes stalling or pausing of protein synthesis. Subsequent work, quantifying the abundance of translation initiation complexes and the extent of translational run-off in the presence of lincomycin, indicated that chlorophyll availability had little influence on either translation initiation or elongation (Kim *et al.*, 1994). It was concluded from these studies that binding of chlorophyll to newly synthesized apoproteins rapidly leads to protein stabilization.

There is substantial genetic evidence that nuclear-encoded factors are required for translation of chloroplast mRNAs. As with the factors involved in mediating mRNA stability, these factors are transcript specific, regulating the translation of individual thylakoid proteins. Nuclear mutants of *Chlamydomonas reinhardtii* have been characterized, which are unable to translate chloroplast mRNAs from *psbC* (Rochaix *et al.*, 1989), *psbD* (Kuchka *et al.*, 1988; Rattanachaikunsopon *et al.*, 1999) and *atpA* (Drapier *et al.*, 1992). In each case, the native mRNAs are present at wild-type or slightly elevated amounts. A chloroplast suppressor of the nuclear F34 mutant (Rochaix *et al.*, 1989), deficient in PSII, was shown to be a single base change in the 3′ region of the *psbC* mRNA. The single base change may destabilize a potential stem loop sequence that is a target for a nuclear-encoded translational activator.

Ribonucleobinding proteins have been identified that bind to the 5′ untranslated region of several chloroplast mRNAs (Danon and Mayfield, 1991). One of these proteins, a 47 kDa polypeptide, binds to the *psbA* 5′ untranslated region in a redox dependent manner (Danon and Mayfield, 1994). Binding of the 47 kDa protein requires interaction with a 60 kDa protein, which may be reversibly phosphorylated by an adenosine diphosphate (ADP)-dependent kinase. ADP-dependent phosphorylation of the 60 kDa protein reduces binding of the 47 kDa protein to the 5′ untranslated region of *psbA*. Based on DNA sequence analysis, the 60 kDa protein has been identified as a chloroplast disulfide isomerase (Kim and Mayfield, 1997). Kim and Mayfield (1997) proposed that ferredoxin is able to reduce the 60 kDa disulfide isomerase, which, in turn, will reduce the 47 kDa protein. The reduced 47 kDa protein is then able to activate translation of *psbA* transcripts. Thus, in darkness, the more oxidizing conditions and elevated ADP levels result in decreased translation of *psbA* mRNA. In the light, the more reducing conditions and decreased ADP concentrations will promote translation of *psbA* mRNA. This is very similar to the ADP-dependent phosphorylation events regulating chloroplast transcription, as well as chloroplast enzyme inactivation.

6.3.3 Coordinated translation

Primarily based on studies of *Chlamydomonas reinhardtii* mutants, there is strong indication that the synthesis of certain chloroplast proteins may be coordinated. The lack of expression of an individual subunit will sometimes lead to a decline in the rate of synthesis of a second protein of the same complex. For example, a *Chlamydomonas reinhardtii* mutant, *thm24*, that cannot synthesize the β-subunit of the CF1 complex also exhibits a reduced rate of synthesis of the α-subunit (Drapier *et al.*, 1992; Lemaire and Wollman, 1989). In the case of the PSI reaction center, mutants carrying reading frame-shifts in *psaB* that do not produce PsaB, also do not synthesize PsaA (Girard-Bascou *et al.*, 1987). The converse is not true, since mutants unable to synthesize PsaA still synthesize PsaB (Girard-Bascou *et al.*, 1987; Takahashi *et al.*, 1991), which is then rapidly degraded. In the case of PSII, a mutant lacking D2 exhibits reduced rates of synthesis of both D1 and CP47 (Erickson *et al.*, 1986; de Vitry *et al.*, 1989). In the absence of D1 synthesis, however, D2 protein synthesis is unaffected, although synthesis of CP47 is reduced (Jensen *et al.*, 1986; Bennoun *et al.*, 1986; de Vitry *et al.*, 1989).

The mechanism controlling coordinated synthesis of chloroplast polypeptides is mostly unknown. In at least one case, however, important information about how translation may be controlled has been reported. In the case of the cytochrome b_6f complex, there is a specific drop in the rate of cytochrome *f* synthesis in mutants that fail to assemble the cytochrome b_6f complex correctly. It has been demonstrated that cytochrome *f* is capable of autoregulating its own synthesis by interaction of the C-terminal region of the protein with the 5′ region of the *petA* transcript. As predicted by such a model, mutants lacking the cytochrome *f* C-terminal region display high rates of cytochrome *f* synthesis (Kuras *et al.*, 1995; Choquet *et al.*, 1998). It is likely that similar control mechanisms govern the coordinated synthesis of other chloroplast proteins, but further work is required to demonstrate such control.

6.4 Chloroplast protein import

6.4.1 Import into the stroma

The majority of chloroplast proteins are encoded by nuclear genes and synthesized in the cytosol as precursor proteins. The precursor proteins are then post-translationally transported to their final destination in the chloroplast (reviewed by Keegstra and Cline, 1999). The steps involved in this import process are diagrammatically represented in Figure 6.2. The cytosolic precursors are typically larger than the corresponding chloroplast localized protein. The increased size of the precursor protein is due to an N-terminal transit peptide, containing stromal-targeting information, which is cleaved off by a protease following transport into the chloroplast. Precursor proteins that are destined for the thylakoid lumen have a bipartite transit sequence, which contains both a cleavable stromal-targeting domain and an additional cleavable lumenal-targeting domain. The two domains of the bipartite transit sequence are sequentially removed by processing proteases. The stromal-targeting domain is removed in the stroma by the stromal-processing pro-

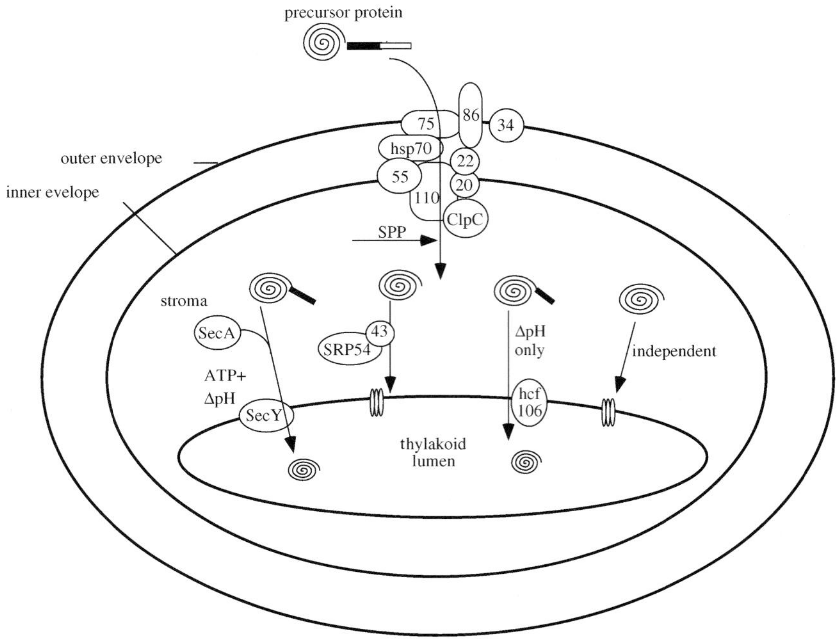

Figure 6.2 Schematic representation of the chloroplast protein import system and the pathway for routing to the various suborganelle compartments. The chloroplast import apparatus is located at the chloroplast envelope. Nucleus encoded pre-proteins are synthesized in the cytosol and transferred across the chloroplast envelope. The pre-protein is then processed by the stromal-processing protease (SPP). A second presequence targets proteins to the thylakoid lumen (see text for further details). Abbreviations: hsp, heat shock protein; ATP, adenosine triphosphate.

tease. Following transfer into the lumen, the second domain is removed by a processing protease in the lumen. An area that is still very poorly understood is the nature of the information contained in the targeting sequences that directs the precursor to the correct organelle and intra-organelle compartment. The stromal-targeting domains of different precursor proteins vary in length from 30 to more than 100 amino-acids and share little common sequence similarity. The lumenal-targeting domains contain predominantly hydrophobic amino-acids.

A combination of biochemical and molecular genetic approaches has led to recent progress in identifying components of the chloroplast envelope and thylakoid membrane translocation apparatus (reviewed by Keegstra and Cline, 1999). The complex responsible for translocating precursor proteins into the stroma must mediate transfer across both the outer and inner chloroplast membranes. Three proteins of 86, 75 and 34 kDa, associated with the outer chloroplast envelope membrane, have been identified as components of the translocation apparatus. The 86 and 34 kDa proteins are both guanosine triphosphate (GTP)-binding proteins (Hirsch *et al.*, 1994; Kessler *et al.*, 1994). The 75 kDa protein has been postulated to form the channel through which the precursor protein is transferred (Schnell *et al.*, 1994; Tranel *et al.*, 1995). Four additional proteins of the chloroplast translocation apparatus are associated with the inner envelope membrane. Interestingly, one of these inner envelope membrane proteins contains an iron-sulfur complex (Caliebe *et al.*, 1997).

In addition to the outer and inner envelope membrane proteins, several molecular chaperones are required for protein import. A heat shock protein, hsp70, has been localized to the intermembrane space between the outer and inner envelope membranes (Schnell *et al.*, 1994). It has been suggested that this hsp70 helps to maintain the precursor protein in an unfolded state. The translocation apparatus also contains the ClpC protein, located in the stroma. ClpC might work by repeatedly binding and releasing the precursor protein, working rather like a ratchet to pull the precursor protein into the chloroplast. The involvement of a stromal hsp70 in protein import is unknown. Recently, a mutant of *Arabidopsis* was isolated carrying a deletion in a gene encoding a component of the chloroplast import machinery (Jarvis *et al.*, 1998). The deleted protein was a small GTP-binding protein of 33 kDa, with high sequence similarity to the 34 kDa GTP-binding protein of the outer envelope membrane. Deletion of the 33 kDa protein resulted in impaired chloroplast protein import but was not lethal, possibly because of the presence of the similar 34 kDa protein. The mutation impaired chloroplast import most severely at an early stage of leaf development. This coincided with the time when the 33 kDa protein was much more highly expressed than the 34 kDa protein.

6.4.2 Targeting to thylakoids

Proteins associated with the thylakoid membrane are first imported into the stroma, and subsequently targeted to or across the thylakoid membrane. Four separate pathways or mechanisms are now known to exist for targeting distinct precursor proteins to the thylakoid (reviewed by Keegstra and Cline, 1999). These have been termed the Sec pathway, the signal recognition particle (SRP)-like pathway, the ΔpH-pathway and the independent pathway. Each pathway differs in the energy requirement for transport and in the protein composition of the import/targeting machinery. The Sec and SRP-like pathways require ATP and GTP, respectively, and additionally require a *trans*-thylakoid ΔpH gradient. The ΔpH-dependent pathway requires only a *trans*-thylakoid pH gradient, whereas the independent pathway requires no external energy source.

The 33 kDa OEC protein, PsaF, cytochrome *f* and PC are targeted by the

Sec pathway, so named because of its similarity to the bacterial Sec system for protein export (Keegstra and Cline, 1999). The bacterial Sec system consists of an azide sensitive ATPase, named SecA, together with the trimeric SecY/E/G membrane proteins. The chloroplast homologue of SecA (cpSecA) and SecY (cpSecY) have been characterized biochemically and cDNAs isolated. Maize high chlorophyll fluorescence mutants have been isolated, which have deletions in the genes encoding cpSecA and cpSecY (Voelker *et al.*, 1997; Roy and Barkan, 1998). Both mutations are lethal, indicating the importance of these two gene products *in vivo*.

Integration of some specific proteins into the thylakoid membrane involves proteins with similarity to components of bacterial and endoplasmic reticulum SRPs. Following transfer into the stroma, light-harvesting complex protein (LHCP) is assembled into a 200 kDa complex containing cpSRP54 and cpSRP43 (Li *et al.*, 1995). Reconstitution of LHCP with cpSRP54/43 is sufficient for formation of the 200 kDa complex, but additional unidentified stromal components are required for integration into the thylakoid membrane (Schuenemann *et al.*, 1998). Using an *in vitro* translation system, cpSRP54, but not cpSRP43 or SecA, has been shown to cross-link to the D1 protein as it is synthesized by ribosomes (Nilsson *et al.*, 1999). This suggests that the chloroplast SRP system might also be involved in targeting chloroplast-encoded subunits to the thylakoid. In this latter role, cpSRP54 appears to function in the co-translational targeting of plastid-encoded proteins, in contrast to its role in post-translation targeting of LHCP.

Several integral thylakoid proteins, including PsbW, PsbX and CF0-II, have no energy or protein requirements for membrane integration, suggesting that integration may occur by a spontaneous process. Finally, several lumenal proteins use only the ΔpH gradient as the energy source for transport from the stroma to the lumen. Two maize mutants, *hcf106* and *tha4* (Voelker and Barkan, 1995; Roy and Barkan, 1998), have been isolated that are selectively defective in the ΔpH-dependent import pathway. The Hcf106 and Tha4 proteins are homologues of a bacterial export pathway involved in biogenesis of periplasmic redox proteins.

6.5 Assembly of photosynthetic protein complexes

Genetic transformation of cyanobacteria and chloroplast genomes has made it possible to use gene deletion to study the requirement of specific polypeptides for the assembly of thylakoid protein complexes (reviewed by Vermaas, 1993; Golbeck, 1992; Wollman *et al.*, 1999). Deletion of *psbA* (Nilsson *et al.*, 1990), *psbD* or *psbB* genes (Vermaas *et al.*, 1988) leads to a lack of assembly of the remainder of the major PSII polypeptides. In the absence of CP43, due to interruption of *psbC*, the remainder of the PSII complex is still able to assemble and is capable of normal charge separation and electron flow (Rögner, *et al.*, 1991). Deletion of either of the PSI reaction center genes, *psaA* and *psaB*, prevents the assembly of the remainder of the PSI reaction center in *Synechocystis* PCC 6803 (Smart *et al.*, 1991). Deletion of *psaD* (Chitnis *et al.*, 1989a), *psaE* (Chitnis *et al.*, 1989b), *psaF* (Chitnis *et al.*, 1991) and *psaL* (Chitnis *et al.*, 1993) genes still allows photoheterotrophic growth of the mutants, implying that PSI still assembles in these deletion mutants, although all subunits may not assemble normally in the absence of others (Cohen *et al.*, 1993). In *Anabaena variabilis* ATCC 29413 (Mannan *et al.*, 1991), deletion of the *psaC* gene does not disrupt the assembly of PSI, although in *Chlamydomonas reinhardtii* similar disruption of *psaC* in chloroplasts results in destabilization of the remainder of the PSI complex (Takahashi *et al.*, 1991). The individual subunits of the b_6f complex all show a concerted accumulation *in vivo*. Similarly, without either the α, β or ε subunits (Lemaire and Wollman, 1989; Robertson *et al.*, 1989, 1990) the remainder of the ATPase complex does not assemble. Again, rapid degradation of unassembled subunits is thought to occur. Taken together, these studies suggest that there must be concerted synthesis of the majority of the thylakoid proteins, for each complex, in order for correct assembly to occur.

6.5.1 Non-structural protein factors required for assembly

The involvement of special proteins required for the assembly of thylakoid membrane proteins has become apparent in recent years. Such proteins obviously include the components of the import apparatus described above. However, additional proteins also appear to operate on post-translational steps, in many cases serving to catalyze attachment or insertion of nonproteinaceous co-factors. At least four nuclear genes (*Ccs1*–*Ccs4*) and one chloroplast gene (*ccsA*) are involved in attachment of *c*-type hemes to cytochrome *f* and cytochrome c_6 (Xie *et al.*, 1998). Structural models predict that the *Ccs* products form a transmembrane complex, with the site for *c*-heme attachment activity on the lumenal side. Similarly, attachment of heme to the cytochrome b_6 also requires multiple nucleus-encoded proteins in *Chlamydomonas reinhardtii* (Kuras *et al.*, 1998).

Two genes, termed *ycf3* and *ycf4*, have been implicated in the assembly of the PSI complex. Deletion of *ycf3* and *ycf4* in *Chlamydomonas reinhardtii*, or *ycf3* in tobacco, leads to a complete loss of all PSI proteins (Boudreau *et al.*, 1997). In *Synechocystis* 6803, deletion of *ycf4* results in a 50% reduction of PSI. Both Ycf3 and Ycf4 are associated with the thylakoid membrane, but neither polypeptide is tightly associated with known thylakoid membrane protein complexes. In *Chlamydomonas reinhardtii*, it has been shown that chloroplast gene transcription, and translation of PsaA and PsaB, proceeds normally. The function of Ycf3/4, therefore, appears to be as assembly factors regulating post-translation events of PSI biogenesis. It is possible that Ycf3/4 function in insertion or assembly of co-factors into the PSI complex. Such a protein has recently been identified in cyanobacteria. Deletion of *rubA*, a gene encoding a rubredoxin, results in the accumulation of a low level of PSI complexes that lack the iron-sulfur cluster, F_X. It is speculated that RubA is involved in F_X iron-sulfur cluster formation or insertion (Shen, G. and Bryant, DA, pers. comm.). A high chlorophyll fluorescence mutant of *Arabidopsis*, HCF136, lacks PSII due to a T-DNA insertion in the nuclear genome (Meurer *et al.*, 1998). The wild-type copy of the interrupted gene has been cloned and characterized, and shown to encode a protein that is located as a peripheral membrane protein in the lumen of unstacked thylakoids. It is speculated that this protein may be involved in assembly of PSII in unstacked membranes.

6.5.2 Proteolytic turnover of protein complexes

Based on the above discussion, there is clearly rapid, and sometimes selective, post-translational degradation of unassembled or damaged thylakoid-membrane proteins. These post-translational events may be important in determining the steady-state level of thylakoid proteins, and perhaps function as a proofreading mechanism to ensure that misassembled complexes do not accumulate. In PSII, the turnover rate of the D1 protein is significantly faster than other protein components (Mattoo *et al.*, 1984). Similar behavior has been reported for the cytochrome b_6f complex, where turnover of cytochrome b_6 is slower than other components of the complex (Bruce and Malkin, 1991). These results suggest that individual proteins can be selectively degraded, resynthesized and reinserted into a functional complex. The selective degradation and resynthesis of D1 is important because it becomes damaged by oxygen radicals produced during electron transfer.

Rapid degradation of unassembled thylakoid proteins suggests the presence of chloroplast proteases (reviewed by Vierstra, 1993; Adam, 1996). Proteases involved in post-translational modification of chloroplast proteins following import into the chloroplast were discussed above. Protease enriched fractions have been isolated from chloroplasts (Liu and Jagendorf, 1984, 1986; Malek *et al.*, 1984; Adamska *et al.*, 1996; Sokolenko *et al.*, 1997). Further characterization of pea chloroplast proteases has revealed a neutral endopeptidase and three amino peptidases (Liu and Jagendorf, 1986). The endopeptidase was inhibited by dithiothreitol (Liu and Jagendorf, 1986), suggesting it may be under redox control. A protease purified from *Chlamydomonas reinhardtii y-1* membranes degrades newly synthesized Cab proteins and other unidentified thylakoid membrane proteins (Hoober and Hughs, 1992). Inhibitor studies indicate that *Chlamydomonas reinhardtii* protease was different from commonly found proteases and may contain an active histidyl moiety (Hoober and Hughs, 1992). A prolyl endopeptidase, which has been isolated from spinach thylakoids, specifically degrades the 18 kDa polypeptide of the oxygen-evolving complex (Kuwabara, 1992). More recently, chloroplast homologues of several bacterial proteases have been identified. Amongst these is the ATP-dependent ClpP protease and FtsH. The FtsH protein has been shown to be involved in degradation of unassembled Reiske protein (Lindhal *et al.*, 1996; Ostersetzer and Adam, 1997). In the future, it will be important to understand the regulation of chloroplast proteolytic activity. It is very probable that many steps in the response of thylakoid membrane protein composition to changing environmental growth conditions are mediated through post-translational events. The recent characterization of an ATP-dependent protease that degrades light-harvesting complex II (LHCII) during high-light acclimation supports this view (Yang *et al.*, 1998).

6.6 Control of chloroplast division

The division of chloroplasts is under the control of multiple nuclear genes. This has been elegantly demonstrated through the isolation of *Arabidopsis* mutants altered in the accumulation and replication of chloroplasts (*arc* mutants) (reviewed by Pyke, 1999). The most dramatic of these mutants is *arc6*, which contains only two large chloroplasts per mesophyll cell (Pyke *et al.*, 1994). The *arc3* and *arc5* mutants accumulate approximately 15 enlarged mesophyll chloroplasts per mesophyll cell. The mutant plants are reported to have a nearly normal phenotype. It appears that the chloroplast to mesophyll cell volume is maintained in the mutant leaves, and that this compensates for the reduced chloroplast number. An *Arabidopsis* gene has been identified that shows high similarity to the *ftsZ* gene involved in bacterial cell division. The *Arabidopsis FtsZ* gene is important for chloroplast division because transgenic plants expressing antisense *FtsZ* constructs have fewer enlarged chloroplasts, similar to the *arc6* mutant (Osteryoung *et al.*, 1998). Surprisingly, no differences in the photosynthetic apparatus were reported (although no detailed investigation appears to have been carried out). Why the total number of plastids per plant cell is so precisely regulated throughout leaf development is an intriguing question, warranting further studies.

6.7 Environmental light effects

As new leaves emerge into a plant canopy, they are subject to vastly changing light environments. In monocotyledons, leaves emerge at the top of the canopy and subsequently become progressively shaded as new leaves emerge above them. At the opposite extreme, seedlings germinating in an established canopy receive only a fraction of the irradiance that is received by mature plants. As the seedling grows toward the top of the canopy, it will be exposed to a range of light levels. Plants have evolved a number of adaptive responses that allow the photosynthetic apparatus to acclimate to changing light environment in the long-term, and to cope with increases in light levels that are potentially damaging.

Leaves of most species respond to increases in growth irradiance by increasing the capacity for linear electron transport per chlorophyll (Evans, 1996). Such acclimatory responses involve increases in the concentration of PSI, PSII, cytochrome b_6f and ATP synthase complexes relative to the antennae chlorophyll–protein complexes (Anderson, 1986). Enhancement of the far-red region of the spectrum, as would occur due to shading by vegetation, results in an increase in the ratio of PSII:PSI complexes (Chow *et al.*, 1990; McKiernan and Baker, 1991; Kim J.H. *et al.*, 1993). Changes can also occur in the stoichiometry of light-harvesting antenna complexes and reaction centers (Anderson, 1986). It was found that intensity changes alone do not elicit such a change in reaction center stoichiometry or size of light-harvesting complex per reaction center in pea leaves (Lee and Whitmarsh, 1989). Such light-induced changes in the stoichiometry of thylakoid components have been implicated in the optimization of photosynthetic performance (Evans, 1996; Percy, 1998) and are important, but mechanistically poorly understood, acclimatory responses.

The few studies that have investigated the molecular basis for photosystem stoichiometry adjustments have indicated that light quality can regulate chloroplast gene expression (Glick *et al.*, 1986; Deng *et al.*, 1989). Germinating spinach seedlings adapted to red light show a ten fold decrease in *psaA* mRNA levels that correlate with decreased levels of PSI polypeptides (Deng *et al.*, 1989). Run-on transcription assays demonstrate that the decrease in mRNA from *psaA* under red-light growth conditions is a result of a decrease in *psaA* transcription activity and *psaA* transcript stability (Deng *et al.*, 1989). The redox state of the PQ pool has been implicated as a signal for the control of the rate of transcription of chloroplast genes (Pfannschmidt *et al.*, 1999). Such a mechanism might provide a sensitive indicator of imbalance in the stoichiometry of PSII and PSI activity under different light environments. In addition, control of chloroplast transcription and translation by ADP-dependent phosphorylation and thioredoxin, described above, may play a role in light acclimation. Finally, protein turnover must also play a role in light acclimation. An ATP-dependent protease has been shown to degrade LHCII when the antenna size of PSII is decreased in plants following a shift from a low to high light environment (Yang *et al.*, 1998). The acclimation response showed a lag of 2 days, presumably the time required for expression or activation of the protease.

6.7.1 Photoprotective molecules and mechanisms

In the natural environment, leaves can be exposed to light levels that exceed the rate of photosynthesis (Long *et al.*, 1994). Under these conditions, photosynthesis can generate excessive amounts of reactive compounds that can cause oxidative damage. Plants have evolved a range of mechanisms to dissipate excess light energy and/or remove toxic by-products generated by overexcitation of the photosynthetic apparatus (Foyer and Mullineaux, 1994; Owens 1996; Noctor and Foyer, 1998; Niyogi, 1999). Even with such a wide range of protective mechanisms, however, plants still show reduced productivity when exposed to excess light, particularly when combined with temperature extremes that limit photosynthetic activity.

6.7.1.1 Carotenoids

Carotenoids serve as accessory light-harvesting pigments and also fulfil a photoprotective role in the antenna (Owens, 1996; Niyogi, 1999). Excess light conditions can be potentially damaging to the photosynthetic apparatus if the energy is not dissipated in a harmless form. When excited, chlorophyll can undergo a process called 'intersystem conversion', where the excited singlet state forms a triplet state. The triplet state of chlorophyll is long-lived and may transfer its energy to oxygen. Carotenoids are able to quench the energy from potentially damaging excited molecules and dissipate the energy in a harmless form (reviewed by Niyogi, 1999). Certain carotenoids, termed xanthophylls, are also able to quench excitation energy from singlet chlorophyll by a phenomena known as 'non-photochemical' quenching (Demmig-Adams, 1990; Owens, 1996; Niyogi, 1999). Under high light conditions, violaxanthin is rapidly converted to zeaxanthin by violaxanthin de-epoxidase associated with the thylakoid membrane lumen. Zeaxanthin is thought to have an energy level below chlorophyll, and can thus act as an excited state quencher. Although the mechanism of quenching is not known in detail, the xanthophyll cycle, as it is now known, is thought to play an important role in regulating the dissipation of excitation energy in antenna complexes, before this energy can be used to initiate photochemical processes that could lead to photo-oxidative damage in high light. A genetic approach to the identification of mutants defective in non-photochemical quenching processes is beginning to yield interesting insights into the precise components required for non-photochemical quenching (Niyogi, 1999). One mutant of *Arabidopsis*, *npq1*, is unable to convert violaxanthin to zeaxanthin due to a recessive mutation in violaxanthin de-epoxidase (Niyogi *et al.*, 1998). In the *npq1* mutant, thermal dissipation of excitation energy is almost completely inhibited, and leaves are more susceptible to photoinhibition under high light conditions. Interestingly, when acclimated to high light, growth of *npq1* plants is similar to wild-type, suggesting that other photoprotective processes can compensate for a defect in violaxanthin de-epoxidase.

6.7.1.2 Catalases, peroxidases and antioxidant metabolites

Over excitation of PSII and PSI can lead to the production of superoxide O_2^- and singlet oxygen ($^1O_2$), which can potentially lead to extensive damage to the photosynthetic apparatus. Moreover, H_2O_2 formed from O_2^- may lead to the formation of $^{\cdot}OH$, which is among the most reactive radicals in nature. Active oxygen species may initiate cascade reactions that result in the production of other destructive oxidative species, such as lipid peroxides. Plants have evolved several mechanisms to deal with potential photo-oxidative damage and its consequences. These mechanisms include antioxidant systems involving carotenoids, α-tocopherol, ascorbate, glutathione, and the scavenging enzymes superoxide dismutase, ascorbate peroxidase and catalase. The antioxidant enzymes need to act in a synchronous manner to effectively remove O_2^- and H_2O_2 (Noctor and Foyer, 1998). Superoxide produced in chloroplasts and peroxisomes is rapidly converted to H_2O_2 by superoxide dismutase. However, H_2O_2 itself is reactive and must be dealt with by catalase and/or peroxidases. In plants, the best defined pathway for removal of H_2O_2 is the ascorbate-glutathione cycle (reviewed by Noctor and Foyer, 1998). In this cycle, ascorbate peroxidase uses two molecules of ascorbate to reduce H_2O_2 to water, with the concomitant production of two molecules of monodehydroascorbate. Monodehydroascorbate can rapidly disproportionate to ascorbate and dehydroascorbate. Enzymes are present in several cell compartments to regenerate ascorbate from monodehydroascorbate and dehydroascorbate. Dehydroascorbate reductase converts dehydroascorbate to ascorbate using glutathione, generating glutathione disulfide. Glutathione disulfide is subsequently re-reduced by glutathione reductase using NADPH. Thus, the ascorbate-glutathione pathway uses NAD(P)H to reduce H_2O_2 to H_2O, without the consumption of ascorbate or glutathione.

6.7.1.3 Tocopherols

Although all forms of tocopherols and tocotrienols possess antioxidant properties *in vitro*, α-tocopherol (vitamin E) is the most bioactive of all tocopherols *in vivo* and is the predominant form in thylakoid membranes (Foyer *et al.*, 1994). It has been estimated from *in vitro* experiments that one α-tocopherol molecule is capable of quenching as many as 120 $^1O_2$ molecules before being degraded (Fryer, 1992). In addition to serving as antioxidants, tocopherols also function to stabilize polyunsaturated fatty acids within lipid bilayers (Fryer, 1992). The interaction between the tocopherol phytyl tail and polyunsaturated fatty acids has been shown to protect membrane lipids from lipoxygenase attack and oxidative damage.

The best characterized function of tocopherols in biological membranes is to act as recyclable chain reaction terminators of polyunsaturated fatty acid free radicals generated by lipid oxidation (Fryer, 1992). Polyunsaturated fatty acid radicals are re-reduced by donation of a hydrogen atom from the ring hydroxyl of tocopherols, resulting in a tocopherol radical. The tocopherol radical is recycled back to the corresponding tocopherol by interaction with either ascorbate or glutathione, by yet to be identified mechanism(s) (Padh, 1990). This regeneration chain is very rapid and allows each tocopherol to quench a large number of polyunsaturated fatty acid free radicals before being degraded.

6.7.2 Repair of photodamage

The most susceptible site of photodamage is the PSII reaction center. Photoinhibition of PSII is a constant problem, and repair of PSII is an important photoprotective mechanism. The repair of photosystem II involves the selective removal and degradation of damaged D1 protein. Newly synthesized D1 is then reincorporated to form a functional complex (Van Wijk *et al.*, 1995). The selective replacement of D1 to repair damaged PSII complexes has been referred to as the 'damage and repair cycle'. In this model, the damaged PSII complexes migrate to the unstacked region of the thylakoid membrane, where damaged D1 is degraded and D1 replacement occurs. Following incorporation of new D1, the PSII complex migrates back to the granal stack region.

The acceptor side co-factors of PSI have a potential low enough to transfer electrons to oxygen and generate O_2^-. Thus, PSI is potentially a highly susceptible target for photoinhibition. The subunit composition on the PSI acceptor side must help protect PSI from oxygen. More work is required to understand whether damage to PSI is of physiological significance. There are reports that PSI may be susceptible to photoinhibition under chilling stress conditions. The selective replacement of damaged PSI complexes *in vivo* has not been reported.

References

Adam, Z. (1996) Protein stability and degradation in chloroplasts. *Plant Molecular Biology*, **32** 773–783.

Adamska, I., Lindhal, M., Roobol-Boza, B. and Andersson, B. (1996) Degradation of the light-stress protein is mediated by an ATP-independent, serine-type protease under low-light conditions. *European Journal of Biochemistry*, **236** 591–599.

Anderson, J.M. (1986) Photoregulation of the composition, function and structure of thylakoid membranes. *Annual Review of Plant Physiology*, **37** 93–136.

Bennoun, P., Spierer-Herz, M., Erickson, J., Girard-Bascou, J., Pierre, Y., Delosme, M. and Rochaix, J-D. (1986) Characterization of photosystem II mutants of *Chlamydomonas reinhardtii* lacking the *psbA* gene. *Plant Molecular Biology*, **6** 151–160.

Boudreau, E., Takahashi, Y., Lemieu, C., Turmel, M. and Rochaix, J-D. (1997) The chloroplast *ycf3* and *ycf4* open reading frames are required for the accumulation of the photosystem I complex. *EMBO Journal*, **16** 6095–6104.

Bruce, B.D. and Malkin, R. (1991) Biosynthesis of the chloroplast b_6f complex: studies in a photosynthetic mutant of *Lemna*. *Plant Cell*, **3** 203–212.

Bruick, R.K. and Mayfield, S.P. (1998) Processing of the *psbA* 5′ untranslated region in *Chlamydomonas reinhardtii* depends upon factors mediating ribosome association. *Journal of Cell Biology*, **143** 1145–1153.

Caliebe, A., Grimm, R., Kaiser, G. Lubeck, J., Soll, J. and Heins, L. (1997) The chloroplastic protein import machinary contains a Reiske-type iron-sulfur cluster and a mononuclear iron-binding protein. *EMBO Journal*, **16** 7342–7350.

Chitnis, P.R., Purvis, D. and Nelson, N. (1991) Molecular cloning and targeted mutagenesis of the gene, *psaF*, encoding subunit III of photosystem I from the cyanobacterium, *Synechocystis* sp. PCC 6803. *Journal of Biological Chemistry*, **266** 20146–20151.

Chitnis, P.R., Reilly, P.A. and Nelson, N. (1989a) Insertional inactivation of the gene encoding subunit II of photosystem I from the cyanobacterium, *Synechocystis* sp. PCC 6803. *Journal of Biological Chemistry*, **264** 18381–18385.

Chitnis, P.R., Reilly, P.A., Miedel, M.C. and Nelson, N. (1989b) Structure and targeted mutagenesis of the gene encoding 8 kDa subunit of photosystem I from the cyanobacterium, *Synechocystis* sp. PCC 6803. *Journal of Biological Chemistry*, **264** 18374–18380.

Chitnis, V.P., Xu, Q., Yu, L., Golbeck, J.H., Nakamoto, H., Xie, D.L. and Chitnis, P.R. (1993) Targeted inactivation of the gene, *psaL*, encoding a subunit of photosystem I of the cyanobacterium, *Synechocystis* sp. PCC 6803. *Journal of Biological Chemistry*, **268** 11678–11684.

Choquet, Y., Stern, D.B., Wostrikoff, K., Kuras, R., Girard-Bascou, J. and Wollman, F-A. (1998) Translation of cytochrome *f* is autoregulated through the 5′ untranslated region of *petA* mRNA in *Chlamydomonas* chloroplasts. *Proceedings of the National Academy of Sciences of the United States of America*, **95** 4380–4385.

Chow, W.S., Goodchild, D.J., Miller, C. and Anderson, J.M. (1990) The influence of high levels of brief or prolonged supplementary far-red illumination during growth on the photosynthetic characteristics, composition and morphology of *Pisum sativum* chloroplasts. *Plant, Cell and Environment*, **13** 135–145.

Christopher, D.A. and Mullet, J.E. (1994) Separate photosensory pathways co-regulate blue light/ultraviolet-A-activated *psbD-psbC* transcription and light-induced D2 and CP43 degradation in barley (*Hordeum vulgare*) chloroplasts. *Plant Physiology*, **104** 1119–1129.

Christopher, D.A., Kim, M. and Mullet, J.E. (1992) A novel light-regulated promoter is conserved in cereal and dicot chloroplasts. *Plant Cell*, **4** 785–798.

Cohen, Y., Chitnis, V.P., Nechushtai, R. and Chitnis, P.R. (1993) Stable assembly of PsaE into cyanobacterial photosynthetic membranes is dependent on the presence of other accessory subunits of photosystem I. *Plant Molecular Biology*, **23** 895–900.

Danon, A. and Mayfield, S.P. (1991) Light-regulated translational activators: identification of chloroplast gene specific mRNA binding proteins. *EMBO Journal*, **10** 3993–4001.

Danon, A. and Mayfield, S.P. (1994) Light-regulated translation of chloroplast messenger RNAs through redox potential. *Science*, **266** 1717–1719.

De Vitry, C., Olive, J., Drapier, D., Recouvreur, M. and Wollman, F-A. (1989) Post-translational events leading to the assembly of photosystem II protein complex: a study using photosynthesis mutants from *Chlamydomonas reinhardtii*. *Journal of Cell Biology*, **109** 991–1006.

Demmig-Adams, B. (1990) Carotenoids and photoprotection in plants: a role for xanthophyll zeaxanthin. *Biochimica et Biophysica Acta*, **1020** 1–24.

Deng, X-W. and Gruissem, W. (1987) Control of plastid gene expression during development: the limited role of transcriptional regulation. *Cell*, **49** 379–387.

Deng, X-W., Tonkyn, J.C., Peter, G.F., Thornber, J.P. and Gruissem, W. (1989) Post-transcriptional control of plastid mRNA accumulation during adaptation of chloroplasts to different light quality environments. *Plant Cell*, **1** 645–654.

Diner, B.A. and Babcock, G.T. (1996) Structure, dynamics and energy conversion efficiency in photosystem II, in *Oxygenic Photosynthesis: the Light Reactions* (eds. D. Ort and C. Yocum), Kluwer Academic Publishers, Dordrecht, The Netherlands, pp. 41–58.

Drapier, D., Girard-Bascou, J., and Wollman, F-A. (1992) Evidence for nuclear control of expression of the *atpA* and *atpB* chloroplast genes in *Chlamydomonas*. *Plant Cell*, **4** 283–295.

Eichacker, L.A., Soll, J., Lauterbach, P., Rüdiger, W., Klein, R.R. and Mullet, J.E. (1990) *In vitro* synthesis of chlorophyll *a* in the dark triggers accumulation of chlorophyll *a* apoproteins in barley etioplasts. *Journal of Biological Chemistry*, **265** 13566–13571.

Erickson, J.M., Rahire, M., Malnoe, P., Girard-Bascou, J., Pierre, Y., Bennoun, P. and Rochaix, J-D. (1986) Lack of the D2 protein in a *Chlamydomonas reinhardtii psbD* mutant affects photosystem II stability and D1 expression. *EMBO Journal*, **5** 1745–1754.

Evans, J.R. (1996) Developmental constraints on photosynthesis: effects of light and nutrition, in *Photosynthesis and the Environment*, (ed. N.R. Baker), Kluwer Academic Publishers, Dordrecht, The Netherlands, pp. 281–304.

Foyer, C.H. and Mullineaux, P. (1994) *Causes of Photooxidative Stress and Amelioration of Defense Systems in Plants*, CRC Press, Baton Rouge, LA, USA.

Foyer, C.H., Descourvieres, P. and Kunert, K.J. (1994) Protection against oxygen radicals: an important defense mechanism studied in transgenic plants. *Plant, Cell and Environment*, **17** 507–523.

Fryer, M.J. (1992) The antioxidant effects of thylakoid vitamin E (α-tocopherol) *Plant, Cell and Environment*, **15** 381–392.

Gamble, P.E. and Mullet, J.E. (1989) Blue light regulates the accumulation of two *psbD-psbC* transcripts in barley chloroplasts. *EMBO Journal*, **8** 2785–2794.

Gamble, P.E., Sexton, T.B. and Mullet, J.E. (1988) Light-dependent changes in *psbD* and *psbC* transcripts of barley chloroplasts: accumulation of two transcripts maintains *psbD* and *psbC* translation capability in mature chloroplasts. *EMBO Journal*, **7** 1289–1297.

Girard-Bascou, J., Choquet, Y., Schneider, M., Delosme, M. and Dron, M. (1987) Characterization of a chloroplast mutation in the *psaA2* gene of *Chlamydomonas reinhardtii*. *Current Genetics*, **12** 489–495.

Glick, R.E., McCauley, S.W., Gruissem, W. and Melis, A. (1986) Light quality regulates expression of chloroplast genes and assembly of photosynthetic membrane complexes. *Proceedings of the National Academy of Sciences of the United States of America*, **83** 161–170.

Golbeck, J.H. (1992) Structure and function of photosystem I. *Annual Review of Plant Physiology and Plant Molecular Biology*, **43** 293–324.

Gruissem, W. (1989) Chloroplast RNA: transcription and processing, in *The Biochemistry of Plants*, (ed. A. Marcu), Vol. 15, Academic Press, New York, USA, pp. 151–191.

Gruissem, W. and Tonkyn, J.C. (1993) Control mechanisms of plastid gene expressions. *Critical Reviews in Plant Sciences*, **12** 19–55.

Haley, J. and Bogorad, L. (1989) A 4 kDa maize chloroplast polypeptide associated with the cytochrome b_6f complex: subunit 5, encoded by the chloroplast *petE* gene. *Proceedings of the National Academy of Sciences of the United States of America*, **86** 1534–1538.

Haley, J. and Bogorad, L. (1990) Alternative promoters are used for genes within maize chloroplast polycistronic transcription units. *Plant Cell*, **2** 323–333.

Hird, S.M., Webber, A.N., Wilson, R., Dyer, T.A. and Gray, J.C. (1991) Differential expression of the chloroplast genes for the 47 kDa chlorophyll *a* protein and the 10 kDa phosphoprotein during chloroplast development in wheat. *Current Genetics*, **19** 199–206.

Hirsch, S., Muckel, E., Heemeyer, F., von Heijne, G. and Soll, J. (1994) A receptor component of the chloroplast protein translocation machinery. *Science*, **266** 1989–1992.

Hoober, J.K. and Hughs, M.J. (1992) Purification and characterization of a membrane-bound protease from *Chlamydomonas reinhardtii*. *Plant Physiology*, **99** 932–937.

Jarvis, P., Chen, L.J., Li, H.M., Pete, C.A., Fankhauser, C. and Chory, J. (1998) An *Arabidopsis* mutant defective in the plastid general protein import apparatus. *Science*, **282** 100–103.

Jensen, K.H., Herrin, D.L., Plumley, G. and Schmidt, G.W. (1986) Biogenesis of photosystem II complexes: transcriptional, translational and post-translational regulation. *Journal of Cell Biology*, **103** 1315–1325.

Keegstra, K. and Cline, K. (1999) Protein import and routing systems of chloroplasts. *Plant Cell*, **11** 557–570.

Kessler, F., Blobel, G., Patel, H.A. and Schnell, D.J. (1994) Indentification of two GTP-binding proteins in the chloroplast import machinary. *Science*, **266** 1035–1039.

Kim, J.H., Glick, R.E. and Melis, A. (1993) Dynamics of photosystem stoichiometry adjustment by light quality in chloroplasts. *Plant Physiology*, **102** 181–190.

Kim, J.M. and Mayfield, S.P. (1997) Protein disulfide isomerase as a regulator of chloroplast translational activation. *Science*, **278** 1954–1957.

Kim, J.M., Klein, P.G. and Mullet, J.E. (1991) Ribosomes pause at specific sites during synthesis of membrane-bound chloroplast reaction center protein D1. *Journal of Biological Chemistry*, **266** 14931–14938.

Kim, J., Eichacker, L.A., Rudiger, W. and Mullet, J.E. (1994) Chlorophyll regulates accumulation of the plastid-encoded chlorophyll proteins, P700 and D1, by increasing apoprotein stability. *Plant Physiology*, **104** 907–916.

Kim, M.Y. and Mullet, J.E. (1995) Identification of a sequence specific DNA-binding factor required for transcription of the barley chloroplast blue light-responsive *psbD-psbC* promoter. *Plant Cell*, **7** 1445–1457.

Kim, M.Y., Christopher, D.A. and Mullet, J.E. (1993) Direct evidence for selective modulation of *psbA*, *rpoA*, *rbcL* and 16S RNA stability during barley chloroplast development. *Plant Molecular Biology*, **22** 447–463.

Kim, M., Christopher D.A., and Mullet, J.E. (1999) ADP-dependent phosphorylation regulates association of a DNA-binding complex with the barley chloroplast psbD blue-light-responsive promoter. *Plant Physiology*, **119** 663–670.

Klaff, P. and Gruissem, W. (1991) Changes in chloroplast mRNA stability during leaf development. *Plant Cell*, **3** 517–529.

Klein, R.R. and Mullet, J.E. (1986) Regulation of chloroplast-encoded chlorophyll-binding protein translation during higher plant chloroplast biogenesis. *Journal of Biological Chemistry*, **261** 11138–11145.

Klein, R.R. and Mullet, J.E. (1987) Control of gene expression during higher plant chloroplast biogenesis: protein synthesis and transcript levels of *psbA*, *psaA-psaB* and *rbcL* in dark-grown and illuminated barley seedlings. *Journal of Biological Chemistry*, **262** 4341–4348.

Klein, R.R. and Mullet, J.E. (1990) Light-induced transcription of chloroplast genes: *psbA* transcription is differentially enhanced in illuminated barley. *Journal of Biological Chemistry*, **265** 1895–1902.

Klein, R.R., Mason, H. and Mullet, J.E. (1987) Light-regulated translation of chloroplast proteins: transcripts of *psaA-psaB*, *psbA* and *rbcL* are associated with polysomes in dark-grown and illuminated barley seedlings. *Journal of Cell Biology*, **106** 289–302.

Krauss, N., Schubert, W-D., Klukas, O., Fromme, P., Witt, H.T. and Saenger, W. (1996) Photosystem I at 4Å resolution: a joint reaction center and core antenna system. *Nature Structural Biology*, **3** 965–973.

Kreuz, K., Dehesh, K. and Apel, K. (1986) The light-dependent accumulation of the P700 chlorophyll *a* protein of the photosystem I reaction center in barley. *European Journal of Biochemistry*, **159** 459–467.

Kuchka, M.R., Mayfield, S.P. and Rochaix, J-D. (1988) Nuclear mutations specifically affect the synthesis and/or degradation of the chloroplast-encoded D2 polypeptide of photosystem II in *Chlamydomonas reinhardtii*. *EMBO Journal*, **7** 319–324.

Kuchka, M.R., Goldschmidt-Clermont, M., Van Dillewijn, J. and Rochaix, J-D. (1989) Mutation at the *Chlamydomonas* nuclear NAC2 locus specifically affects stability of the chloroplast *psbD* transcript encoding polypeptide D2 of PSII. *Cell*, **58** 869–876.

Kudla, J., Hayes, R. and Gruissem, W. (1996) Polyadenylation accelerates degradation of chloroplast mRNA. *EMBO Journal*, **15** 7137–7146.

Kuras, R., Wollman, F-A. and Joliot, P. (1995) Conversion of cytochrome *f* to a soluble form *in vivo* in *Chlamydomonas reinhardtii*. *Biochemistry*, **34** 7468–7475.

Kuras, R., De Vitry, C., Choquet, Y., Girard-Bascou, J., Culler, D., Buschlen, S., Merchant, S. and Wollman, F-A. (1998) Molecular genetic identification of a pathway for heme binding to cytochrome b_6. *Journal of Biological Chemistry*, **272** 32427–32435.

Kuwabara, T. (1992) Characterization of a prolyl endopeptidase from spinach thylakoids. *FEBS Letters*, **300** 127–130.

Lee, H., Bingham, S.E. and Webber, A.N. (1996) Function of 3′ non-coding sequences and stop codon usage in expression of the chloroplast *psaB* gene in *Chlamydomonas reinhardtii*. *Plant Molecular Biology*, **31** 337–354.

Lee, W.-J. and Whitmarsh, J. (1989) Photosynthetic apparatus of pea thylakoid membranes: response to growth light intensity. *Plant Physiology*, **89** 932–940.

Lemaire, C. and Wollman, F-A. (1989) The chloroplast ATP synthase in *Chlamydomonas reinhardtii*. II. Biochemical studies on its biogenesis using mutants defective in photophosphorylation. *Journal of Biological Chemistry*, **264** 10235–10242.

Li, X., Henry, R., Yuan, J., Cline, K. and Hoffman, N.E. (1995) A chloroplast homologue of the signal recognition particle subunit, SRP54, is involved in the post-translational integration of a protein into thylakoid membranes. *Proceedings of the National Academy of Sciences of the United States of America*, **92** 3789–3793.

Lindhal, M., Tabak, S., Cseke, L., Pichersky, B., Andersson, B. and Adam, Z. (1996) Identification, characterization and molecular cloning of a homologue of the bacterial FtsH protease in chloroplasts of higher plants. *Journal of Biological Chemistry*, **271** 29329–29334.

Link, G. (1988) Photocontrol of plastid gene expression. *Plant, Cell and Environment*, **11** 329–338.

Lisitsky, I., Klaff, P. and Schuster, G. (1996) Addition of poly (A)-rich sequences to endonucleolytic cleavage sites in the degradation of spinach chloroplast mRNA. *Proceedings of the National Academy of Sciences of the United States of America*, **93** 13398–13403.

Liu, X-Q. and Jagendorf, A.T. (1984) ATP-dependent proteolysis in pea chloroplasts. *FEBS Letters*, **166** 248–252.

Liu, X-Q. and Jagendorf, A.T. (1986) Neutral peptidases in the stroma of pea chloroplasts. *Plant Physiology*, **81** 603–608.

Long, S.P., Humphries, S. and Falkowski, P.G. (1994) Photoinhibition of photosynthesis in nature. *Annual Review of Plant Physiology and Plant Molecular Biology*, **45** 633–662.

Malek, K., Bogorad, L., Ayers, A. and Goldberg, A.L. (1984) Newly synthesized proteins are degraded by ATP stimulated proteolytic process in isolated pea chloroplasts. *FEBS Letters*, **166** 253–255.

Maliga, P. (1998) Two plastid RNA polymerases of higher plants: an evolving story. *Trends in Plant Science*, **3** 4–6.

Mannan, R.M., Whitmarsh, J., Nyman, P. and Pakrasi, H.B. (1991) Directed mutagenesis of an iron sulfur protein of the photosystem I complex in the filamentous cyanobacterium, *Anabaena variabilis* ATCC 29413. *Proceedings of the National Academy of Sciences of the United States of America*, **88** 10168–10172.

Mattoo, A.K., Hoffman-Falk, H., Marder, J.B. and Edelman, E. (1984) Regulation of protein metabolism: coupling of photosynthetic electron transport to *in vivo* degradation of the rapidly metabolized 32 kDa protein of the chloroplast membranes. *Proceedings of the National Academy of Sciences of the United States of America*, **81** 1380–1384.

McKiernan, M. and Baker, N.R. (1991) Adaptation to shade of the light-harvesting apparatus in *Silene dioica*. *Plant, Cell and Environment*, **14** 205–212.

Meurer, J., Plucken, H., Kowallik, K.V. and Westhoff, P. (1998) A nuclear-encoded protein of prokaryotic origin is essential for the stability of photosystem II in *Arabidopsis thaliana. EMBO Journal*, **17** 5286–5297.

Monod, C., Goldschmidt-Clermont, M. and Rochaix, J-D. (1992) Accumulation of chloroplast *psbB* RNA requires a nuclear factor in *Chlamydomonas reinhardtii. Molecular and General Genetics*, **231** 449–459.

Mullet, J.E. (1988) Chloroplast development and gene expression. *Annual Review of Plant Physiology and Plant Molecular Biology*, **39** 475–502.

Mullet, J.E. (1993) Dynamic regulation of chloroplast transcription. *Plant Physiology*, **103** 309–313.

Mullet, J.E. and Klein, R.R. (1987) Transcription and RNA stability are important determinants of higher plant chloroplast RNA levels. *EMBO Journal*, **6** 1571–1579.

Nickelsen, J., Van-Dillewijn, J., Rahire, M. and Rochaix, J-D. (1994) Determinants for stability of the chloroplast psbD RNA are located in its short leader region in *Chlamydomonas reinhardtii. EMBO Journal*, **13** 3182–3191.

Nilsson, F., Andersson, B. and Jansson, C. (1990) Photosystem II characteristics of a constructed *Synechocystis* 6803 mutant lacking synthesis of the D1 polypeptide. *Plant Molecular Biology*, **14** 1051–1054.

Nilsson, R., Brunner, J., Hoffman, N.E. and van Wijk, K.L. (1999) Interactions of ribosome nascent chain complexes of the chloroplast-encoded D1 thylakoid membrane protein with cpSRP54. *EMBO Journal*, **18** 733–742.

Niyogi, K.K. (1999) Photoprotection revisited: genetic and molecular approaches. *Annual Review of Plant Physiology and Plant Molecular Biology*, **50** 333–359.

Niyogi, K., Grossman, A. and Björkman, O. (1998) *Arabidopsis* mutants define a central role for the xanthophyll cycle in the regulation of photosynthetic energy conversion. *Plant Cell*, **10** 1121–1134.

Noctor, G. and Foyer, C.H. (1998) Ascorbate and glutathione: keeping active oxygen under control. *Annual Review of Plant Physiology and Plant Molecular Biology*, **49** 249–279.

Ostersetzer, O. and Adam, Z. (1997) Light-stimulated degradation of an unassembled Rieske FeS protein by a thylakoid-bound protease: the possible role of the FtsH protease. *Plant Cell*, **9** 957–965.

Osteryoung, K.W., Stokes, K.D., Rutherford, S.M., Percival, A.L. and Lee, W.Y. (1998) Chloroplast division in higher plants requires members of two functionally divergent gene families with homology to bacterial *ftsZ. Plant Cell*, **10** 1991–2004.

Owens, T.G. (1996) Processing of excitation energy by antenna pigments, in *Photosynthesis and the Environment*, (ed. N.R. Baker), Kluwer Academic Publishers, Dordrecht, The Netherlands, pp. 1–23.

Padh, H. (1990) Cellular functions of ascorbic acid. *Biochemistry and Cell Biology*, **68** 1166–1173.

Percy, R.W. (1998) Acclimation to sun and shade, in *Photosynthesis, A Comprehensive Treatise*, (ed. A.S. Raghavendra,), Cambridge University Press, Cambridge, UK, pp. 250–263.

Pfannschmidt, T., Nilsson, A. and Allen, J.F. (1999) Photosynthetic control of chloroplast gene expression. *Nature*, **397** 625–628.

Pyke, K. (1999) Plastid division and development. *Plant Cell*, **11** 549–556.

Pyke, K.A., Rutherford, S.M., Robertson, E.J. and Leech, R.M. (1994) Arc6, a fertile *Arabidopsis* mutant with only two mesophyll cell chloroplasts. *Plant Physiology*, **106** 1169–1177.

Rapp, J.C., Baumgartner, B.J. and Mullet, J. (1992) Quantitative analysis of transcription and RNA levels of 15 barley chloroplast genes: transcription rates and mRNA levels vary over 300-fold; predicted mRNA stabilities vary 30 fold. *Journal of Biological Chemistry*, **267** 21404–21411.

Rattanachaikunsopon, P., Rosch, C. and Kuchka, M.R. (1999) Cloning and characterization of the nuclear AC115 gene of *Chlamydomonas reinhardtii. Plant Molecular Biology*, **39** 1–10.

Robertson, D., Woessner, J.P., Gillham, N.W. and Boynton, J.E. (1989) Molecular characterization of two point mutants in the chloroplast *atpB* gene of the green alga, *Chlamydomonas reinhardtii*, defective in assembly of the ATP synthase complex. *Journal of Biological Chemistry*, **264** 2331–2337.

Robertson, D., Boynton, J.E. and Gillham, N.W. (1990) Co-transcription of the wild-type chloroplast *atpE* gene encoding the CF_1/CF_0 epsilon subunit with the 3′ half of the *rps7* gene in *Chlamydomonas reinhardtii* and characterization of frameshift mutations in *atpE*. *Molecular and General Genetics*, **221** 155–163.

Rochaix, J-D. (1992) Post-transcriptional steps in the expression of chloroplast genes. *Annual Review of Cell Biology*, **8** 1–28.

Rochaix, J-D., Kuchka, M., Mayfield, S., Schirmer-Rahire, M., Girard-Bascou, J. and Bennoun, P. (1989) Nuclear and chloroplast mutations affect the synthesis or stability of the chloroplast *psbC* gene product in *Chlamydomonas reinhardtii*. *EMBO Journal*, **8** 1013–1021.

Rögner, M., Chisholm, D.A. and Diner, B.A. (1991) Site-directed mutagenesis of the *psbC* gene of photosystem II: isolation and functional characterization of CP43-less photosystem II core complexes. *Biochemistry*, **30** 5387–5395.

Roy, L.M. and Barkan, A. (1998) A SecY homologue is required for the elaboration of the chloroplast thylakoid membrane and for normal chloroplast gene expression. *Journal of Cell Biology*, **141** 385–395.

Ruffle, S.V. and Sayre, R.T. (1998) Functional analysis of photosystem II, in *Molecular Biology of Chloroplasts and Mitochondria in Chlamydomonas*, (eds. J-D. Rochaix, M. Goldschmidt-Clermont and S. Merchant), Kluwer Academic Publishers, Dordrecht, The Netherlands, pp. 287–322.

Sakamoto, W., Kindle, K.L. and Stern, D.B. (1993) *In vivo* analysis of *Chlamydomonas* chloroplast *petD* gene expression using stable transformation of β-glucuronidase translational fusions. *Proceedings of the National Academy of Sciences of the United States of America*, **90** 497–501.

Salvador, M.L., Klein, U. and Bogorad, L. (1993) 5′ sequences are important positive and negative determinants of the longevity of *Chlamydomonas* chloroplast gene transcripts. *Proceedings of the National Academy of Sciences of the United States of America*, **90** 1556–1560.

Schnell, D.J., Kessler, F. and Blobel, G. (1994) Isolation of components of the chloroplast import machinery. *Science*, **266** 1007–1012.

Schuenemann, D., Gupta, S., Persello-Cartieaux, F., Klimyuk, V.I., Jones, J.D.G., Nussaume, L. and Hoffman, N.E. (1998) A novel signal recognition particle targets light-harvesting proteins to the thylakoid membranes. *Proceedings of the National Academy of Sciences of the United States of America*, **95** 10312–10316.

Sexton, T.B., Christopher, D.A. and Mullet, J.E. (1990) Light-induced switch in barley *psbD-psbC* promoter utilization: a novel mechanism regulating chloroplast gene expression. *EMBO Journal*, **9** 4485–4494.

Sieburth, L.E., Berry-Lowe, S. and Schmidt, G.W. (1991) Chloroplast RNA stability in *Chlamydomonas*: rapid degradation of *psbB* and *psbC* transcripts in two nuclear mutants. *Plant Cell*, **3** 175–189.

Smart, L.B, Anderson, S.L. and McIntosh, L. (1991) Targeted genetic inactivation of the photosystem I reaction center in the cyanobacterium, *Synechocystis* sp. PCC 6803. *EMBO Journal*, **10** 3289–3296.

Sokolenko, A., Altschmied, L. and Herrmann, R.G. (1997) Sodium dodecyl sulfate-stable proteases in chloroplasts. *Plant Physiology*, **115** 827–832.

Stern, D.B. and Gruissem, W. (1987) Control of plastid gene expression: 3′ inverted repeats act as mRNA processing and stabilizing elements but do not terminate transcription. *Cell*, **51** 1145–1157.

Stern, D.B. and Kindle, K.L. (1993) 3′ end maturation of the *Chlamydomonas reinhardtii* chloroplast *atpB* mRNA is a two-step process. *Molecular and Cellular Biology*, **13** 2277–2285.

Takahashi, Y., Goldschmidt-Clermont, M., Soen, S-Y., Franzen, L.G. and Rochaix, J-D. (1991) Directed chloroplast transformation in *Chlamydomonas reinhardtii*: insertional inactivation of the *psaC* gene encoding the iron sulfur protein destabilizes photosystem I. *EMBO Journal*, **10** 2033–2040.

Tan, S. and Troxler, R.F. (1999) Characterization of two chloroplast RNA polymerase sigma factors from *Zea mays*: photoregulation and differential expression. *Proceedings of the National Academy of Sciences of the United States of America*, **96** 5316–5321.

Tiller, K. and Link, G. (1993) Sigma-like transcription factors from mustard (*Sinapis alba* L.) etioplast are similar in size to, but functionally distinct from, their chloroplast counterparts. *Plant Molecular Biology*, **21** 503–513.

Tranel, P.J., Froehlich, J., Goyal, A. and Keegstra, K. (1995) A component of the chloroplastic protein import apparatus is targeted to the outer envelope membrane via a novel pathway. *EMBO Journal*, **14** 2436–2446.

Van Wijk, K.J., Bingsmark, S., Aro, E.M. and Andersson, B. (1995) *In vitro* synthesis and assembly of photosystem II core proteins: the D1 protein can be incorporated into photosystem II in isolated chloroplasts and thylakoids. *Journal of Biological Chemistry*, **270** 25685–25695.

Vermaas, W. (1993) Molecular biological approaches to analyze photosystem II structure and function. *Annual Review of Plant Physiology and Plant Molecular Biology*, **44** 457–481.

Vermaas, W.F.J., Ikeuchi, M. and Inoue, Y. (1988) Protein composition of the photosystem II core complex in genetically engineered mutants of the cyanobacterium, *Synechocystis* sp. PCC 6803. *Photosynthesis Research*, **17** 97–113.

Vierstra, R.D. (1993) Protein degradation in plants. *Annual Review of Plant Physiology and Plant Molecular Biology*, **44** 385–410.

Voelker, R. and Barkan, A. (1995) Two nuclear mutations disrupt distinct pathways for targeting proteins to the chloroplast thylakoid. *EMBO Journal*, **14** 3905–3914.

Voelker, R., Mendel-Hartvig, J. and Barkan, A. (1997) Transposon-disruption of a maize nuclear gene, *tha1*, encoding a chloroplast SecA homologue: *in vivo* role of cp-SecA in thylakoid protein targeting. *Genetics*, **145** 467–478.

Wollman, F-A., Minai, L. and Nechushtai, R. (1999) The biogenesis and assembly of photosynthetic proteins in thylakoid membranes. *Biochemica Biophysica Acta*, **1411** 21–85.

Webber, A.N. and Bingham, S.E. (1998) Structure and function of photosystem I, in *Molecular Biology of Chloroplasts and Mitochondria in Chlamydomonas*, (eds. J-D. Rochaix, M. Goldschmidt-Clermont and S. Merchant), Kluwer Academic Publishers, Dordrecht, The Netherlands, pp. 323–348.

Xie, Z.Y., Culler, D., Dreyfuss, B.W., Kuras, R., Wollman, F-A., Girard-Bascou, J. and Merchant, S. (1998) Genetic analysis of chloroplast *c*-type cytochrome assembly in *Chlamydomonas reinhardtii*: one chloroplast locus and at least four nuclear loci are required for heme attachment. *Genetics*, **148** 681–692.

Yang, D.H., Webster, J., Adam, Z., Lindahl, M. and Andersson, B. (1998) Induction of acclimative proteolysis of the light-harvesting chlorophyll *a*/*b* protein of photosystem II in response to elevated light intensities. *Plant Physiology*, **118** 827–834.

7 Nitrogen utilisation and the photosynthetic system

Niels P.R. Anten, Kouki Hikosaka and Tadaki Hirose

7.1 Introduction

On a worldwide scale, nitrogen is one of the factors most limiting to plant growth. This is evident from the increase in plant production that is observed when nitrogen fertiliser is added (e.g. Kachi and Hirose, 1983; Lee *et al.*, 1983). Even under non-limiting conditions, nitrogen acquisition involves considerable carbon costs for a plant. These costs involve both the construction and maintenance of a root system and the adenosine triphosphate (ATP) requirements for the uptake process itself (van der Werf *et al.*, 1989). Plants are, therefore, restricted in the amount of nitrogen which they can allocate to their leaves.

A large fraction of the nitrogen in leaves is incorporated into proteins associated with the photosynthetic process (Evans, 1989b; Evans and Seemann, 1989), and there is generally a strong positive correlation between the photosynthetic rates and nitrogen contents of leaves (e.g. Osman and Milthorpe, 1971; Nátr, 1975; Field and Mooney, 1986; Hirose and Werger, 1987a; Evans, 1989a). Photosynthesis, in turn, provides the structural substrates for growth and reproduction, which are primary determinants of plant fitness.

In view of the limited availability of nitrogen and the importance of photosynthesis for growth and reproduction, the hypothesis that plants should maximise the photosynthetic carbon gain per unit of nitrogen (PNUE) (Mooney and Gulmon, 1979) is used as a starting point for this study. This hypothesis is based on the theory of optimisation, i.e. natural selection favours characteristics that will maximise fitness within the constraints imposed by the environment. As in most optimisation studies, gross simplifications are made by assuming photosynthesis to be a measure of fitness and by taking nitrogen availability to be the only constraint imposed by the environment. Any prediction, thus derived, will be a consequence of these assumptions (Parker and Maynard Smith, 1990). However, it is not claimed here that all plants maximise their PNUE; rather, the optimisation theory is used as a null hypothesis to analyse the adaptiveness of nitrogen utilisation for photosynthesis.

In this review, two questions are addressed: 1) do plants utilise their nitrogen optimally in such a way as to maximise PNUE? and 2) can the optimisation theory predict the differences in nitrogen utilisation between plants with different characteristics or between plants growing under different conditions? Firstly, the photosynthesis–nitrogen relationship is analysed and the way in which nitrogen is partitioned among the different components of the photosynthetic system is discussed. Then models which analyse the optimal nitrogen partitioning and the optimal leaf nitrogen content that maximises daily carbon gain of a leaf per unit nitrogen (leaf-PNUE) are reviewed. Secondly, scaling-up from leaves to plants, it is shown that an optimal pattern of nitrogen allocation exists among the leaves of plants, which maximises whole plant daily photosynthesis per unit of nitrogen in the canopy (canopy-PNUE). Calculated optimal patterns of nitrogen partitioning within and among leaves are compared to actual data. Differences between optimal and actual distribution patterns are discussed, and the optimisation theory and the way in which it has been applied is critically assessed. Finally, biological mechanisms that potentially may explain how optimal nitrogen partitioning could be achieved are reviewed.

7.2 Photosynthesis–nitrogen relationships at leaf level

7.2.1 Photosynthesis–nitrogen relationship: physiological basis

The relationship between net leaf photosynthesis (P_n) and light incident on a leaf (I_L) is generally fitted with a non-rectangular hyperbola:

$$P_n = \frac{(P_{max} - \phi_L I_L) - \sqrt{(P_{max} - \phi_L I_L)^2 - 4\phi_L \theta P_{max} I_L}}{2\theta} - R_d \qquad (1)$$

where, P_{max} is the light saturated rate of photosynthesis, ϕ_L the apparent quantum yield, θ the curvature factor and R_d the dark respiration (Marshall and Biscoe, 1980). To understand the physiological basis of the photosynthesis–nitrogen relationship, it is important to analyse the extent to which each of these four parameters is related to the nitrogen content per unit leaf area (N_{area}) (Hirose and Werger, 1987a).

There is generally a strong positive correlation between P_{max} and N_{area} (see reviews by Nátr, 1975; Field and Mooney, 1986). A large proportion of the nitrogen in leaves is incorporated in components which limit photosynthesis at high light availability, i.e. mainly ribulose-1,5-bisphosphate carboxylase (RuBPCase) (von Caemmerer and Farquhar, 1981; Woodrow and Berry, 1988). The P_{max} to N_{area} relationship has often been found to be linear (e.g. Hirose and Werger, 1987a; Pons *et al.*, 1989; Schieving *et al.*, 1992a,b; Anten *et al.*, 1995a, 1998a). However, saturating P_{max} to N_{area} relationships have also been found (e.g. Lugg and Sinclair, 1981; Evans, 1983; Hikosaka *et al.*, 1999b). Curvilinearity at high N_{area} can be attributed to the fact that CO_2 levels at the sites of carboxylation tend to decrease with increasing P_{max} because of restrictions imposed by the resistances for CO_2 diffusion from the ambient air to the intercellular space (Makino *et al.*, 1988) and/or from the cell walls to the stroma (Evans, 1983; Evans and Terashima, 1988).

The P_{max} achieved per unit of N_{area} (potential photosynthetic nitrogen-use efficiency, PPNUE) may differ substantially between species. Plants with the C_4 photosynthetic pathway generally have higher PPNUEs than those with the C_3 photosynthetic pathway (Brown, 1978; Sage and Pearcy, 1987; Anten *et al.*, 1995a), though this does not necessarily hold true in cases where C_3 and C_4 plants are compared under similar natural conditions (Anten *et al.*, 1998a). Large variations in PPNUE are also observed among C_3 species (Field and Mooney, 1986; Evans, 1989a). PPNUEs are generally relatively high: in plants which commonly grow in high light conditions (Chazdon and Field, 1987) or high nutrient availability (Poorter *et al.*, 1990); in herbaceous as opposed to woody plants (Field and Mooney, 1986; Hikosaka *et al.*, 1998); in species with short leaf-life spans (Reich *et al.*, 1991); and in deciduous as compared to evergreen trees (Reich *et al.*, 1995). In general, high PPNUEs are associated with rapid growth and high competitiveness in favourable growth environments and low PPNUEs with adverse growth environments and stress tolerance.

Field and Mooney (1986) listed a number of factors which could explain differences in PPNUE between species: 1) permeability of cell walls, which determine the CO_2 concentration at the sites of carboxylation, and which thus influence the efficiency with which photosynthetic enzymes fix CO_2; 2) difference in fractional allocation of nitrogen to photosynthetically active and non-active compounds; 3) efficiency of nitrogen partitioning among components of the photosynthetic system; and 4) specific activity of photosynthetic enzymes. The first point might explain the low PPNUEs in evergreen sclerophylls, where thick cell walls impose large resistances to CO_2 diffusion (Medina, 1981). Several more recent studies (Lloyd *et al.*, 1992; Epron *et al.*, 1995) have found CO_2 levels in the chloroplasts to be lower in plants with low than in plants with high PPNUEs. Lloyd and co-workers (1992) also found species with a high PPNUE to have higher levels of RuBPCase than those with a low PPNUE. Vu and Yelenosky (1988) measured the specific activity, the Michaelis–Menten constant, and the activation state of RuBP-Case in *Citrus sinensis*, an evergreen tree with a low PPNUE, and found them to be similar to those that have been measured for herbaceous species. More recently, Hikosaka and co-workers (1998) conducted a study in which they compared all aspects of PPNUE between *Chenopodium album*, a herbaceous plant, and *Quercus myrsinaefolia*, an evergreen tree. They found that *C. album* had a twofold higher PPNUE than *Q. myrsinaefolia*. This difference was caused by higher chloroplast CO_2 levels, increased allocation of nitrogen to photosynthetically active compounds and a higher specific activity of enzymes, but not by a change in the partitioning of nitrogen among photosynthetic components. In a comparative study with ten species with different PPNUEs, Poorter and Evans (1998) showed that high PPNUE in low-light grown plants resulted from a higher relative allocation to photosynthetic components, while in high-light grown plants it resulted more from a greater specific activity of RuBPCase. Apparently, high cell wall resistances and low fractional allocation of nitrogen to photosynthetic components tend to be more important in explaining low PPNUE than the other two factors. These factors are also most closely associated with such factors as leaf longevity and stress tolerance (Field and Mooney, 1986).

Dark respiration (R_d) is generally observed to be strongly correlated with N_{area} (Hirose and Werger, 1987a; Pons *et al.*, 1989; Schieving *et al.*, 1992a,b; Anten *et al.*, 1995a, 1998a), Dark respiration is related to the maintenance of ion gradients across cell membranes and the rates of protein turnover (Penning de Vries, 1975). These costs should be expected to increase with the protein levels, and thus with the N_{area} of leaves. Respiration is also regulated by the concentration of assimilates in the leaf (Azcón-Bieto *et al.*, 1983; Lambers, 1985), which depends on the photosynthetic rate and thus on the nitrogen content of a leaf. However, this dependency of respiration on carbohydrate supply does not exist in all plants (Azcón-Bieto *et al.*, 1983). For example, Noguchi and Terashima (1997) showed that respiration in the sun species, *Spinacia oleracea*, was limited by carbohydrate supply, while in the shade species, *Alocasia odora*, it was not. Thus, mechanisms of respiration might be different between shade and sun species.

The initial slope of the photosynthetic light response curve (ϕ_L) is the product of the quantum yield on the basis of absorbed light (ϕ_{abs}) and leaf absorbance ($\phi_L = \phi_{abs} \times$ absorbance). Measurements on a wide variety of C_3 plants have indicated that there is little variation in ϕ_{abs} when it is measured on unstressed leaves at the same temperature and CO_2 level (Björkman and Demmig, 1987). Leaf absorbance, on the other hand, is a hyperbolically increasing function of leaf chlorophyll content (chl) (Gabrielsen, 1948), and ϕ_L can, therefore, be expected to be similarly related to chlorophyll (Evans, 1993b). Leaf chlorophyll levels increase proportionally with N_{area} when plants are grown at the same light availability (Terashima and Evans, 1988), and under these conditions a positive relationship between ϕ_L and N_{area} should be expected. However, when reductions in N_{area} are associated with decreasing light availability, leaf chlorophyll content remains relatively constant leading to increased chl/N ratios (e.g. Terashima and Evans, 1988; Evans, 1993a). Under such conditions, correlations between ϕ_L and N_{area} are generally found to be very weak (Pons *et al.*, 1989; Schieving *et al.*, 1992b; Anten *et al.*, 1995a, 1998a; Hikosaka *et al.*, 1999b).

A number of studies have found negative but weak correlations between the curvature factor of the photosynthetic light response curve (θ) and N_{area} (Hirose and Werger, 1987a; Schieving *et al.*, 1992b; Hikosaka *et al.*, 1999b), while other studies (Pons *et al.*, 1989; Anten *et al.*, 1995a, 1998a) found no correlations. A possible reason for the existence of a negative relationship between θ and N_{area} might be that θ correlates negatively with self-shading within a leaf, which increases with leaf chlorophyll content and thus with N_{area} (Leverenz, 1987). However, as noted above, leaf chlorophyll content is relatively constant when differences in N_{area} result from differences in light availability. Moreover, in well-acclimated leaves, the negative effect of self-shading on θ tends to be mitigated (Terashima and Hikosaka, 1995).

In summary, P_{max} and R_d exhibit strong positive correlations with N_{area}, while θ and ϕ_L are correlated with N_{area} weakly or not at all. The positive effect of N_{area} on leaf photosynthesis is, therefore, stronger at high than at low light availability (Gulmon and Chu, 1981; Hirose and Werger, 1987a). This is clearly illustrated in Figure 7.1.

7.2.2 *Nitrogen partitioning among photosynthetic components: acclimation of the photosynthetic apparatus*

About half of the nitrogen in the photosynthetic system (photosynthetic N, hereafter) is allocated to thylakoid proteins, while the other half is allocated

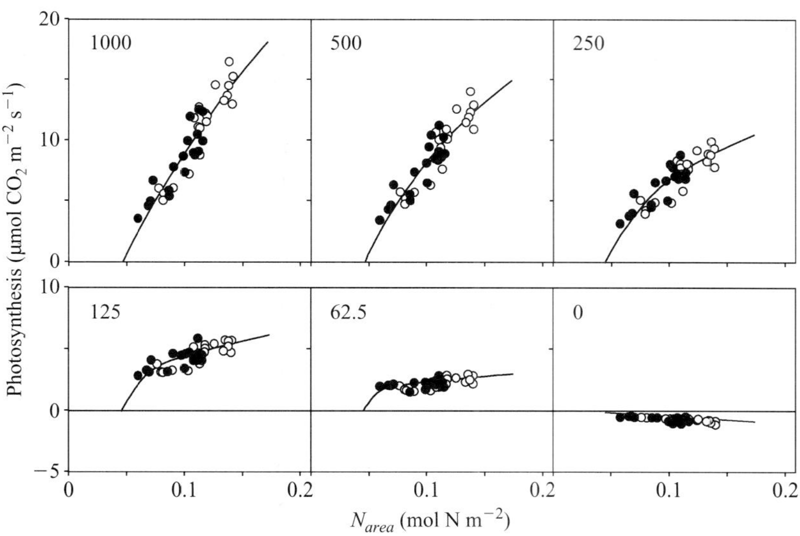

Figure 7.1 Dependence of net leaf photosynthesis of *Solidago altissima* on leaf nitrogen per unit area (N_{area}) under different photon flux densities of 1000, 500, 250, 125, 62.5 and 0 (μmol m^{-2} s^{-1}). (Redrawn with permission from Hirose and Werger, 1987a.)

to Calvin cycle enzymes (Evans and Seemann, 1989). Incoming quanta are absorbed by the antenna chlorophylls and then passed on to the reaction centres of photosystems I and II (PSI and PSII, respectively). Light stimulates these reaction centres to release electrons. Electrons released from PSII are transferred to the reaction centre of PSI via a number of electron carriers, such as plastoquinone, cytochrome b/f complex and plastocyanine. Electrons provided by the photolysis of water reduce the oxidised reaction centre of PSII. Electrons released from PSI are used for the reduction of nicotinanide adenine dinucleotide phosphate (NADP) to NADPH. The photolysis of water and the transport of protons by plastoquinone from outside of the thylakoid membranes (stroma) results in an accumulation of protons inside thylakoid membranes (thylakoid lumen). The resulting proton gradient between the stroma and the thylakoid lumen provides the energy for the formation of ATP. ATP and NADPH are subsequently utilised in the Calvin cycle for CO_2 fixation. Here, CO_2 is first bound to ribulose-1,5-bisphosphate (RuBP) forming 3-phosphoglyceric acid (PGA), a reaction which is catalysed by the enzyme RuBPCase. Through a number of steps, RuBP is then regenerated from PGA and this process utilises ATP and NADPH. Triose-phosphate is simultaneously formed, exported from the chloroplasts and utilised for the synthesis of substances.

The organisation of the photosynthetic system may vary depending on the environmental conditions to which leaves are exposed. In this respect, changes that occur in response to differences in growth irradiance (i.e. sun–shade acclimation) have been particularly well documented. As growth irradiance increases, the following changes occur: 1) an increase in the amounts of Calvin cycle enzymes, electron carriers and ATPase relative to leaf chlorophyll content; 2) an increase in the ratio of PSII to leaf chlorophyll content, though the ratio of PSI to leaf chlorophyll content does not change; and 3) an increase in the ratio of chlorophyll a to chlorophyll b (Boardman, 1977; Leong and Anderson, 1984; Evans, 1987; Terashima and Evans, 1988). Changes in the organisation of photosynthetic components, similar to those associated with shade acclimation, have also been observed to occur during age-related senescence (Friedrich and Huffaker, 1980; Jenkins and Woolhouse, 1981; Makino *et al.*, 1983). However, Hikosaka (1996) demonstrated that these changes only occur when senescing leaves are simultaneously shaded (e.g. by younger leaves in a canopy), indicating sun–shade acclimation to be strictly associated with light availability.

The relationship between sun–shade acclimation and leaf-PNUE can be understood when both the roles in the photosynthetic process and the nitrogen costs of each of the photosynthetic components are taken into account. Light saturated photosynthesis (P_{max}) increases with the amount of Calvin cycle enzymes, electron carriers and ATPase (Evans, 1987; Terashima and Evans, 1988; Hikosaka and Terashima, 1996), and each of these substances is costly in terms of nitrogen (Terashima and Evans, 1988; Evans and Seemann, 1989). On the other hand, a large amount of chlorophyll contributes to the photosynthetic rate at lower light intensity, due to the positive correlation between the apparent quantum yield (ϕ_L) and leaf chlorophyll level. All chlorophyll in leaves is bound to chl–protein complexes, which are major nitrogenous compounds (Evans and Seemann, 1989). Consequently, there is a trade-off between nitrogen allocation to either P_{max}-related proteins or chl-protein complexes (Evans, 1989b). In low light conditions, a relatively large investment in chlorophyll should be expected to be more advantageous in terms of leaf-PNUE, while a greater investment of nitrogen in Calvin cycle enzymes is expected to enhance leaf-PNUE in high light conditions.

Hikosaka and Terashima (1995) developed this idea further and constructed a comprehensive model to predict the optimal nitrogen partitioning among photosynthetic components that maximises leaf-PNUE. In this model, photosynthetic components were categorised into five functional groups: Group I, RuBPCase; Group II, electron carriers, ATPase, and Calvin cycle enzymes other than RuBPCase; Group III, core complex of photosystem II (PSII core); Group IV, core complex of photosystem I and light-harvesting chl–protein complex I (PSI); Group V, light-harvesting chl–protein complex II (LHCII). The nitrogen cost for each of the groups was calculated from previously published data (see Table 2 in Hikosaka and Terashima, 1995). P_{max} depends on the amounts of Groups I, II and III and, in order to maximise PPNUE, nitrogen allocation to each group was assumed to be such that P_{max} was co-limited by the amounts in each of them (von Caemmerer and Farquhar, 1981). All chlorophyll molecules are associated with chl–protein complexes, either Group III, IV, or V, and the amounts in these groups determine ϕ_L. Nitrogen partitioning among chl–protein complexes was assumed to occur as follows. Allocation to PSII is determined by the requirement for co-limitation of P_{max} (i.e. by groups I, II and III). The amounts of LHCII and PSI are then adjusted so that light absorption by PSI and PSII is balanced. Nitrogen costs of leaf chlorophyll level differ among Group III, IV and V. LHCII is the most efficient because it is only associated with light-harvesting, while the core complexes also play an important role in electron transport and are, therefore, bound to the relatively nitrogenous proteins of the photosystems. Consequently, under low light conditions, when electron transport is not high, a greater relative allocation of nitrogen to LHCII rather than to the PSII core should be expected to enhance leaf-PNUE. This explains the relatively low chlorophyll a/b ratio, commonly observed in shade-acclimated leaves, because chlorophyll b is predominantly bound to LHCII.

Hikosaka and Terashima (1996) applied this model to plants of a shade-intolerant (*Chenopodium album*) and a shade-tolerant (*Alocasia odora*) species, grown under various light intensities, to see whether actual plants optimise nitrogen allocation among photosynthetic components. Figure 7.2 shows daily carbon gain for different light levels plotted as a function of the fraction of photosynthetic nitrogen that is invested in RuBPCase (i.e. Group I). For each light level, there was an optimal allocation of photosynthetic nitrogen to RuBPCase and this optimum increased with increasing growth irradiance. Differences between the optimal and the actually measured nitrogen partitioning were surprisingly small, except for the *C. album* plants grown at the lowest light (5% of daylight). These results suggest that plants are able to adjust the partitioning of their photosynthetic nitrogen optimally to different light environments, although sun plants might be unable to do so in very low light conditions.

Apart from the trade-off outlined above, there is another trade-off between nitrogen allocated to components which limit either the carboxylation or the regeneration of RuBP. An analysis of this trade-off might be particularly important to understand how plants may acclimate optimally to increasing CO_2 levels in the air. The amount of RuBPCase (Group I in the model of Hikosaka and Terashima, 1995) determines the capacity of RuBP carboxylation, while the amounts of electron carriers, ATPase and Calvin cycle enzymes other than RuBPCase (Group II), and PSII core proteins (Group III) determine the capacity of RuBP regeneration. As noted above, all these compounds contain large amounts of nitrogen. In most C_3 plants, carboxylation and regeneration of RuBP are considered to co-limit the photosynthetic

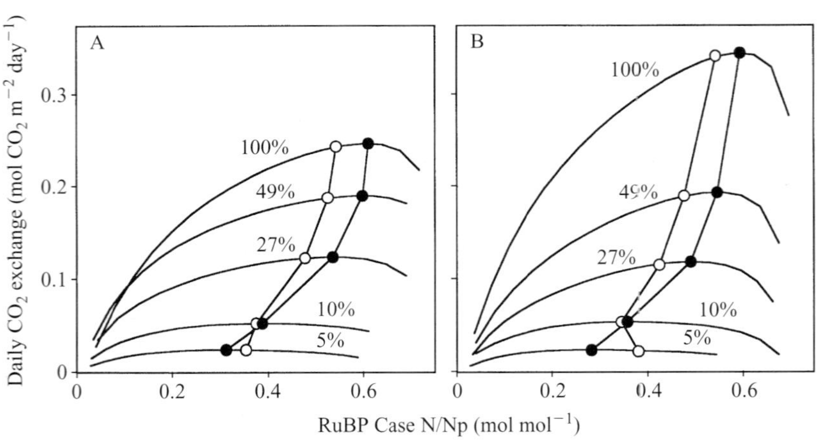

Figure 7.2 Effects of nitrogen partitioning on daily CO_2 exchange rate in A) *Alocasia odora*; and B) *Chenopodium album*. Nitrogen partitioning is expressed as a fraction of the nitrogen in photosynthetic components (N/Np) allocated to ribulose-1,5-biphosphate carboxylase (RuBPCase). Open and closed circles denote actual and optimal patterns of nitrogen partitioning, respectively. Percentage values given above each line denote light conditions (% of full sunlight). (Redrawn with permission from Hikosaka and Terashima, 1996.)

rate at the present CO_2 level (von Caemmerer and Farquhar, 1981; Wullschleger, 1993) and, as mentioned previously, this results in a maximisation of PPNUE. However, at elevated CO_2, the necessity for RuBPCase decreases, and RuBP regeneration will limit photosynthesis. The reverse happens at low CO_2 levels, and RuBP carboxylation will tend to limit photosynthesis (Farquhar *et al.*, 1980). Therefore, at elevated CO_2 levels, a reduction in the amount of RuBPCase relative to the amounts of other photosynthetic components can be expected to enhance leaf-PNUE (Hogan *et al.*, 1991; Sage, 1994; Webber *et al.*, 1994). Two theoretical studies (Medlyn, 1996; Hikosaka and Hirose, 1998) supported this idea and predicted that, plants growing at high CO_2 can maximise their leaf-PNUE by reducing the fractional allocation of nitrogen to RuBPCase and that this optimisation may result in approximately 15% higher leaf-PNUE, depending on the temperature. In support of these predictions, transgenic plants with reduced levels of RuBPCase were found to have higher PNUEs than wild-type plants at elevated CO_2 (Makino *et al.*, 1997). On the other hand, Sage (1994) concluded that in most C_3 species nitrogen partitioning among photosynthetic components is hardly affected by CO_2 levels. Apparently, most C_3 plants are unable to optimally acclimate their photosynthetic apparatus to high CO_2.

Another factor that may influence nitrogen partitioning is temperature. The temperature dependencies of RuBP carboxylation and regeneration differ from each other. Therefore, the CO_2 concentration point at which co-limitation of photosynthesis by these two processes occurs will also vary with temperature. However, leaves may be able to re-establish co-limitation of photosynthesis by changes in nitrogen partitioning between the components that limit RuBP carboxylation and regeneration. Hikosaka (1997) developed a model with which the optimal nitrogen partitioning at different temperatures could be calculated. He demonstrated that this optimisation resulted in a substantial increase in leaf-PNUE at different temperatures. Interestingly, predicted changes in nitrogen partitioning appeared to account for the differences in temperature dependence of photosynthesis, which have been observed between plants grown at different temperatures (Berry and Björkman, 1980). Furthermore, Hikosaka and co-workers (1999a) showed for the evergreen tree, *Quercus myrsinaefolia*, that optimal shifts in the capacities for regeneration and carboxylation of RuBP occurred with changes in growth temperature. However, it remained to be determined whether these shifts were accompanied by changes in nitrogen partitioning.

7.2.3 Optimisation of absolute nitrogen content per unit leaf area

Hirose (1984) showed that, if the relationship between daily carbon gain and N_{area} is curvilinear, there is an optimal N_{area} that maximises daily carbon gain per unit nitrogen (leaf-PNUE) (Figure 7.3). Hirose and Werger (1987a)

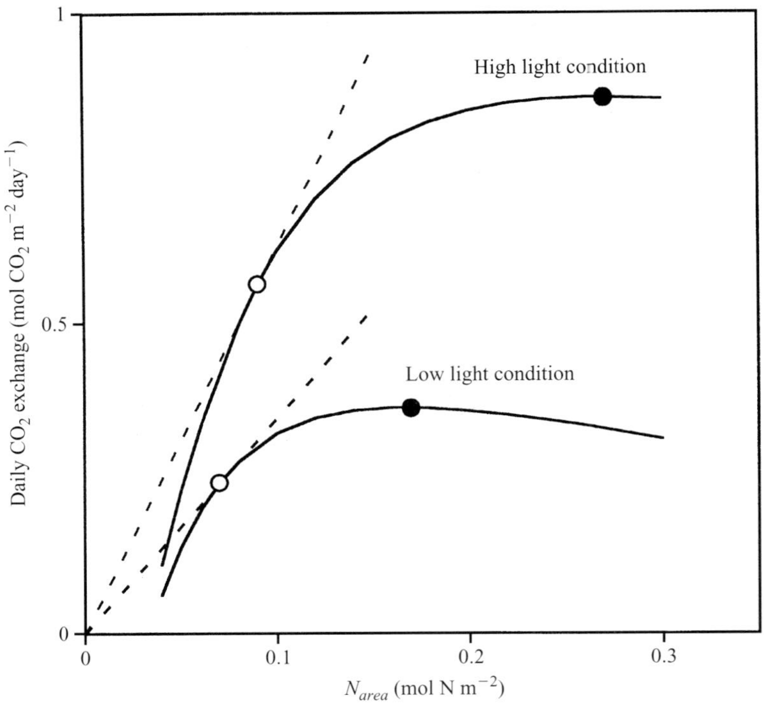

Figure 7.3 Daily CO_2 exchange *vs* leaf nitrogen per unit area (N_{area}) assuming optimal partitioning of nitrogen among components of the photosynthetic system. Open circle denotes the nitrogen content that maximises daily CO_2 exchange per unit nitrogen (note that it is a tangent from the origin to the curve). Closed circle denotes the nitrogen content that maximises daily CO_2 exchange per unit leaf area. (Calculated after Hikosaka and Terashima, 1995.)

applied this concept to *Solidago altissima* plants from dense stands. The optimal N_{area} values increased with light availability. This result can easily be understood when considering the fact that the correlation between photosynthesis and N_{area} is much stronger at high light than at low light (Figure 7.1). Interestingly, actually measured N_{area} values were very close to the predicted optima (Hirose and Werger, 1987a).

Apart from the optimal N_{area} that maximises leaf-PNUE, it is possible to define another optimum: namely, the N_{area} that maximises photosynthesis per unit leaf area (Figure 7.3) (Hikosaka and Terashima, 1995). It can be shown that the N_{area} that maximises leaf-PNUE should be lower than the one that maximises photosynthesis per unit area (Figure 7.3) (Hikosaka and Terashima, 1995; Anten *et al.*, 1996). Hikosaka and Terashima (1995) argued that actual nitrogen contents in leaves should vary between these two optima. If nitrogen availability is low, N_{area} should tend towards the value that maximises leaf-PNUE. Conversely, if nitrogen availability is high, and leaf area production is constrained by a factor other than nitrogen, e.g. availability of carbon, N_{area} values should be expected to be close to the value that maximises photosynthesis per unit area. Actual data appear to support this idea. At high nitrogen availability, N_{area} values in *Spinacia oleracea* were close to the values that maximise photosynthesis per unit area, while at low nitrogen availability they were close to the values that maximise leaf-PNUE (Terashima and Evans, 1988). In trees of the species *Tetrorchidium rubrivenium*, growing at extremely low light availability (3% of natural daylight) in the understorey of a tropical rain forest, N_{area} values were very close to the values that maximise photosynthesis per unit area (Anten *et al.*, 1996). On the other hand, at very low light availabilities, shade-intolerant plants tend to have N_{area} values which are greater than the optima for maximisation of photosynthesis per unit area (Hikosaka and Terashima, 1996; Hikosaka *et al.*, 1999b). Such nitrogen contents can be considered to be non-adaptive (Hikosaka and Terashima, 1995), and indicate an inability of shade-intolerant plants to optimise their photosynthetic system under very low light conditions.

7.3 Optimal nitrogen distribution

Until now, this chapter has focused on the optimal utilisation of leaf nitrogen for maximisation of leaf-PNUE. However, maximisation of nitrogen-use efficiency at the individual leaf level does not automatically imply that nitrogen-use efficiency of an entire plant, consisting of a number of leaves, is also maximised. The main reason for this is that different leaves on a plant do not function independently from one another. The nitrogen allocated to one leaf cannot simultaneously be allocated to other leaves. Moreover, leaves in a canopy shade one another and, as a result, some leaves on a plant receive more light than others. This section looks at the optimal nitrogen allocation patterns among leaves of plants that maximise whole plant canopy photosynthesis per unit nitrogen (canopy-PNUE).

In dense stands of vegetation, light intensity (photon flux density, PFD) decreases from the top towards the bottom of the canopy. The distribution of light in a canopy is often approximated using Beer's law (Monsi and Saeki, 1953):

$$I(h) = I_o e^{-K_L L(h)} \tag{2}$$

where, $I(h)$ and I_o are the PFDs in a horizontal plane at depth h and above the canopy, and $L(h)$ the cumulative leaf area index (LAI) from the top of the canopy to depth h. The parameter K_L, the extinction coefficient for light, determines the rate at which I declines with increasing L and is determined mainly by the geometric arrangement of leaves in the canopy. With the vertical components of sky radiation predominating, horizontally projected leaves intercept more light than vertically inclined ones, and K_L will therefore increase with decreasing leaf inclination (Monsi and Saeki, 1953).

The photosynthetic benefits of high N_{area} are greater at high than at low light availability (see Figure 7.1). In view of this, it can be argued that plants which grow in non-uniform light climates may enhance their photosynthetic performance by adjusting N_{area} levels in such a way that leaves in the more shady environments have low and those in the more illuminated environments have high N_{area} values (Mooney and Gulmon, 1979). Building on this concept, Field (1983) formalised the condition that defines the optimal nitrogen distribution in a stand of vegetation for maximisation of canopy-PNUE. In a canopy with a fixed total amount of leaf nitrogen (N_T), whole canopy carbon gain will be maximised when the change in daily carbon gain (P_{day}) with changing N_{area} is equal in every microsite in the canopy:

$$\delta P_{day}/\delta N_{area} = \lambda \tag{3}$$

where λ is a Lagrange multiplier. Since $\delta P_{day}/\delta N_{area}$ is a negative function of N_{area} and a positive function of light availability, this condition predicts that N_{area} should decrease with increasing depth and decreasing light availability in the canopy.

Equation (3) only gives the condition under which carbon gain is maximised, without any explicit indication of the form of the optimal nitrogen distribution. Hirose and Werger (1987b) took a different approach by calculating canopy photosynthesis as a function of a set of predetermined nitrogen distribution patterns. They found nitrogen distribution in the canopy to be exponential and calculated canopy photosynthesis as a function of a nitrogen allocation coefficient (K_a). This parameter, K_a, was defined such that the steepness of the nitrogen distribution (i.e. the difference between N_{area} at the top and bottom of the canopy) increased with K_a, and that $K_a = 0$ represented the uniform distribution in which N_{area} of all the leaves equals the mean. The K_a at which canopy photosynthesis was maximised was assumed to be equivalent to the optimal distribution. This model was applied to stands of *Solidago altissima*, a herbaceous dicotyledonous perennial. They calculated that optimal K_a values increased with increasing leaf area indices and predicted that nitrogen distribution should, therefore, be steeper in dense than in open stands.

Other studies (Farquhar, 1989; Sellers *et al.*, 1992; Anten *et al.*, 1995a,b) have treated the problem of optimal nitrogen distribution analytically. The advantage of analytical mathematics is that it is possible to formalise the optimal nitrogen distribution in relation to certain leaf, canopy or environmental characteristics (Anten, 1995). The latter is important to predict how nitrogen distribution and related photosynthetic nitrogen-use efficiency may differ between stands of plants with different characteristics or how they may be affected by changes in the environment. The disadvantage of the analytical approach is that analytical solutions can only be derived by means of gross simplifications, ignoring the complexity of real photosynthetic systems.

Farquhar (1989) applied the condition defined by Field (1983), Equation (3), to analyse the distribution of nitrogen within a single leaf. Within leaves there is a light gradient, with the top layers receiving more light than the bottom layers. Farquhar showed that the nitrogen distribution in a leaf is optimal if the associated photosynthetic capacities (P_{max}) in each leaf layer are proportional to the amount of irradiance absorbed. This idea was extended to the distribution of nitrogen among leaves in a canopy (Sellers *et al.*, 1992; Anten *et al.*, 1995a) and it was shown that nitrogen is optimally distributed if P_{max} is a linear function of the relative incident PFD (I_L/I_o) in the canopy:

$$P_{max} = \alpha I_L/I_o \qquad (4)$$

where, α is a constant. From this condition, Anten and co-workers (1995a) derived an analytical equation for the optimal nitrogen distribution:

$$N_{area}(h) = \frac{K_L(N_T - N_b L_T)}{(1 - e^{-K_L L_T})} e^{-K_L L(h)} + N_b \qquad (5)$$

where, K_L is the extinction coefficient for light, N_T the total canopy nitrogen, and L_T and $L(h)$ the total LAI and the cumulative LAI above a given point in the canopy, respectively. The parameter N_b is the amount of nitrogen in a leaf which is invested in structural components and which can, therefore, not be reallocated. From Equation (5), it can be shown that the steepness of the optimal nitrogen distribution increases with K_L and N_T and that it decreases with N_b. Surprisingly, parameters describing leaf physiological characteristics and the PFD at top of the canopy (I_o) are absent from Equation (5). Thus, nitrogen distribution in canopies is primarily determined: by structural characteristics that determine the distribution of relative PFD in the canopy (i.e. K_L); by the availability of nitrogen to the canopy (N_T); and by the amount of nitrogen in leaves which is invested in structural characteristics (N_b). It is independent of leaf physiological characteristics and the absolute amount of light which plants receive.

One aspect not yet discussed is the actual photosynthetic benefit of optimising the nitrogen distribution. Reallocation of nitrogen is costly in terms of energy. For example, the costs of transporting amino-acids from one leaf to another have been estimated to be about 3 ATPs per molecule of amino-acid (Radin and Elmore, 1980). Thus, optimisation of nitrogen distribution is only of benefit to plants if it results in an increase in photosynthesis that is greater than the associated costs. Accurate cost–benefit analyses of nitrogen distribution have not yet been conducted. Various studies (Hirose and Werger, 1987a; Pons *et al.*, 1989; Schieving *et al.*, 1992a,b; Anten *et al.*, 1995a,1998a), however, have calculated that the optimisation of the nitrogen distribution results in a considerable (20–40%) increase in canopy photosynthesis over a uniform nitrogen distribution, indicating that optimisation of the nitrogen distribution is indeed beneficial to plants. The relative photosynthetic benefit of the optimal nitrogen distribution has also been found to increase both with the leaf area index (Hirose and Werger, 1987b; Schieving *et al.*, 1992a) and the extinction coefficient (K_L) (Anten *et al.*, 1995a) of a stand. This supports the notion that stands with high LAI or K_L values should have steeper nitrogen distributions than those with low LAI or K_L.

7.4 Actual patterns of nitrogen: a critical test of the optimisation theory

7.4.1 Mono-species stands

The previous section showed how the optimisation theory produced specific predictions with regard to the optimal nitrogen distribution which maximises canopy-PNUE. However, the potential of the optimisation theory to explain the growth patterns of plants can only be tested by comparing these predictions with observed patterns of nitrogen distribution.

Non-uniform patterns of nitrogen distribution (i.e. N_{area} declining with depth in the canopy), strongly differing from the uniform distribution, have been found in mono-specific stands of a large variety of species, including both perennials and annuals and both agricultural crops and wild plants (Hirose and Werger, 1987b; Charles-Edwards *et al.*, 1987; Hirose *et al.*, 1988, 1989; Pons *et al.*, 1989; Lemaire *et al.*, 1991; Werger and Hirose, 1991; Schieving *et al.*, 1992a,b; Anten *et al.*, 1995a,b, 1998a,b; Hollinger, 1996; Anten, 1997; Del Pozo and Dennett, 1999). In all cases, nitrogen distribution resulted in a considerable increase in canopy photosynthesis as compared to the uniform distribution, ranging from 6% in *Nothofagus fusca*, a broad leaved evergreen (Hollinger, 1996), to almost 30% in *Glycine max*, a leguminous crop (Anten *et al.*, 1995a). Nevertheless, in all stands studied actual patterns of nitrogen distribution did not correspond to the optimal distribution, in that they were always more uniform. Canopy photosynthetic rates calculated for the actual distributions were 4–15% lower than the predicted maximum rates (Hirose and Werger, 1987b; Pons *et al.*, 1989; Schieving *et al.*, 1992a,b; Anten *et al.*, 1995a, 1998a).

The prediction that nitrogen distribution in stands is predominantly determined by structural characteristics, such as the LAI and the extinction coefficient for light (K_L), and that it is mostly independent of leaf physiological characteristics, was tested in a number of experiments. In the case of

Lysimachia vulgaris (Hirose *et al.*, 1988) and *Carex acutiformis* (Schieving *et al.*, 1992b), nitrogen distribution was found to be steeper in stands with high LAI than in stands with low LAI. On the other hand, in *Vicea faba* there was no difference in nitrogen distribution between stands with low and high LAI (Del Pozo and Dennett, 1999). However, the low LAI stand had a higher K_L value and nitrogen distribution was strongly correlated with the distribution of relative PFD in the canopy.

Anten and co-workers (1995a,b) compared patterns of nitrogen distribution between stands of species with either the C_3 or C_4 photosynthetic pathway and between stands of mono-cotyledonous and dicotyledonous species. The C_4 plants had considerably higher leaf photosynthetic capacities (P_{max}) for a given N_{area} (PPNUE) than the C_3 plants, which was consistent with previous findings (see Section 7.2.1). However, in spite of this difference in PPNUE, nitrogen distributions were very similar between the C_3 and C_4 species. Dicotyledons had more horizontally projected leaves, formed stands with considerably higher K_L and exhibited much steeper nitrogen distributions (i.e. greater variation in nitrogen with depth of canopy) in the stands than monocotyledons (Figure 7.4). A field experiment by Anten and co-workers (1998a) with C_3 and C_4 tropical savanna grasses with contrasting canopy structures yielded very similar results. These results strongly support the prediction derived from the optimisation theory, that differences in canopy structure had large effects on nitrogen distribution while the effects of large differences in leaf physiology (e.g. C_3 *vs* the C_4 photosynthetic pathway) were negligible.

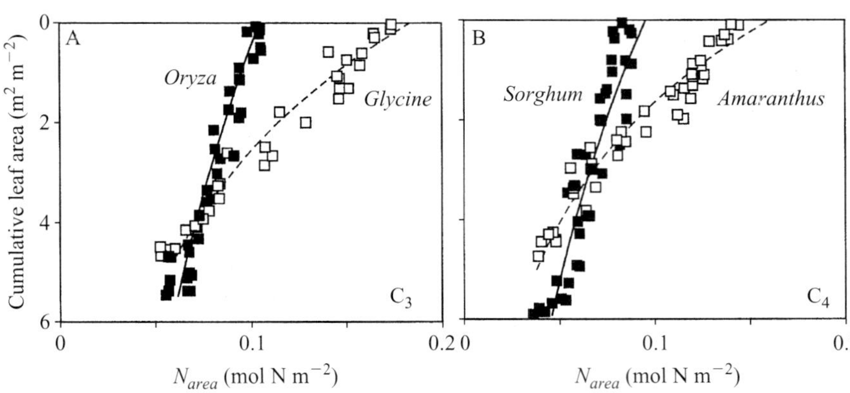

Figure 7.4 Nitrogen per unit area (N_{area}) in dense stands of: A) *Glycine max* (C_3 dicotyledonous) and *Oryza sativa* (C_3 monocotyledonous); and B) *Amaranthus cruentus* (C_4 dicotyledonous) and *Sorghum bicolor* (C_4 monocotyledonous) as a function of the cumulative leaf area index. (Redrawn with permission from Anten *et al.*, 1995a.)

To further test the capability of the optimisation theory to predict differences in nitrogen distribution between stands of species with different characteristics, the nitrogen allocation coefficient (K_n) defined by Anten and co-workers (1995a) was plotted as a function of the extinction coefficient for light (K_L) (Figure 7.5). The coefficient, K_n, is defined in such a way that the non-uniformity of the nitrogen distribution increases with K_n, while a $K_n = K_L$ is equivalent to the optimal distribution. Data from a large number of species were included in Figure 7.5, including monocotyledons and dicotyledons as annuals and perennials. There was a strong and significantly positive correlation between K_n and K_L ($r^2 = 0.59$), indicating that much of the variation in nitrogen distribution in the stands of these species was explained by differences in light distribution. This supports the prediction that structural characteristics are an important factor in determining nitrogen distribution. On the other hand, the slope of the regression of K_n against K_L was considerably smaller than one, indicating that K_n for each of the stands

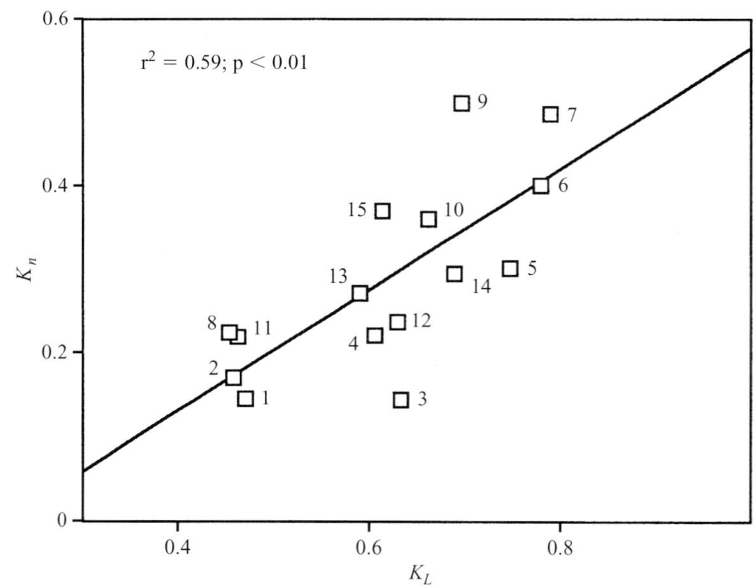

Figure 7.5 The relationship between the coefficient of nitrogen allocation (K_n) (see Equation 8 in Anten *et al.*, 1995a) and the extinction coefficient for light (K_L) (see Equation 2 in this chapter). The solid line represents the linear regression. Numbers indicate: 1,2 *Oryza sativa* high and low nitrogen availability; 3,4 *Sorghum bicolor* high and low nitrogen availability; 5 *Glycine max*; 6,7 *Amaranthus cruentus* high and low nitrogen availability (1–7, Anten *et al.*, 1995b); 8 *Leersia hexandra*; 9 *Paspalum fasciculatum*; 10 *Hymenachne amplexicaulis* (8–10, Anten, *et al.*, 1998a); 11 *Oryza sativa*; 12 *Pennisetum americanum* (11 and 12, Anten 1997); 13 *Carex acutiformis* (Schieving *et al.*, 1992b); 14 *Lysimachia vulgaris* (Hirose *et al.*, 1988); and 15 *Solidago altissima* (Hirose and Werger, 1987b).

was smaller than K_L and that, as noted before, actual nitrogen distributions were more uniform than the optimal ones predicted.

7.4.2 Individual plants in mono-species stands

In the studies discussed in the previous section, height differences between plants in the stand were not taken into consideration. This implied that all plants in a stand were assumed to experience the same light environment. This assumption made it possible to describe nitrogen distribution, which is an individual plant-based characteristic, at stand level. However, in many stands of vegetation there are large variations in plant height (Ford, 1975; Harper, 1977). In such stands, taller plants are able to project their leaves in higher, more illuminated regions of the canopy, while shorter plants grow in the shade of their taller neighbours.

The optimisation theory predicts that nitrogen distribution in plants is determined by differences in PFD between their leaves, and that it is independent of the absolute PFD level that they receive. If this prediction holds true, patterns of nitrogen distribution in plants of different heights in a stand would be expected to be similarly related to the distribution of relative incident PFD (Anten and Werger, 1996). In accordance with this prediction, Lemaire and co-workers (1991) found the nitrogen distribution in shoots of different heights in a stand of *Medicago sativa* to be closely related to the relative PFD. At any layer in the canopy, leaves of different shoots had very similar N_{area}. On the other hand, in a stand of *Xanthium canadense*, nitrogen distribution patterns differed considerably between plants of different heights (Figure 7.6) (Anten *et al.*, 1998b). The nitrogen distribution plotted as a function of relative PFD was steeper in shorter than in taller plants. Interestingly, N_{area} of leaves of shorter plants were always higher than those of leaves from taller plants at the same layer in the canopy. As a result, nitrogen distributions in individual plants were steeper and closer to optimal than the average nitrogen distribution in the stand (Figure 7.6).

7.4.3 Multi-species stands

So far, the focus has been on single-species stands, in which every individual could be assumed to have basically the same physiological and structural characteristics. This assumption will certainly not hold for multi-species stands, where such traits may differ considerably among individuals of different species. Many stands of vegetation consist of a range of species differing in plant height. In such stands, the taller species will be able to project their leaves towards the top of the canopy, where light availability is high, while the shorter species are growing in the shade. The taller species generally have physiological and structural traits that are very different from those of the shorter species (Hirose and Werger, 1994, 1995; Turner and

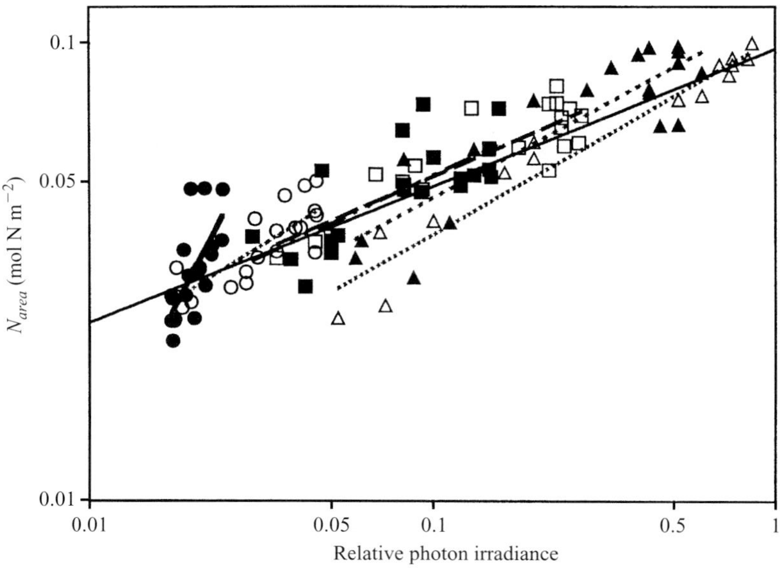

Figure 7.6 Nitrogen per unit leaf area (N_{area}) as a function of relative incident photon iradiance for plants from different height classes in a dense stand of *Xanthium canadense*. Median height and average photon irradiance on the top-most leaf for each height class (hc) are: hc 1 (thick lines with closed circles), 32.5 cm and 2.6%; hc 2 (chain with open circles), 47.5 cm and 4.7%; hc 3 (long broken line with closed squares), 77.5 cm and 17.3%; hc 4 (medium broken line with open squares), 87.5 cm and 29.3%; hc 5 (short broken line with closed triangles), 97.5 cm and 64.6%; and hc 6 (dotted line with open triangles), 102.5 cm and 80.8%. (Redrawn with permission from Anten *et al.*, 1998b.)

Knapp, 1996; Anten and Hirose, 1999). The question arises whether nitrogen in such stands is optimally distributed. If, as predicted by the optimisation theory, the level of PFD indeed strictly determines nitrogen distribution, then taller species should have considerably higher N_{area} values than shorter species. On the other hand, it can also be argued that growth of the taller species is limited by nitrogen, while growth of the shorter ones is limited by light. In that case, taller species would be expected to exhibit high PNUEs and, therefore, relatively low N_{area} values, while shorter species would have low PNUEs and relatively high N_{area} values (Hirose and Werger, 1994).

Hirose and Werger (1994) analysed nitrogen distribution in a multi-species stand in a wet meadow. The stand consisted of four relatively tall grass species that dominated the upper and middle layers of the canopy and a number of shorter forb species that were found in the lower layers. They found N_{area} values of the taller grass species to be much higher than those of the shorter species and, when the data for all species were combined, the relationship between N_{area} and relative PFD was more or less linear, which was consistent with the prediction derived from the optimisation theory. Interestingly, they also found the PNUEs of the shorter species to be similar or even somewhat higher than those of the taller species. It can, therefore, be concluded that growth of the shorter species was not only limited by light but also by nitrogen availability (Hirose and Werger, 1994).

7.5 Mechanisms of nitrogen distribution

7.5.1 Control factors: is nitrogen distribution controlled by light distribution or leaf age?

The optimisation theory indicates the relative PFD is the predominant factor controlling nitrogen distribution in canopies. However, in stands of many plants and in those of erect dicotyledonous plants in particular, there is a negative correlation between leaf age and light availability. Young leaves are formed at the top of the canopy, shading the older leaves. This confounding of leaf age with light availability makes it difficult to determine whether non-uniform patterns of nitrogen distribution form in direct response to the formation of a light gradient in the canopy or whether they are the result of age-related senescence. Mooney and co-workers (1981) showed that N_{area} decreased with leaf age, even though older leaves were only minimally shaded, and suggested that nitrogen distribution is developmentally linked to the leaf ageing process. On the other hand, in *Lysimachia vulgaris* the nitrogen distribution was steeper in a dense stand than in an open stand, suggesting differences in light availability to be the most important factor in the control of nitrogen distribution (Hirose *et al.*, 1988). Hirose and co-workers (1989) obtained similar results in stands of *Carex acutiformis*. This species forms long erect leaves from basal meristems, with the oldest leaf parts receiving more light than the younger ones. They found N_{area} values to be higher in the older than in the younger leaf parts. Similarly, Ackerly (1992) found N_{area} in the vine species, *Synogonium podophyllum*, which exhibits little correlation between leaf age and the distribution of PFD, to be more strongly determined by PFD than by leaf age. Nitrogen distribution in shoots of different heights in a stand of the erect dicotyledonous perennial, *Medicago sativa*, was also predominantly determined by PFD (Lemaire *et al.*, 1991).

More convincing evidence for PFD being the major factor controlling nitrogen distribution comes from experiments in which light climate was experimentally manipulated. In two grass species, *Carex acutiformis* and *Brachypodium pinnatum*, gradients in N_{area} did not develop when leaves were evenly illuminated. When a light gradient was imposed, matching N_{area}

gradients were found in *C. acutiformis* but not in *B. pinnatum* (Pons *et al.*, 1993). In the case of *B. pinnatum*, some environmental factors other than light, possibly the spectral quality of light, might contribute to the nitrogen distribution observed in dense stands of this species. Hikosaka and co-workers (1994) subjected a vine (*Ipomoea tricolor*) to reverse canopy-type shading (progressively shading of younger leaves), and found the more illuminated older leaves to have higher N_{area} than the more shaded younger leaves. Other experiments have shown that shading of individual leaves resulted in a reduction in N_{area} in herbaceous species (Pons and Pearcy, 1994; Pons and Jordi, 1998) but not in a woody evergreen species (Pons and Jordi, 1998).

The studies discussed above strongly indicate that, although nitrogen reallocation is to some extent regulated by the leaf ageing process, most plants are highly flexible in adjusting their nitrogen distribution to changes in the light climate. This flexibility appears to be particularly strong in plants that grow in highly unpredictable light climates, such as vines, or plants, such as many grasses, which normally experience positive correlations between leaf age and light availability. On the other hand, plants whose leaves are generally exposed to relatively stable light climates, e.g. shade-tolerant understorey evergreens that produce plagiotropic canopies in which there is little self-shading, are probably much less flexible.

7.5.2 Physiological mechanisms: which mechanisms mediate nitrogen reallocation in response to partial shading?

If plants are indeed highly flexible in redistributing nitrogen in response to changes in the light gradient, how then are they able to 'perceive' differences in light availability between their leaves and what mechanisms regulate this redistribution of nitrogen? One possibility is that senescence and export of nitrogen is directly induced by a reduction in photosynthesis caused by shading. Hensel and co-workers (1993) found a decline in the production of photosynthetic metabolites to result in an increase in *SAG* gene expression, the products of which stimulate nitrogen export and senescence of leaf tissue. However, while the *SAG* gene expression has been found to occur during age-related senescence, it appears to be absent during shade-induced senescence. Moreover, Pons and Bergkotte (1996) found that a reduction in photosynthesis per se did not result in nitrogen export from leaves.

Another possibility is that nitrogen reallocation from shaded to more illuminated leaves is associated with differences in the red to far-red (R/FR) ratio of the light to which leaves are exposed. Leaves selectively absorb more red than far-red light and the R/FR ratio therefore declines as light penetrates through the canopy, with shaded leaves receiving light with a considerably lower R/FR ratio than more illuminated leaves (Monteith, 1965). Low R/FR ratios have been found to induce senescence-related processes, while high R/FR ratios tend to reverse these processes (De Greef *et al.*, 1971). A field experiment with dense stands of *Helianthus annuus*, using filters and FR mirrors, demonstrated that FR enrichment of the light penetrating through the canopy enhanced senescence of leaves lower down in the canopy (Rousseaux *et al.*, 1996). On the other hand, in transgenic plants that lack phytochrome (a family of photoreceptors that mediate the perception of R/FR ratios, see Chapter 5) shading of individual leaves resulted in a reduction in P_{max} similar to that observed in wild-type plants (Smith *et al.*, 1993). Moreover, in experiments where the light distribution across leaves of plants was manipulated by using neutral shading, thus ensuring that all leaves were exposed to light with the same R/FR ratio, nitrogen was still reallocated from shaded to more illuminated leaves in the vine, *Ipomoea tricolor* (Hikosaka *et al.*, 1994) and in the grass, *Carex acutiformis*, though not in another grass, *Brachypodium pinnatum* (Pons *et al.*, 1993). Apparently, differences in R/FR ratio received by leaves might play a role in inducing nitrogen redistribution in some species but not in others.

Finally, compounds transported in the transpiration stream could mediate nitrogen reallocation from shaded leaves. Transpiration is reduced in shaded leaves, mainly because of a reduction in stomatal conductance. Xylem sap is partitioned among leaves according to their rate of transpiration. Any compound transported in the xylem would, therefore, be partitioned in proportion to the light received by leaves, and could be instrumental in signalling partial shading in plants.

Pons and Bergkotte (1996) conducted an experiment in which they manipulated the transpiration rate of individual leaves. A reduced transpiration resulted in a reduction in both leaf nitrogen content (N_{area}) and associated light saturated photosynthesis. Surprisingly, a reduced transpiration also induced a number of other responses typically associated with shade acclimation, i.e. chl/N ratios increased and chlorophyll a/b ratios decreased. These data strongly support the notion that compounds transported in the transpiration stream serve as regulators of nitrogen reallocation from shaded leaves. Cytokinins appear to be suitable candidates (Pons and Bergkotte, 1996). They are produced in the roots and transported in the xylem (Matthyse and Scott, 1982), while they play an important role in delaying leaf senescence (Stoddart and Thomas, 1982). Moreover, low cytokinin levels have been found to cause a decreased transcription of genes encoding proteins of the photosynthetic apparatus (Kusnetsov *et al.*, 1994), which could in turn reduce the sink strength of leaves for amino-acids. However, no one has yet found cytokinin levels to be lower in shaded than in more illuminated leaves, so that it can not be concluded that they are indeed regulators of nitrogen distribution in plants.

7.6 Explaining the differences between optimal and actual nitrogen partitioning

As noted above, the optimisation theory predicts nitrogen distributions that are steeper than the actual nitrogen distributions. Several reasons can be suggested to explain the deviation between actual and optimal. One is that there are shortcomings in the models that have been used to calculate optimal nitrogen distribution.

Models for optimal nitrogen distribution generally assume the relationship between light saturated photosynthesis (P_{max}) and leaf nitrogen content (N_{area}) to be linear. In order to calculate the optimal nitrogen distribution, this linear relationship was extrapolated to N_{area} values that were considerably higher than the range for which P_{max} was measured. However, as discussed in Section 7.2.1, there is an indication that at high N_{area} the P_{max} to N_{area} relationship becomes curvilinear. Extrapolation of linear P_{max} to N_{area} relationships may, therefore, lead to an overestimation of the photosynthetic benefit of high N_{area} in the top of the canopy.

Most studies on optimal nitrogen distribution assume the only factors that limit photosynthesis to be light and nitrogen. However, a number of other environmental factors, such as photoinhibition, water stress, wind damage and herbivory, may negatively affect the photosynthetic carbon gain of leaves. In leaves that are exposed to high irradiance, photoinhibition may lead to reduced carbon gain (Björkman *et al.*, 1988; Gamon and Pearcy, 1989). In stands of vegetation, photoinhibition is more likely to occur in higher, more illuminated leaves, thus reducing photosynthesis per unit nitrogen of these leaves. Leaves with high nitrogen content tend to be preferred by herbivores (Mooney and Gulmon, 1979; White 1984), which means that leaves higher in the canopy would suffer more from herbivory than those lower down. Wind damage is also more likely to occur in leaves which are higher up in the canopy. Under such conditions, it might be favourable to have more uniform nitrogen distribution in the canopy, partly to reduce the loss of nitrogen and partly so that leaves lower down, which become exposed after higher leaves are removed, have a higher P_{max} (Stockhoff, 1994).

In most models of optimal nitrogen distribution, light distribution in the canopy is calculated by using Beer's law, Equation (2). This is inaccurate, mainly because it assumes that all leaves in a given horizontal layer intercept the same amount of light. Leaves in light flecks receive more light than predicted by Beer's law. Moreover, interception of direct beam irradiance depends on both the leaf inclination angle and azimuth orientation, both of which differ considerably between leaves (e.g. Goudriaan, 1977, see Chapter 9). However, De Pury and Farquhar (1997) found the optimal nitrogen distribution calculated using a more realistic light model to be steeper and to differ more strongly from the actual distribution than the optimal distribution calculated with Beer's law. The use of an overly simplified light model should thus be discarded as a possible explanation of the difference between actual and optimal nitrogen distribution.

The second reason why there is a difference between the optimal and actual distributions could be that the optimisation theory is flawed. Several authors (Chen *et al.*, 1993; Kull and Jarvis, 1995; Grindlay, 1997) have indicated that the concept that plants should have maximisation of photosynthesis as their 'objective' is highly anthropomorphic, and that it is difficult to imagine how optimisation of the nitrogen distribution may be achieved in plants. Optimisation requires that plants adjust the nitrogen distribution to the light climate which they experience. While plants may have 'learned' to respond to the average conditions in which they have evolved, it is hard to imagine how they are able to respond to actual conditions that may differ substantially from long-term averages (Chen *et al.*, 1993). Another difficulty that has been raised is that there seem to be no realistic biological mechanisms which could allow an optimal nitrogen distribution to develop. However, it has been pointed out above that a number of mechanisms have been proposed that could regulate the development of such a distribution.

Chen and co-workers (1993) proposed the coordination theory as an alternative to the optimisation theory to explain patterns of nitrogen distribution in canopies. They assumed that photosynthesis is limited by either the RuBPCase-limited rate of carboxylation (W_c) or the electron transport limited rate of carboxylation (W_j). Both processes are positively related to N_{area}, while W_j also depends on the PFD level (Farquhar *et al.*, 1980). Chen and co-workers (1993) assumed that at a given light level there is a unique N_{area} at which photosynthesis is co-limited by W_j and W_c. If N_{area} is greater than this value there is a surplus of nitrogen and if it is smaller there is a deficit. For each layer in a canopy, the model calculates surplus or deficits of nitrogen at hourly intervals from dawn to dusk. Nitrogen is then reallocated between layers in the canopy according to their daily average surplus or deficit. The nitrogen distribution patterns thus estimated were slightly less steep than the optimal distribution.

However, the coordination theory as presented by Chen and co-workers (1993) can be criticised. Firstly, the assumption that for any given light availability there is a unique nitrogen content at which W_j and W_c co-limit photosynthesis ignores the effects of nitrogen partitioning among components of the photosynthetic system. Hikosaka and Terashima (1995) showed that at a given PFD, preferential allocation of nitrogen to components limiting either the apparent quantum yield or light saturated photosynthesis will allow co-limitation of photosynthesis to occur for a whole range of N_{area} values (see Figure 3 in Hikosaka and Terashima, 1995). Secondly, it is hard to imagine biological mechanisms allowing leaves to keep track of the exact surplus or deficit of nitrogen which they accumulate over a day. Finally, natural selection should be expected to favour optimisation rather than coordination because optimisation gives higher fitness.

A third reason that could explain differences between optimal and actual nitrogen distributions could be that, while the optimisation theory is in itself correct, it has not been correctly formulated. In most studies, the optimal nitrogen distribution has been defined as the distribution that maximises photosynthesis of a stand of vegetation. One of the assumptions here is that the optimal characteristics of a population reflect the optimal characteristics of individuals. This holds true if a population is a summation of very similar individuals. For many stands of vegetation this does not appear to be the case (see Sections 7.4.2 and 7.4.3 and Chapter 10). A second assumption is that the optimum for one individual is independent of the activity of its neighbours. This is the condition for simple optimisation (Parker and Maynard Smith, 1990) and seems unlikely in situations where individuals compete for light and resources.

The ability of a plant to compete for light depends on its leaf area development. In this context, it is important to note that there is, at least in theory, a trade-off between the area of a leaf and its nitrogen content per unit area (N_{area}). N_{area} is equal to the quotient of the nitrogen content per unit mass (N_{mass}) and the specific leaf area (SLA), i.e. $N_{area} = N_{mass}/\mathrm{SLA}$. A high N_{area} will thus result from a high N_{mass}, a low SLA or both. Changes in N_{area} in response to light availability are regulated, at least in part, through changes in SLA (e.g. Pons and Pearcy, 1994). Sun leaves are thicker and have lower SLA than shade leaves (see Dijkstra, 1989). It can thus be argued that changes in the nitrogen distribution in plants are most probably directly associated with changes in their leaf area distribution, which will, in turn, affect the capacity of plants to capture the light that would otherwise be available to their neighbours. The important consequence of this is that nitrogen distribution in an individual affects not only its own photosynthesis but also that of its neighbours.

In cases where changes in a trait affects not only an individual's fitness but also that of its neighbours, a competitive rather than a simple optimisation criterion should be applied (Iwasa *et al.*, 1984; Parker and Maynard Smith, 1990). Schieving (1998) analysed nitrogen distributions in stands and, assuming a trade-off between N_{area} and SLA, proposed the following competitive optimisation criterion: the nitrogen distribution in a stand is evolutionarily stable if no change in the distribution of N_{area} and associated change in leaf area distribution of an individual can increase its photosynthesis above the average photosynthesis of individuals in the stand. By applying this criterion, he showed that the optimal nitrogen distribution is not evolutionarily stable and that the evolutionarily stable nitrogen distribution is more uniform and probably closer to the actual distribution than the optimal one. However, the analysis of Schieving (1998) is based on a number of simplifying assumptions. The most important one is that he takes the leaf area of individuals and neighbours to be evenly mixed in space, which is most probably not the case. The important point which should be stressed, however, is that competitive optimisation is conceptually more correct than simple optimisation when analysing the performance of plants which compete in stands.

7.7 Optimal leaf area index

The present review focuses mainly on nitrogen partitioning; it being either among the photosynthetic components within a leaf or among leaves of a plant. However, photosynthetic performance per unit nitrogen of a plant may also be determined by its leaf area. With more leaf area, plants are able to intercept more light and thus enhance photosynthesis. On the other hand, at a given total amount of leaf nitrogen an increase in leaf area also implies that N_{area} and associated P_{max} values of leaves are reduced. Consequently, there is probably an optimal leaf area that maximises photosynthesis per unit nitrogen.

Anten and co-workers (1995b) applied this concept to stands of vegetation. Using an analytical model, they showed that for a given amount of canopy nitrogen (N_T), there is an optimal leaf area index (LAI^{op}) and therefore an optimal average leaf nitrogen content (N_{av}^{op}, $N_{av}^{op} = N_T/\mathrm{LAI}^{op}$) that maximises canopy-PNUE. They further derived that LAI^{op} increases and that therefore N_{av}^{op} decreases with the ratio of P_{max} to N_{area} (i.e. PPNUE) and that LAI^{op} decreases and that therefore N_{av}^{op} increases with the extinction coefficient for light (K_L). From this, they predicted that plants with high PPNUE should produce relatively high amounts of leaf area per unit nitrogen, while plants with horizontal leaves, forming stands with high K_L, should produce comparatively low amounts of leaf area per unit nitrogen. In accordance with this prediction, C_4 species were found to have higher PPNUEs and produced more leaf area for a given amount of nitrogen than C_3 species, while dicotyledonous plants had more horizontally projected leaves, formed stands with higher K_L and formed lower amounts of leaf area per unit nitrogen than monocotyledonous plants (Anten *et al.*, 1995b, 1998a).

Actual LAIs, however, were always greater and actual nitrogen contents always lower than the optimal values. Photosynthesis for the actual LAI was 2–20% lower than the maximum values (Schieving *et al.*, 1992a; Anten *et al.*, 1995b, 1998a). Similar arguments to those proposed to explain the difference between actual and optimal nitrogen distribution may apply here as well. For example, extrapolation of a linear P_{max} to N_{area} relationship might have led to an overestimation of photosynthesis at high nitrogen contents, while greater susceptibility of high nitrogen leaves to herbivory may have favoured selection for lower than optimal nitrogen contents. More importantly, however, the optimisation criterion which was used to determine the optimal LAI—maximisation of whole stand canopy photosynthesis—ignored the fact that stands consist of several plants competing with each other for light. Assume that a plant in a stand with optimal LAI increases its leaf area, this plant will capture a larger fraction of the available light, which would otherwise be used by its neighbours. If this would increase its photosynthesis above the average photosynthesis of plants in the stand, the optimal LAI would not be evolutionarily stable and the evolutionarily stable LAI (ES-LAI) would then be greater than optimal LAI (Schieving, 1998). It is further important to note that the problem of ES-LAI is very similar to the problem of evolutionarily stable nitrogen (ES-N) distribution, discussed above. In fact they cannot be viewed independently from each other. In the case of ES-N distribution, changes in the SLA result in changes in the leaf area of individuals, and thus in changes in the LAI.

7.8 Conclusions

At the leaf level, plants generally respond to reduced light by allocating more photosynthetic nitrogen to photosynthetic components that limit apparent quantum yield and less to P_{max} related proteins. In response to changing temperatures, plants are probably able to adjust the nitrogen partitioning among components that limit either RuBPCase or RuBP regeneration, such that co-limitation of P_{max} by these two processes is maintained. At the whole-plant level, nitrogen distribution is generally observed to follow the light environment in the canopy. Differences in nitrogen distribution between stands of different species are largely determined by differences in structural traits of the canopy that determine the light climate therein and much less by physiological characteristics of the leaf, while most plants are very flexible in adjusting their nitrogen distribution patterns to changes in the light climate. Finally, leaf area per unit nitrogen appears to increase with potential leaf-PNUE, and seems to be lower for plants with horizontal leaves forming stands with high extinction coefficients for light. All these trends were well predicted by models for optimal nitrogen utilisation in plants. This gives a strong indication that the assumptions regarding the adaptive significance of nitrogen utilisation were well founded; that is, maximisation of carbon gain per unit of available nitrogen is probably an important driving force in determining nitrogen utilisation in plants.

One predicted trend, which was not observed in most plants, was a reduction in the fractional allocation of nitrogen to RuBPCase in response to increased CO_2 levels in the air, even though this was predicted to significantly enhance leaf-PNUE. This, however, is probably due to the fact that the CO_2 levels, under which plants have evolved, have been relatively constant for at least the last 160,000 years (Barnola *et al.*, 1987).

While most of the general trends in nitrogen partitioning were well predicted by the optimisation theory, predicted values did not exactly match actual observations. The optimal nitrogen distributions in plants were always steeper than the measured ones, while optimal LAIs were considerably lower than actual LAIs. One reason for these deviations could lie with the complexity of the photosynthetic system. Models are, by definition, simplifications of reality, and the deviations between predictions and observations could have resulted from the fact that it is not possible to include every single detail of the photosynthetic system in a predictive model. From this perspective, it can be argued that models should be expected to predict general trends correctly rather than exact values. However, it is also possible that the deviations between predictions and actual data are the result of more fundamental shortcomings in the studies on optimal nitrogen utilisation in plants. Firstly, some of the assumptions about the photosynthetic system and its relation to the environment could have been unrealistic. For example, the P_{max} to N_{area} relationship is probably not linear at high nitrogen contents, and factors other than light and nitrogen, e.g. water availability, herbivory and photoinhibition, may impose limitations on the photosynthetic performance of leaves. Consequently, it is very likely that the optimisation models overestimated PNUE at high N_{area}. Secondly, the application of simple optimisation to analyse optimal nitrogen distribution and optimal leaf areas was incorrect because it ignores the fact that the leaf area of a plant, which is also determined by its nitrogen distribution, affects the performance of its neighbours. A competitive optimisation criterion would, therefore, be more appropriate to analyse optimal nitrogen utilisation at the whole-plant level.

Acknowledgements

The authors acknowledge support from the National Science Foundation grant IBN 9604030 (to N.P.R.A.) and from Monbusho grant-in-aid no. 09041142 (to T.H.). Thanks are also due to Reiko Yambe for preparing the figures.

References

Ackerly, D.D. (1992) Light, leaf age and nitrogen concentration in a tropical vine. *Oecologia*, **89** 596–600.

Anten, N.P.R. (1995) *Canopy Structure and Patterns of Leaf Nitrogen Distribution in Relation to Carbon Gain*, Ph.D. thesis, Utrecht University, Utrecht, The Netherlands.

Anten, N.P.R. (1997) Modelling canopy photosynthesis using parameters determined from simple non-destructive measurements. *Ecological Research*, **12** 77–88.

Anten, N.P.R. and Hirose, T. (1999) Interspecific differences in above-ground growth patterns result in spatial and temporal partitioning of light among species in a tall-grass meadow. *Journal of Ecology*, **87** 583–597.

Anten, N.P.R. and Werger, M.J.A. (1996) Canopy structure and nitrogen distribution in dominant and subordinate plants in a dense stand of *Amaranthus dubius* (L.) with a size hierarchy of individuals. *Oecologia*, **105** 30–37.

Anten, N.P.R., Schieving, F. and Werger, M.J.A. (1995a) Patterns of light and nitrogen distribution in relation to whole canopy carbon gain in C_3 and C_4 mono- and dicotyledonous species. *Oecologia*, **101** 504–513.

Anten, N.P.R., Schieving, F., Medina, E., Werger, M.J.A. and Schuffelen, P. (1995b) Optimal leaf area indices in C_3 and C_4 mono- and dicotyledonous species at low and high nitrogen availability. *Physiologia Plantarum*, **95** 541–550.

Anten, N.P.R., Hernandez, R. and Medina, E. (1996) The photosynthetic capacity and leaf nitrogen concentration as related to light regime in shade leaves of a montane tropical forest tree, *Tetrorchidium rubrivenium*. *Functional Ecology*, **10** 491–500.

Anten, N.P.R., Werger, M.J.A. and Medina, E. (1998a) Nitrogen distribution and leaf area indices in relation to photosynthetic nitrogen-use efficiency in savana grasses. *Plant Ecology*, **138** 63–75.

Anten, N.P.R., Miyazawa, K., Hikosaka, K., Nagashima, N. and Hirose, T. (1998b) Leaf nitrogen distribution in relation to leaf age and photon flux density in dominant and subordinate plants in dense stands of a dicotyledonous herb. *Oecologia*, **113** 314–324.

Azcón-Bieto, J., Lambers, H. and Day, D.A. (1983) Effect of photosynthesis and carbohydrate status on respiratory rates and the involvement of the alternative pathway in leaf respiration. *Plant Physiology*, **72** 598–603.

Barnola, J.M., Raynaud, D., Korotkevich, Y.S. and Lorius, C.D. (1987) Vostok ice pore provides a 160,000 year record of atmospheric CO_2. *Nature*, **329** 408–414.

Berry, J.A. and Björkman, O. (1980) Photosynthetic response and adaptation to temperature in higher plants. *Annual Reviews of Plant Physiology*, **31** 491–543.

Björkman, O. and Demming, B. (1987) Photon yield of O_2 evolution and chlorophyll fluorescence characteristics at 77 K among vascular plants of diverse origins. *Planta*, **170** 489–504.

Björkman O., Demming, B. and Andrew, T.J. (1988) Mangrove photosynthesis: response to high irradiance stress. *Australian Journal of Plant Physiology*, **15** 43–62.

Boardman, N.K. (1977) Comparative photosynthesis of sun and shade plants. *Annual Reviews of Plant Physiology*, **28** 355–377.

Brown, R.H. (1978) A difference in N use efficiency in C_3 and C_4 plants and its implications in adaptation and evolution. *Crop Science*, **18** 93–98.

Charles-Edwards, D.A., Stutzel, H. Ferraris, R. and Beech, D.F. (1987) An analysis of spatial variation in the nitrogen content of leaves from different horizons within a canopy. *Annals of Botany*, **60** 421–426.

Chazdon, R.L. and Field, C. (1987) Determinants of photosynthetic capacity in six rain forest *Piper* species. *Oecologia*, **73** 222–230.

Chen, J.L., Reynolds, J.F., Harley, P.C. and Tenhunen, J.D. (1993) Coordination theory of leaf nitrogen distribution in a canopy. *Oecologia*, **93** 63–69.

De Greef, J., Butler, W.L. and Roth, T.F. (1971) Control of senescence in *Marchantia* by phytochrome. *Plant Physiology*, **48** 407–412.

De Pury, D.G.G. and Farquhar, G.D. (1997) Simple scaling of photosynthesis from leaves to canopies without the errors of big-leaf models. *Plant, Cell and Environment*, **20** 537–557.

Del Pozo, A. and Dennett, M.D. (1999) Analysis of the distribution of light, leaf nitrogen and photosynthesis within the canopy of *Vicia faba* (L.) at two contrasting plant densities. *Australian Journal of Agricultural Research*, **50** 183–189.

Dijkstra, P. (1989) Cause and effect of differences in specific leaf area, in *Causes and Consequences of Variation in Growth Rate and Productivity of Higher Plants* (eds. H. Lambers, M.L. Cambridge, H. Konings and T.L. Pons), SPB Academic Publishing, The Hague, The Netherlands, pp. 125–140.

Epron, D., Godard, D., Cornic, G. and Genty, B. (1995) Limitation of net CO_2 assimilation rate by internal resistances to CO_2 transfer in the leaves of two tree species *Fagus sylvatica* (L.) and *Castanea sativa* (Mill.). *Plant, Cell and Environment*, **18** 43–51.

Evans, J.R. (1983) Nitrogen and photosynthesis in the flag leaf of wheat *(Triticum aestivum L.). Plant Physiology*, **72** 297–302.

Evans, J.R. (1987) The relationship between electron transport components and photosynthetic capacity in pea leaves grown at different irradiance. *Australian Journal of Plant Physiology*, **14** 157–170.

Evans, J.R. (1989a) Photosynthesis and nitrogen relationships in leaves of C_3 plants. *Oecologia*, **78** 9–19.

Evans, J.R. (1989b) Photosynthesis: the dependence on nitrogen partitioning, in *Causes and Consequences of Variation in Growth Rate and Productivity of Higher Plants* (eds. H. Lambers, M.L. Cambridge, H. Konings and T.L. Pons), SPB Academic Publishing, The Hague, The Netherlands, pp. 159–170.

Evans, J.R. (1993a) Photosynthetic acclimation and nitrogen partitioning within a lucerne canopy. I. Canopy characteristics. *Australian Journal of Plant Physiology*, **20** 55–67.

Evans, J.R. (1993b) Photosynthetic acclimation and nitrogen partitioning within a lucerne canopy. II. Stability through time and comparison with a theoretical optimum. *Australian Journal of Plant Physiology*, **20** 69–82.

Evans, J.R. and Seemann, J.R. (1989) The allocation of protein nitrogen in the photosynthetic apparatus: costs, consequences and control, in *Photosynthesis* (ed. W.R. Brigs), Alan R. Liss, Inc., New York, USA, pp. 183–205.

Evans, J.R. and Terashima, I. (1988) Photosynthetic characteristics of spinach leaves grown with different nitrogen treatments. *Plant and Cell Physiology*, **29** 157–165.

Farquhar, G.D. (1989) Models of integrated photosynthesis of cells and leaves. *Philosophical Transactions of the Royal Society of London, Series B, Biological Sciences*, **323** 357–367.

Farquhar, G.D., von Caemmerer, S. and Berry, J.A. (1980) A biochemical model of photosynthetic CO_2 assimilation in leaves of C_3 species. *Planta*, **149** 78–90.

Field, C. (1983) Allocating leaf nitrogen for the maximisation of carbon gain: leaf age as a control on the allocation program. *Oecologia*, **56** 341–347.

Field, C. and Mooney, H.A. (1986) The photosynthesis–nitrogen relationship in wild plants, in *On the Economy of Form and Function* (ed. T.J. Givnish), Cambridge University Press, Cambridge, UK, pp. 25–55.

Ford, E.D. (1975) Competition and stand structure in some even-aged plant monocultures. *Journal of Ecology*, **63** 311–333.

Friedrich, J.W. and Huffaker, R.C. (1980) Photosynthesis, leaf resistances and ribulose-1,5-bisphosphate carboxylase degradation in senescing barley leaves. *Plant Physiology*, **65** 1103–1107.

Gabrielsen, E.K. (1948) Effects of different chlorophyll concentrations on photosynthesis in foliage leaves. *Physiologia Plantarum*, **1** 5–37.

Gamon, J.A. and Pearcy, R.W. (1989) Leaf movement stress avoidance and photosynthesis in *Vitis californica. Oecologia*, **79** 475–481.

Goudriaan, J. (1977) *Crop Micrometeorology: a Simulation Study*, PUDOC, Wageningen, The Netherlands.

Grindlay, D.J.C. (1997) Towards an explanation of crop nitrogen demand based on the optimisation of leaf nitrogen per unit leaf area. *Journal of Agricultural Science*, **128** 377–396.

Gulmon, S.L. and Chu, C.C. (1981) The effects of light and nitrogen on photosynthesis, leaf characteristics and dry matter allocation in the chaparral shrub, *Diplacus aurantiacus. Oecologia*, **49** 207–212.

Harper, J.L. (1977) *Population Biology of Plants,* Academic Press, London, UK.

Hensel, L.L., Vojislava, G., Baumgarten, D.A. and Bleecker, A.B. (1993) Development and age-

related processes that influence the longevity and senescence of photosynthetic tissues in *Arabidopsis. The Plant Cell*, **5** 553–564.
Hikosaka, K. (1996) Effects of leaf age, nitrogen nutrition and photon flux density on the organisation of the photosynthetic apparatus in leaves of a vine (*Ipomoea tricolor* Cav.) grown horizontally to avoid mutual shading of leaves. *Planta*, **198** 144–150.
Hikosaka, K. (1997) Modelling optimal temperature acclimation of the photosynthetic apparatus in C_3 plants with respect to nitrogen use. *Annals of Botany*, **80** 721–730.
Hikosaka, K. and Hirose, T. (1998) Leaf and canopy photosynthesis of C_3 plants at elevated CO_2 in relation to optimal partitioning of nitrogen among photosynthetic components: theoretical prediction. *Ecological Modelling*, **106** 247–259.
Hikosaka, K. and Terashima, I. (1995) A model of the acclimation of photosynthesis in the leaves of C_3 plants to sun and shade with respect to nitrogen use. *Plant, Cell and Environment*, **18** 605–618.
Hikosaka, K. and Terashima, I. (1996) Nitrogen partitioning among photosynthetic components and its consequence in sun and shade plants. *Functional Ecology*, **10** 335–343
Hikosaka, K., Terashima, I. and Katoh, S. (1994) Effects of leaf age, nitrogen nutrition and photon flux density on the distribution of nitrogen among leaves of a vine *(Ipomoea tricolor* Cav.) grown horizontally to avoid mutual shading of leaves. *Oecologia*, **97** 451–457.
Hikosaka, K., Hanba, Y.T., Hirose, T. and Terashima, I. (1998) Photosynthetic nitrogen use efficiency in woody and herbaceous plants. *Functional Ecology*, **12** 896–905.
Hikosaka, K., Murakami, A. and Hirose, T. (1999a) Balancing carboxylation and regeneration of ribulose bisphosphate in leaf photosynthesis: temperature acclimation in an evergreen tree, *Quercus myrsinaefolia. Plant, Cell and Environment*, **22** 841–849.
Hikosaka, K., Sudoh, S. and Hirose, T. (1999b) Light acquisition and use by individuals competing in a dense stand of an annual herb, *Xanthium canadense. Oecologia*, **118** 388–396.
Hirose, T. (1984) Nitrogen use efficiency in growth of *Polygonum cuspidatum* Sieb. et Zucc. *Annals of Botany*, **54** 695–704.
Hirose, T. and Werger, M.J.A. (1987a) Nitrogen-use efficiency in instantaneous and daily photosynthesis of leaves in the canopy of a *Solidago altissima* stand. *Physiologia Plantarum*, **70** 215–222.
Hirose, T. and Werger, M.J.A. (1987b) Maximising daily canopy photosynthesis with respect to the leaf nitrogen allocation pattern in the canopy. *Oecologia*, **72** 520–526.
Hirose, T. and Werger, M.J.A. (1994) Photosynthetic capacity and nitrogen partitioning among species in the canopy of a herbaceous plant community. *Oecologia*, **100** 203–212.
Hirose, T. and Werger, M.J.A. (1995) Canopy structure and photon flux partitioning among species in a herbaceous plant conununity. *Ecology*, **76** 466–474.
Hirose, T., Werger, M.J.A., Pons, T.L. and van Rheenen, J.W.A. (1988) Canopy structure and leaf nitrogen distribution in a stand of *Lysimachia vulgaris* as influenced by stand density. *Oecologia*, **77** 145–150.
Hirose, T., Werger, M.J.A. and van Rheenen, J.W.A. (1989) Canopy development and leaf nitrogen distribution in a stand of *Carex acutiformis. Ecology*, **70** 1610–1618.
Hogan, K.P., Smith, A.P. and Ziska, L.H. (1991) Potential effects of elevated CO_2 and changes in temperature on tropical plants. *Plant, Cell and Environment*, **14** 763–778.
Hollinger, D.Y. (1996) Optimality and nitrogen allocation in a tree canopy. *Tree Physiology*, **16** 627–634.
Iwasa, Y., Cohen, D. and Leon, J.A. (1984) Tree height and crown shape, as a result of competitive games. *Journal of Theoretical Biology*, **112** 279–297.
Jenkins, G.I. and Woolhouse, W. (1981) Photosynthetic electron transport during senescence of the primary leaves of *Phaseolus vulgaris* (L.). I. Non-cyclic electron transport. *Journal of Experimental Botany*, **32** 467–478.
Kachi, N. and Hirose, T. (1983) Limiting nutrients for plants growth in coastal sand dune soils. *Journal of Ecology*, **71** 937–944.
Kull, O. and Jarvis, P.G. (1995) The role of nitrogen in a simple scheme to scale-up photosynthesis from leaf to canopy. *Plant, Cell and Environment*, **18** 1174–1182.

Kusnetsov, V.V., Oelmuller, R., Sarwat, M.I., Porfirova, S.A., Cherepneva, G.N., Herrmann, R.G. and Kulaeva, O.N. (1994) Cytokinins, abscisic acid and light effect accumulation of chloroplast proteins in *Lupinus luteus* cotyledons without notable effect on steady-state mRNA levels. *Planta*, **194** 318–327.

Lambers, H. (1985) Respiration in intact plants and tissues: its regulation and dependence on environmental factors, metabolism and invaded organisms, in *Encyclopedia of Plant Physiology. 11. Higher Plant Cell Respiration* (eds. R. Douce and D.A. Day), Springer-Verlag, Berlin, Germany, pp. 418–473.

Lee, J.A., Harmer, R. and Ignachiuk, R. (1983) Nitrogen as a limiting factor in plant communities, in *Nitrogen as an Ecological Factor* (eds. J.A. Lee, S. McNeil and I.H. Rorison), Blackwell, Oxford, UK, pp. 95–112.

Lemaire, G., Onillon, B., Gosse, G., Chartier, M. and Allirand, J.M. (1991) Nitrogen distribution within a lucerne canopy during regrowth: relation with light distribution. *Annals of Botany*, **68** 483–488.

Leong, T. and Anderson, J.M. (1984) Adaptation of the thylakoid membranes of pea chloroplasts to light intensities. I. Study on the distribution of chlorophyll–protein complexes. *Photosynthesis Research*, **5** 105–115.

Leverenz, J.T. (1987) Chlorophyll content and the light response curve of shade-adapted conifer needles. *Physiologia Plantarum*, **71** 20–29.

Lloyd, J., Syverstsen, J.P., Kriedemann, P.E. and Farquhar, G.D. (1992) Low conductances for CO_2 diffusion from stomata to the sites of carboxylation in leaves of woody species. *Plant, Cell and Environment*, **15** 873–899.

Lugg, D.G. and Sinclair, T.R. (1981) Seasonal changes in photosynthesis of field-grown soybean leaflets. 2. Relation to nitrogen content. *Photosynthetica*, **15** 138–144.

Makino, A., Mae, T. and Ohira, K. (1983) Photosynthesis and ribulose-1,5-biphosphate carboxylase in rice leaves. *Plant Physiology*, **73** 1002–1007.

Makino, A., Mae, T. and Ohira, K. (1988) Differences between wheat and rice in the enzymatic properties of ribulose-1,5-biphosphate carboxylase and the relationship to gas exchange. *Planta*, **174** 30–38.

Makino, A., Shimada, T., Takumi, S., Kaneko, K., Matsuoka, M., Shimamoto, K., Nakano, H., Miyao-Tokutomi, M., Mae, T. and Yamamoto, N. (1997) Does decrease in ribulose-1,5-biphosphate carboxylase by 'antisense' *rbc* S lead to a higher nitrogen-use efficiency of photosynthesis under conditions of saturating CO_2 and light in rice? *Plant Physiology*, **114** 483–491.

Marshall, B. and Biscoe, P.V. (1980) A model for C_3 leaves describing the dependence of net photosynthesis on irradiance. *Journal of Experimental Botany*, **31** 29–39.

Matthysse, A.G. and Scott, T.K. (1982) Function of hormones at the whole plants level of organisation, in *Encyclopedia of Plant Physiology*, New Series Vol. 10. Hormonal Regulation of Development, I (eds. A. Pirson and M.H. Zimmerman), Springer-Verlag, Berlin, Germany, pp. 219–235.

Medina, E. (1981) *Nitrogen Content, Leaf Structure and Photosynthesis in Higher Plants*. A Report to the UNEP Study Group on Photosynthesis and Bioproductivity, IVIC, Caracas, Venezuela.

Medlyn, B.E. (1996) The optimal allocation of nitrogen within the C_3 photosynthetic system at elevated CO_2. *Australian Journal of Plant Physiology*, **23** 593–603.

Monsi, M. and Saeki, T. (1953) Über den Lichtfaktor in den Pflanzengesellschaften und seine Bedeutung fur die Stoffproduktion. *Japanese Journal of Botany*, **14** 22–52.

Monteith, J.L. (1965) Light distribution and photosynthesis in field crops. *Annals of Botany*, **29** 17–38.

Mooney, H.A. and Gulmon, S.L. (1979) Environmental and evolutionary constraints on the photosynthetic characteristics of higher plants, in *Topics in Plant Population Biology* (eds. O.T. Solbrig, S. Jain, G.B. Johnson and P.H. Raven), Columbia University Press, New York, USA, pp. 316–337.

Mooney, H.A., Field, C., Gulmon, S.L. and Bazzaz, F.A. (1981) Photosynthetic capacity in relation to leaf position in desert vs old-field annuals. *Oecologia*, **50** 109–112.

Nátr, L. (1975) Influence of mineral nutrition on photosynthesis and the use of assimilates, in *Photosynthesis and Productivity in Different Environments* (ed. J.P. Cooper), Cambridge University Press, UK, pp. 537–556.

Noguchi, K. and Terashima, I. (1997) Different regulation of leaf respiration between *Spinacia oleracea,* a sun species, and *Alocasia odora,* a shade species. *Physiologia Plantarum*, **101** 1–7.

Osman, A.M. and Milthorpe, F.L. (1971) Photosynthesis of wheat leaves in relation to age, illuminance and nutrient supply. II. Results. *Photosynthetica*, **5** 61–70.

Parker, G.A. and Maynard Smith, J. (1990) Optimality theory in evolutionary biology. *Nature*, **348** 27–33.

Penning de Vries, F.W.T. (1975) The cost of maintenance processes in plant cells. *Annals of Botany*, **39** 77–92.

Pons, T.L. and Bergkotte, M. (1996) Nitrogen allocation in response to partial shading of a plant: possible mechanisms. *Physiologia Plantarum*, **98** 571–577.

Pons, T.L. and Jordi, W. (1998) Induction of leaf senescence and shade acclimation in leaf canopies: variation with leaf longevity, in *Inherent Variation in Plant Growth: Physiological Mechanisms and Ecological Consequences* (eds. H. Lambers, H. Poorter and M.M.I. van Vuuren), Backhuys Publishers, Leiden, The Netherlands, pp. 121–137.

Pons, T.L., and Pearcy, R.W. (1994) Nitrogen reallocation and photosynthetic acclimation in response to partial shading in soybean plants. *Physiologia Plantarum*, **92** 636–644.

Pons, T.L., Schieving, F., Hirose, T. and Werger M.J.A. (1989) Optimization of leaf nitrogen allocation for canopy photosynthesis in *Lysimachia vulgaris*, in *Causes and Consequences of Variation in Growth Rate and Productivity of Higher Plants* (eds. H. Lambers, M.L. Cambridge, H. Konings and T.L. Pons), SPB Academic Publishing, The Hague, The Netherlands, pp. 175–186.

Pons, T.L., Van Rijnbeek, H., Scheurwater, I. and Van der Werf, A. (1993) Importance of the gradient in photosynthetically active radiation in a vegetation stand for leaf nitrogen allocation in two monocotyledons. *Oecologia*, **95** 416–424.

Poorter, H. and Evans, J.R. (1998) Photosynthetic nitrogen-use efficiency of species that differ inherently in specific leaf area. *Oecologia*, **116** 26–37.

Poorter, H., Remkes, C. and Lambers, H. (1990) Carbon and nitrogen economy of 24 wild species differing in relative growth rate. *Plant Physiology*, **94** 621–627.

Radin J.W. and Elmore, C.D. (1980) Concepts of translocation with special reference to assimilation of nitrogen and its movement into fruits, in *Predicting Photosynthesis for Ecosystem Models*, Vol II, CRC Press, Boca Raton, FL, USA, pp. 143–154.

Reich, P.B., Uhl, C., Walters, M.B. and Ellsworth, D.S. (1991) Leaf life span as a determinant of leaf structure among 23 Amazonian tree species. *Oecologia*, **86** 16–24.

Reich, P.B., Kloeppel, B.D., Ellsworth, D.S. and Walters, M.B. (1995) Different photosynthesis-nitrogen relations in deciduous hardwood and evergreen coniferous tree species. *Oecologia*, **104** 24–30.

Rousseaux, M.C., Hall, A.J. and Sanchez, R.A. (1996) Far-red enrichment and photosynthetically active radiation level influence leaf senescence in field-grown sunflower. *Physiologia Plantarum*, **96** 217–224.

Sage, R.F. (1994) Acclimation of photosynthesis to increasing atmospheric CO_2: the gas exchange perspective. *Photosynthesis Research*, **39** 351–368.

Sage, R.F. and Pearcy, R.W. (1987) The nitrogen-use efficiency of C_3 and C_4 plants. II. Leaf nitrogen effects on the gas exchange characteristics of *Chenopodium album* (L.) and *Amaranthus retroflexus* (L.). *Plant Physiology*, **84** 959–963.

Schieving, F. (1998) *Plato's plant*, Ph.D. thesis, Utrecht University, Backhuys Publishers, Leiden, The Netherlands.

Schieving, F., Werger, M.J.A. and Hirose, T. (1992a) Canopy structure, nitrogen distribution and whole canopy photosynthetic carbon gain in growing and flowering stands of tall herbs. *Vegetatio*, **102** 173–182.

Schieving, F., Pons, T.L., Werger, M.J.A. and Hirose, T. (1992b) Vertical distribution of nitrogen in photosynthetic activity at different plant densities in *Carex acutiformis*. *Plant and Soil*, **142** 9–17.

Sellers, P.J., Berry, J.A., Collatz, G.J., Field, C.B. and Mooney, H.A. (1992) Canopy reflectance, photosynthesis and transpiration. III. A reanalysis using improved leaf models and a new canopy integration scheme. *Remote Sensing and Environment*, **42** 187–216.

Smith, H., Samson, G. and Fork, D.C. (1993) Photosynthetic acclimation to shade, probing the role of phytochromes using photomorphogenic mutants of tomato. *Plant, Cell and Environment*, **18** 149–157.

Stockhoff, B.A. (1994) Maximisation of daily canopy photosynthesis: effects of herbivory on optimal nitrogen distribution. *Journal of Theoretical Biology*, **169** 209–220.

Stoddart, J.L. and Thomas, H. (1982) Leaf senescence, in *Encyclopedia of Plant Physiology* New Series, Vol. 14B, *Nucleic Acids and Proteins in Plants*. I. *Proteins* (eds. D. Boulter and B. Parthier), Springer-Verlag, Berlin, Germany, pp. 592–636.

Terashima, I. and Evans, J.R. (1988) Effects of light and nitrogen nutrition on the organisation of the photosynthetic apparatus in spinach. *Plant and Cell Physiology*, **29** 143–155.

Terashima, I. and Hikosaka, K. (1995) Comparative ecophysiology of leaf and canopy photosynthesis. *Plant, Cell and Environment*, **18** 1111–1118.

Turner, C.L. and Knapp, A.K. (1996) Responses of a C_4 grass and three C_3 forbs to variation in nitrogen and light in a tall grass prairie. *Ecology*, **77** 1738–1749.

Van der Werf, A., Lambers, H. and Hirose, T. (1989) Variation in root respiration, causes and consequences for growth, in *Causes and Consequences of Variation in Growth Rate and Productivity of Higher Plants* (eds. H. Lambers, M.L. Cambridge, H. Konings and T.L. Pons), SPB Academic Publishing, The Hague, The Netherlands, pp. 227–240.

Von Caemmerer, S. and Farquhar, G.D. (1981) Some relationships between the biochemistry of photosynthesis and the gas exchange of leaves. *Planta*, **153** 376–387.

Vu, J.C.V. and Yelenosky, G. (1988) Water deficit and associated changes in some photosynthetic parameters in leaves of 'Valencia' orange *(Citrus sinensis* (L.) Osbeck). *Plant Physiology*, **88** 375–378.

Webber, A.N., Nie, G.Y. and Long, S.P. (1994) Acclimation of photosynthetic proteins to rising atmospheric CO_2. *Photosynthesis Research*, **39** 413–425.

Werger, M.J.A. and Hirose, T. (1991) Leaf nitrogen distribution and whole canopy photosynthetic carbon gain in herbaceous stands. *Vegetatio*, **97** 11–20.

White, T.C.R. (1984) The abundance of invertebrate herbivores in relation to the availability of nitrogen in stressed food plants. *Oecologia*, **63** 90–105.

Woodrow, I.E. and Berry, J.A. (1988) Enzymatic regulation of photosynthetic CO_2 fixation in C_3 plants. *Annuals Reviews of Plant Physiology*, **39** 533–594.

Wullschleger, S.D. (1993) Biochemical limitations to carbon assimilation in C_3 plants: a retrospective analysis of the A/C_i curves from 109 species. *Journal of Experimental Botany*, **44** 907–920.

8 Modelling interactions in mixed-plant communities: light, water and carbon dioxide

J.S. Wallace and A. Verhoef

8.1 Introduction

'Plant canopies modify their own microclimate'. This was the opening statement by McNaughton (1989) in the forerunner to this book. He then went on to review how the energy balance of a canopy could influence the atmospheric boundary layer on a regional scale. Starting with the same statement, 'plant canopies modify their own microclimate', this chapter instead looks inwards and at a much smaller scale to present a summary of how this happens within plant canopies and, in particular, within mixed plant communities. At the time McNaughton wrote his chapter, most descriptions of radiation interception and energy transfer in plant canopies dealt with monocultures, and their ability to describe interactions within plant mixtures was quite limited. However, these interactions are vital to the overall performance of plant communities and the success or failure of the component species. Fortunately, considerable progress has been made in the last decade, and it is now possible to describe quantitatively many of the key ways in which plants can affect the microclimate of their neighbours.

The primary way in which soil and plants interact above ground is by altering each other's radiation environment. Plants shade the ground and mixtures of plants can shade each other. The first essential feature of any model of a canopy of mixed species is the ability to quantify the amount of radiation intercepted by each component of the system. This is vital, since radiation is the source of energy for photosynthesis and for subsequent partitioning into latent and sensible heat. Radiation is also the primary determinant of the temperature of the plants and the soil, which influences plant development, photosynthesis and respiration.

Plants are also affected by air temperature, humidity, wind speed and CO_2 concentration. Many types of vegetation have incomplete ground cover for all or part of the season and, in this case, the above ground microclimate is influenced by the soil surface. Plants frequently grow naturally, or are deliberately grown, in mixtures, e.g. savannahs, rainforests, intercropping or agroforestry. Here, neighbouring plants can also affect each other's microclimate. A second essential feature of a mixed canopy model is, therefore, the ability to quantify the effect any component of the system has on the temperature, humidity and CO_2 concentration of the air around its neighbours. It is also important to quantify the amount of water evaporated and CO_2 fixed by each component of the system, especially where there may be competition for light and/or limited water supplies in the soil.

Below-ground interactions may be at least as important as those above ground, but there is only scope in the present chapter to deal in detail with the above-ground component. Most of the calculations presented here do not explicitly account for below-ground factors (e.g. water and nutrient supply). Where the influence of below-ground factors is recognised, it is via their effect on the above-ground physical or physiological behaviour of the soil surface or plants. However, the final section of the present chapter returns to this point to indicate how and where above-ground interaction models need to be coupled with appropriate below-ground models.

The remainder of the chapter is structured in five sections. Section 8.2 describes the main types of light interaction models for plant canopies, and includes brief descriptions of theory, some recent new developments and comparisons of the outputs of the different models for a range of canopy mixtures. This is followed by Section 8.3 on modelling the component and total fluxes of heat and water vapour. Section 8.3 begins with a single-source model, moves to dual-source models and finally extends to the most general case of a multi-source model. The current approaches to describing aerodynamic transfer in these models is then discussed, and the section is completed by showing how CO_2 can also be coupled into the multi-source modelling framework. Section 8.4 focuses on microclimate interactions, using the models described to illustrate how neighbouring plants can affect the temperature, humidity and CO_2 concentration around their neighbours. Section 8.5 shows some of the potential effects that the previous microclimate interactions could have on water use, plant growth and development. Finally, Section 8.6 summarises the remaining key issues in multi-source modelling and, in particular, the need to develop and couple above-ground models with their below-ground equivalents.

8.2 Light interactions

Two basic approaches have been used to describe light interception in plant communities, and these are mainly applicable to binary mixtures with horizontally uniform canopies. The most physically precise approach is to use quite complex models, which divide the canopy into many layers, calculate the radiation absorption of each layer and integrate to obtain the total light absorption of the constituent species. For example, Rimmington (1984) developed a model based on the quantity and optical properties of the foliage of the two constituent species. Where the components occupy different vertical strata, details of the amounts of foliage of each species present in each layer are required. Another model of this type has been described by Sinoquet and co-workers (Sinoquet *et al.*, 1990; Sinoquet and Bonhomme, 1991, 1992), who developed a rigorous multi-layer radiation transfer model for horizontally well-mixed intercrops. In this model, the canopy is divided into cells within which the radiation interception probabilities are related to leaf area density, leaf angle distribution and the foliage dispersion of each of the two component species. The total short-wave radiation intercepted by each of the two species in a mixture is obtained by integrating across all the cells in the mixed canopy. This model has also been extended to allow for long-wave radiation exchanges (Tournebize *et al.*, 1996).

The second and alternative approach to modelling light interception in mixed plant communities is to effectively treat the canopy as a semi-transparent medium, through which light is attenuated according to Beer's law. This is the approach which has been used very successfully in uniform single-species canopies (e.g. Monteith and Unsworth, 1990). Spitters (1990) used this approach to develop a model of light competition in a two-species canopy based on the total leaf area index, extinction coefficient and height of each species. The method developed by Spitters has been used in a number of two-species models, for example, those described by Kropff and van Laar (1992) and Kiniry and co-workers (1992). The ERIN model (Evaporation and Radiation Interception by Neighbouring plants) described by Wallace (1997) is also based on the more simplistic approach used by Spitters, but extends this to a canopy containing any number of species. More details of the ERIN model are described below, together with comparisons with the multi-layer radiation transfer model described by Sinoquet and co-workers (1990). To explain the derivation of the ERIN model, light interception in spatially homogeneous canopies is first described. The simple way in which this model deals with spatial heterogeneity is introduced later.

8.2.1 Spatially homogeneous canopies

Much progress has been made in describing light interception in single-species canopies by assuming that the canopy is a homogeneous medium within which the foliage is randomly distributed. In this case, the fraction of the incident solar radiation, f, intercepted by a canopy of leaf area index, L, is given using Beer's Law:

$$f = 1 - e^{-KL} \tag{1}$$

where, K is the extinction coefficient (e.g. Monteith and Unsworth, 1990). The amount of radiation absorbed by the canopy is given by $(1 - \rho)f$, where ρ is the canopy albedo. However, throughout this section the fraction of total solar radiation intercepted will be used. This has been shown by Russell and co-workers (1989) to be very similar to the fractional absorption of photosynthetically active radiation (PAR) for a wide range of leaf extinction coefficients and leaf areas. It should, therefore, be possible to use values of fractional solar radiation interception to estimate the absorption of PAR in multi-species canopies.

To make progress towards a description of multi-species canopies, it is useful to first consider mixtures of two species, as in an intercrop or agroforestry system. Taking an extreme condition, where Species 1 is very much taller than Species 2 (i.e. all its foliage occurs above Species 2), then the fraction of the incident radiation intercepted by the first canopy is given by Equation (1) as:

$$f_1^d = 1 - \mathrm{e}^{-K_1 L_1} \tag{2}$$

where, the superscript d refers to the condition that Species 1 is dominant, and K_1 and L_1 are the extinction coefficient and leaf area index of the first canopy, respectively. The equivalent fraction of incident radiation intercepted by the second canopy is:

$$f_2^s = \mathrm{e}^{-K_1 L_1}(1 - \mathrm{e}^{-K_2 L_2}) \tag{3}$$

where, the superscript s refers to the condition that Species 2 is subordinate, and K_2 and L_2 are the extinction coefficient and leaf area index of the second canopy, respectively. The second or opposite extreme is where Species 2 dominates, in which case:

$$f_2^d = 1 - \mathrm{e}^{-K_2 L_2} \tag{4}$$

and

$$f_1^s = \mathrm{e}^{-K_2 L_2}(1 - \mathrm{e}^{-K_1 L_1}) \tag{5}$$

When the two species are of comparable height, their canopies will overlap and so their fractional light interception will be somewhere between these two extremes, and largely determined by the relative heights of the two species (e.g. Sinoquet and Caldwell, 1995). The simplest way to interpolate between the extreme values is as follows:

$$f_1 = F f_1^d + (1 - F) f_1^s \tag{6}$$

and

$$f_2 = (1 - F) f_2^d + F f_2^s \tag{7}$$

where, F is a scaling factor between 0 and 1, which is a function of the heights, h_1 and h_2, of the two species. This factor, F, is an expression of the dominance of one species over another, and the exact description of its dependence on h_1 and h_2 will depend on the detailed canopy architecture of the two species involved. However, Wallace (1997) proposed a simple empirical function, which had the correct symmetry and limiting conditions in a two-species mixture:

$$F = \frac{h_1{}^2}{h_1{}^2 + h_2{}^2} \tag{8}$$

The fraction of incident radiation intercepted by each component of a two-species mixture can, therefore, be calculated from Equations (2) to (8), without any explicit knowledge of foliage quantities in different layers, if the height, leaf area index and extinction coefficient of the constituent species are known. However, there is an implicit assumption in the above equations that the leaf area density of each species is constant with height. The consequences of this assumption are discussed in Section 8.2.2.

Figure 8.1 shows light interception values derived using the above

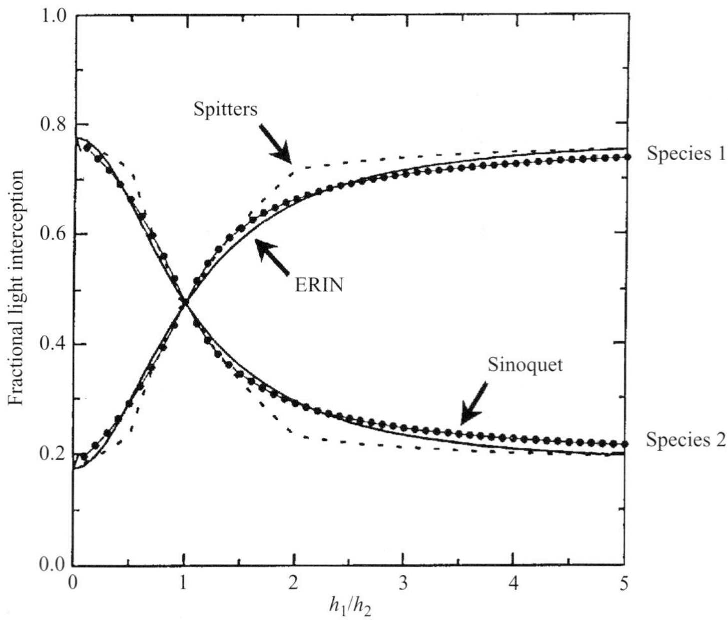

Figure 8.1 The fractional light interception of two species in a mixture as a function of their relative heights (h_1/h_2). Light interception values derived using the ERIN model (——) are compared with the values given by the formula proposed by Spitters (1990) (– – –) and the radiation balance model described by Sinoquet and co-workers (1990) (●–●–●). In this example, each species has the same leaf area index (3.0) and extinction coefficient (0.5).

equations (ERIN model) for a mixture of two species, each with a leaf area index of 3 and an extinction coefficient of 0.5. When Species 1 is completely subordinate ($h_1/h_2 \sim 0$), Species 2 intercepts the incident light first and the remainder falls on Species 1. As h_1/h_2 increases, Species 1 becomes more light dominant until, ultimately, it intercepts the incident light first, the remainder falling on Species 2. Figure 8.1 also shows a comparison of the values of fractional light interception of each species calculated here with values obtained using the simple model described by Spitters (1990) and the more complex model described by Sinoquet and co-workers (1990). The comparison with the model described by Spitters is straightforward since, like the current model, it uses an extinction coefficient (K) to represent the light interception efficiency of each canopy. The model described by Sinoquet and co-workers uses leaf angle and leaf and soil reflectance rather than K. However, Wallace (1997) made some comparisons for a single-species, homogeneous canopy and found that identical light interception was predicted using Beer's law (with $K = 0.5$) and with the model described by Sinoquet and co-workers when the leaf angle was 45°, leaf reflectance 0.3 and soil reflectance 0.2. These values of leaf angle, and leaf and soil reflectance are used in the two species mixture shown in Figure 8.1. All three models give very similar results when the ratio of h_1/h_2 is close to 1. Strictly, the model described by Spitters is restricted to the narrow range of h_1/h_2 ratios between 0.5 and 2.0; however, the original equations can be modified to work outside this range, but this leads to a discontinuity in results when $h_1/h_2 = 0.5$ or 2.0 (see Figure 8.1). At this point, the model developed by Spitters is nearly 20% different from that proposed by Sinoquet and co-workers. The ERIN model is never more than a few per cent different from the model proposed by Sinoquet and co-workers throughout the complete range of species heights.

In a canopy with more than two species, Equations (6), (7) and (8) can be extended so that the fraction of light intercepted by each component of a multi-species mixture, f_i, can be calculated. Unfortunately, when there are more than two species, the light interception of species i can only be calculated when all possible combinations of dominance are taken into account. For example, when there are three species, there are six possible arrangements, i.e. (1, 2, 3) (1, 3, 2) (2, 1, 3) (3, 1, 2) (3, 2, 1) and (2, 3, 1). To calculate f_i for n species, all $n!$ combinations must be considered. The method for doing this is described in the Appendix.

The light interception model described so far applies to horizontally uniform mixtures of species, and so further theoretical development is required for use in plant mixtures where one or more of the component species is spatially clumped, either horizontally or vertically.

8.2.2 Spatially heterogeneous canopies

To allow for spatial heterogeneity within or between the species canopies, Sinoquet and Bonhomme (1991) extended the approach of Sinoquet and co-workers (1990) and developed a theoretical description of radiation interception that allows for a range of canopy geometries within a two-species mixture. This model requires leaf angle distributions for each species and within and between species leaf dispersion parameters. There are other light interception models which explicitly deal with spatially heterogeneous canopies, such as those of Jackson and Palmer (1979), Norman and Welles (1983), Gijzen and Goudriaan (1989) and Wang and Jarvis (1990). These models recognise the three-dimensional structure of the plant canopy and use ray tracing techniques to calculate light interception. However, the main disadvantage of all these physically rigorous models is that they require many canopy architecture and light transmission parameters that are rarely available in practice. In the ERIN model, Wallace (1997) uses a simpler empirical approach and considers vertical and horizontal heterogeneity separately. Vertical heterogeneity in the plant canopies is dealt with by using the vertical thickness of each canopy (t_i), and by assuming that the leaf area density is constant between the top and bottom of the canopy. The assumption of constant leaf area density should not often be a serious limitation of the ERIN model, since Sinoquet and Caldwell (1995) have demonstrated that the relative height of species in mixtures dominates their light interception and that the detailed shape of the leaf area density profile is usually of minor importance.

Horizontal heterogeneity is common in many types of vegetation, especially natural plant communities and man made arrangements, such as intercrops and agroforestry. The spatial irregularity of foliage in a horizontal plane can be described in two main ways. Firstly, the canopies of discrete species may be grouped in space, for example, as in row crops or orchards, where there are definite areas in which foliage is present and areas where it is not. Secondly, within the sub-area that is occupied by foliage, the leaves may not be evenly distributed but clumped in areas of high and low leaf area density. To allow for this as simply as possible in the ERIN model, Wallace (1997) dealt with the first of these spatial heterogeneity problems by considering the 'cover', α, of the species, which is a measure of the fraction of ground covered by the foliage of the species as a projection in the direction of the incident radiation. Consider first a dominant species, i, with partial ground cover, α_i. In this case, light interception can be estimated using Beer's law for the fraction of the ground area occupied by the species (α_i), as no light is intercepted in the area ($1 - \alpha_i$) not occupied by this species. If the incident radiation per unit area of ground is I_0, then over the fraction of ground, α_i, the amount of incident radiation is $\alpha_i I_0$. The fraction of this incident light that is intercepted by the dominant, partial cover species, i, is therefore given by:

$$f_i^d = \alpha_i(1 - e^{-K_i L_i^P}) \tag{9}$$

Note that the appropriate leaf area index in this formula is L_i^P, which is the leaf area index expressed over the fraction of ground area occupied by plants (i.e. $L_i^P = L_i/\alpha_i$).

Equation (9) applies when species i is growing on its own or is totally dominant in a mixture, i.e. it is equivalent to f_i^d in Equation (2). The expression for f_i when a species is completely subordinate to neighbouring species (i.e. f_i^s) also needs to be modified when there is only partial cover. In this case, the ith species is dominated by all of the other species and the ground area average fraction of incident light passing through each of the dominant species is given by:

$$1 - f_i^d = (1 - \alpha_i) + e^{-K_i L_i^P} \tag{10}$$

The resultant light interception by the subordinate species is therefore given by:

$$f_i^s = \alpha_i(1 - e^{-K_i L_i^P}) \prod_{j_{j \neq i}}^{N} \{(1 - \alpha_j) + e^{-K_j L_i^P}\} \tag{11}$$

The light interception of a partial ground cover species when it is growing in a binary mixture can now be calculated using Equations (6), (7) and (8), if Equation (9) is used to specify f_i^d, instead of Equation (2), and Equation (11) is used to specify f_i^s, instead of Equation (5). Calculations for mixtures containing more than two species follow the same procedure as described in Section 8.2.1. It is worth noting that, in this treatment of light interception in partial cover canopies, the model calculates the ground area average light interception and there is no specification of how light levels vary horizontally.

Figure 8.2 shows that the ERIN model can be used to calculate how partial ground cover in a canopy can affect light interception in a two-species mixture. In this example, the understorey has a leaf area index of 3, an extinction coefficient of 0.5 and completely covers the ground (i.e. $\alpha_2 = 1$). The upperstorey is twice the height of the understorey, has the same leaf area index and extinction coefficient and has ground cover ranging from 0 to 1. Note that here, in order to demonstrate the effect of displaying a fixed amount of foliage with different degrees of ground cover, as $\alpha_1 \to 0$, L_1 remains constant. Calculations performed using Equations (9) and (11) predict that, as the upperstorey cover decreases it intercepts less light until ultimately it intercepts none at all, when $\alpha_1 = 0$. At the same time, the understorey light interception increases as the upperstorey cover decreases, eventually inter-

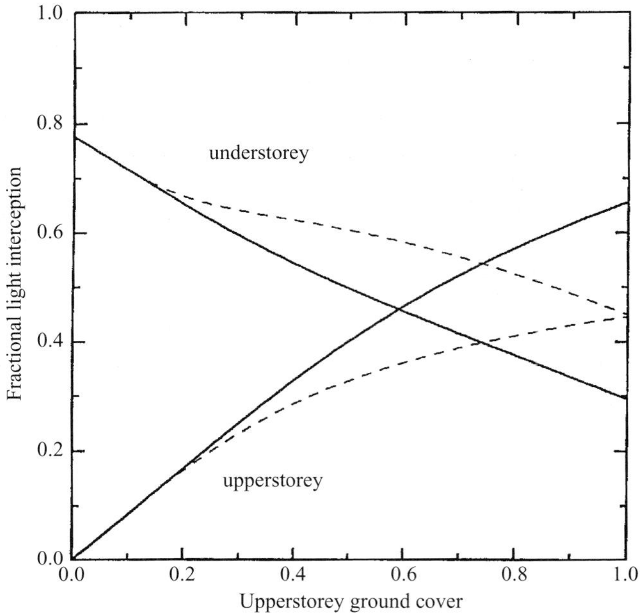

Figure 8.2 The effect of partial ground cover (——) and foliage clumping (– – –) in an upperstorey canopy on the fractional light interception of a two-species mixture. In this example, the upperstorey is twice the height of the understorey, which completely covers the ground and is not clumped. As in Figure 8.1, each species has the same leaf area index (3.0) and extinction coefficient (0.5) (from Wallace, 1997).

cepting the same amount of light as it would if the upperstorey was not present. In the example shown, once the taller upperstorey canopy ground cover is less than 0.6, its fractional light interception is less than the understorey, despite being taller and having the same leaf area index.

Within the ground area where the canopy is present, variations in leaf area density can be represented by a 'clumping factor', which effectively reduces the ability of each unit of leaf area to intercept light (e.g. see Cohen *et al.*, 1995). Wallace (1997) proposed a simple approach to this complex problem by allowing the component species extinction coefficients in the ERIN model to vary with the degree of clumping within the canopy:

$$K_c = K(1 - \delta) \tag{12}$$

where, δ is a canopy clumping coefficient. When there is no clumping within the area where the canopy covers the ground, $\delta = 0$ and $K_c = K$. When $\delta = 1$, the canopy would be 'infinitely' clumped and intercept no light at all; in which case $K_c = 0$. In canopies of intermediate clumping ($0 < \delta < 1$), the

ERIN model simply extrapolates linearly between these two extremes. Figure 8.2 also demonstrates the effect of clumping within a canopy. In this case, the understorey foliage is not clumped and the upperstorey is chosen to have a clumping coefficient of 0.5, which effectively halves its extinction coefficient. This reduces the light interception of the upperstorey at all degrees of ground cover. Conversely, the understorey light interception is increased when the upperstorey foliage is clumped and, in the example shown, even when both species have complete ground cover, the 'advantage' gained by the upperstorey by being twice the height of the understorey is lost when its foliage is clumped by the amount chosen here.

In mixed plant communities, competition for light is a key factor in determining the growth of individual species and, of course, the community as a whole. Wallace (1997) showed how the degree of competition for light by a species growing in a mixture can be quantified by defining a competitive ratio analogous to that used by Willey and Rao (1980) to compare crop yields in intercrops. In the case of light competition in a two-species mixture, the competitive ratio for Species 1 is:

$$\mathfrak{R}_1^L = \frac{f_1/f_1^d}{f_2/f_2^d} \tag{13}$$

and for Species 2 is:

$$\mathfrak{R}_2^L = \frac{f_2/f_2^d}{f_1/f_1^d} \tag{14}$$

These equations express the amount of light a species would intercept in a mixture compared to that which it would intercept on its own relative to the same ratio for its neighbouring species. For example, if two species each had fractional light interceptions of 0.5 when growing on their own and these reduced to 0.4 (Species 1) and 0.2 (Species 2) in a mixture (because say Species 1 was taller than Species 2), then using Equation (13), the competitive ratio for Species 1 would be 2.0. Note that $\mathfrak{R}_2^L$ is simply the reciprocal of $\mathfrak{R}_1^L$, i.e. 0.5 in this example.

Figure 8.3 shows how the values of $\mathfrak{R}_1^L$ change with leaf area of Species 1 and the heights of the two species. In this example, the leaf area of Species 2 is fixed at 1.0 and the extinction coefficient of both species is 0.5. Figure 8.3 demonstrates that, as expected, the competition for light by a species increases with its leaf area and height. It also shows that a high leaf area gives little competitive advantage, unless the foliage is carried at a greater height than the associate species. A major advantage of the formula presented by Wallace (1997) is that it allows light competition to be quantified. This means that within a given plant mixture a species can be said to be, for example, twice ($\mathfrak{R}^L = 2$) or half ($\mathfrak{R}^L = 0.5$) as competitive for light as another species.

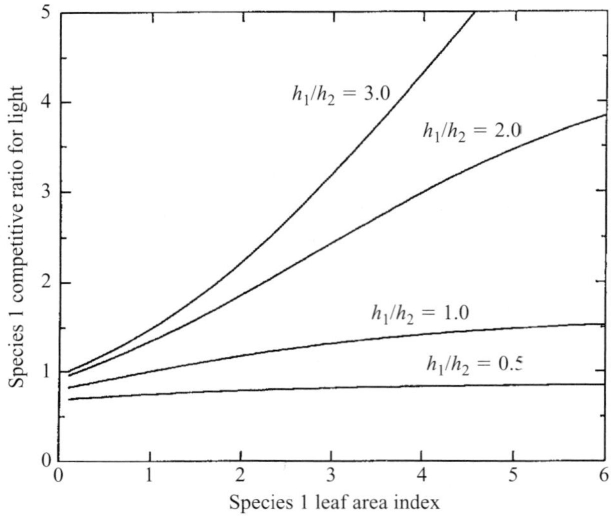

Figure 8.3 The variation in the competitive ratio for light with leaf area index and plant height (h_1 and h_2). The leaf area index of Species 2 is fixed at 1.0 and both species have the same extinction coefficient (0.5) (from Wallace, 1997).

This, therefore, provides the distinct advantage of being able to quantitatively compare light competition in different canopy arrangements.

8.3 Component and total fluxes of heat, water and carbon dioxide

The exchange of momentum, heat, water and CO_2 between vegetation and the atmosphere is normally calculated using what are now called Soil-Vegetation-Atmosphere Transfer schemes (SVATs). In these models, all individual fluxes of water and CO_2 are described, and these are controlled both by the rate of diffusion through the stomata of the vegetation and by diffusion within the plant canopy. The latter is based on diffusion or K-theory, where the flux rate can be described by the product of a concentration (water vapour, heat, CO_2, etc.) gradient and a diffusion coefficient, K_d (Goudriaan, 1977; Norman, 1979). Integration of the diffusion coefficient over a specified path allows the derivation of 'resistances', which can be used to specify fluxes within 'Ohm-type' resistance networks.

Although this approach works well for molecular flow, requiring molecular diffusivities, Raupach (1987) has demonstrated theoretically that the use of this concept to describe turbulent flow within plant canopies can sometimes give imprecise or even incorrect results. In particular, this was found to be the case for tall, sparse forests, where in-canopy 'counter-gradient' fluxes have been observed, which are difficult to describe using diffusion theory (Denmead and Bradley, 1985; Baldocchi and Meyers, 1988). Raupach (1987) developed an alternative Lagrangian theory, which was able to explain these counter-gradient fluxes. However, to predict fluxes this approach requires the source profile of the entity being transferred to be known within the canopy. More recently, McNaughton and Van den Hurk (1995) have combined the K-theory and Lagrangian approaches and have shown that a slightly modified K-theory can still be used for predicting in-canopy fluxes. Shuttleworth and Gurney (1990) also use K-theory within the plant canopy, but point out that it should be recognised that 'this theory is merely a working simplification of a more realistic but complex theoretical model'. In practice, most SVAT models use K-theory and, when compared with the more complex Lagrangian descriptions, K-theory often gives an adequate description of flux transfer, even in tall, sparse canopies (e.g. see Dolman and Wallace, 1991).

8.3.1 From single-source to multi-component models

The energy exchange between an extensive, homogeneous dense canopy and the atmosphere is relatively well understood. In this case, the evaporation can be calculated with the well-known Penman–Monteith equation (Monteith, 1965):

$$\lambda E = \frac{\Delta A + \rho_a c_p D_r / r_a}{\Delta + \gamma(1 + r_s / r_a)} \tag{15}$$

where: A is the available energy (W m^{-2}), i.e. the difference between net radiation, R_n, and soil heat flux, G (W m^{-2}); r_a is the aerodynamic resistance (s m^{-1}) and r_s (s m^{-1}) the surface resistance; D_r (kPa) is the vapour pressure deficit at reference level, z_r (m); Δ (kPa K^{-1}) is the slope of the vapour pressure temperature curve; γ is the psychrometric constant (kPa K^{-1}); ρ_a (kg m^{-3}) is the density of air; and c_p (J kg^{-1} K^{-1}) is the specific heat of air at constant pressure. If evaporation from the soil is assumed to be negligible, then r_s (s m^{-1}) can be equated to the bulk stomatal resistance of the leaves in the canopy. A schematic diagram of the resistance network for evaporation in the single-component approach is presented in Figure 8.4A.

When the surface is sparsely vegetated, a significant percentage of the surface consists of exposed undergrowth or bare soil, so a two-component or sparse canopy model is needed (Figure 8.4B). The single-component approach has, therefore, been extended to explicitly allow for fluxes of heat and water vapour from the soil or understorey (Deardorff, 1978; Shuttleworth and Wallace, 1985; Choudhury and Monteith, 1988; Noilhan and Planton, 1989; Kustas, 1990; Dolman, 1993; Huntingford *et al.*, 1995). Such models introduce more explicit descriptions of aerodynamic transfer through the leaf

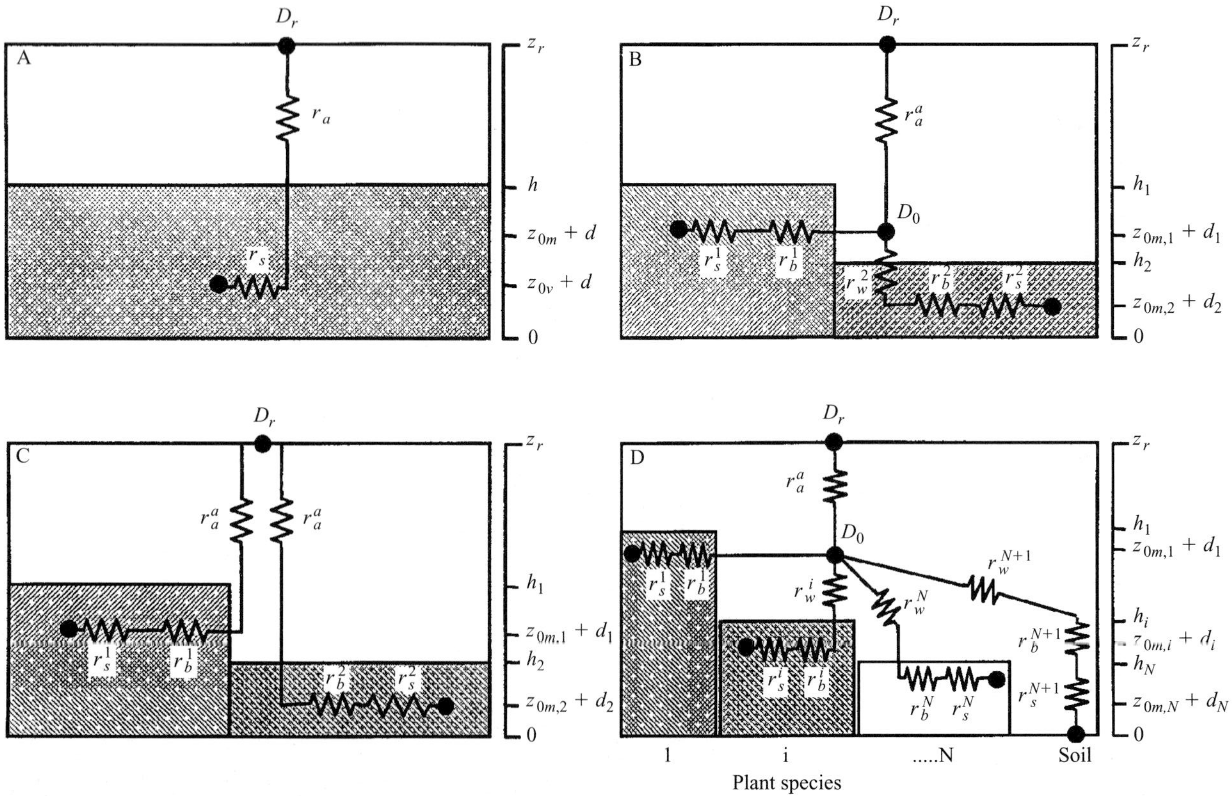

Figure 8.4 Schematic representation of four different soil-vegetation-atmosphere-transfer (SVAT) schemes: A) single source; B) interactive dual source; C) non-interactive dual source; and D) interactive multi-source. The symbols are explained in the text.

boundary layer (via r_b^1, r_b^2, etc.), and they have proved useful in describing a variety of sparse canopies, including crops and natural vegetation (Smith and Barrs, 1988; Lafleur and Rouse, 1990; Wallace *et al.*, 1990; Nichols, 1992; Blyth, 1995; Blyth and Harding; 1995; Van den Hurk *et al.*, 1995; Verhoef, 1995). In some of these two-component models, there is no interaction between the component fluxes as shown in Figure 8.4C, for example, one of the models considered by Huntingford and co-workers (1995). However, there are many situations where fluxes from the substrate can significantly affect the fluxes from the upperstorey canopy. For example, Wallace and co-workers (1990) showed that, in a sparse millet canopy, transpiration from the canopy could be changed by up to 40%, depending on the fluxes of heat and water vapour from the soil below. In tiger bush vegetation in the Sahel, Blyth and Harding (1995) found that fluxes from the vegetation strips exceeded the net radiation input. They concluded that additional energy was being transferred from the hot bare soil to the bushes. To describe the interactions between plants in a mixed canopy it is, therefore, essential to use a model which accounts for interactions between the component fluxes.

The most commonly used two-source model, which allows for interactions between the components, was described by Shuttleworth and Wallace (1985), and is illustrated in Figure 8.4B. Interactions between the upper canopy and the soil or understorey are accounted for by specifying an in-canopy mixing point, where the vapour pressure deficit is D_0 and air temperature is T_0. Fluxes of heat and water vapour from the two components can both affect the values of D_0 and T_0 and, therefore, this is the basic mechanism by which the interaction between the two sources occurs. In contrast, the single-source (Figure 8.4A) and non-interactive two-source (Figure 8.4C) models directly use the reference level vapour pressure deficit, D_r.

Figure 8.4B shows that in the model described by Shuttleworth and Wallace the upper- and understorey fluxes converge at the in-canopy mixing point, which is located at a height $z_1 = z_{0m,1} + d_1$, where $z_{0m,1}$ and d_1 are the roughness length and displacement height for the upper canopy (designated 1), after which the total flux moves into the air above the canopy via the aerodynamic resistance, r_a^a. Each component has its own aerodynamic (subscript b or w) and surface (subscript s) resistances. The aerodynamic resistances are the leaf boundary resistances, r_b^1 and r_b^2, and the within-canopy aerodynamic resistance, r_w^2, arising from transport through the air-layer between the source height of the understorey ($z_2 = z_{0m,2} + d_2$) and the in-canopy mixing point, z_1. If the understorey is bare soil, $d_2 = 0$. The calculation of aerodynamic parameters (such as z_{0m} and d) and resistances is discussed in Sections 8.3.3 and 8.3.4.

In mixed-plant communities, the surface may be composed of any number of plant species (N) and the soil. In order to describe the energy partitioning for such a system, the ERIN model (Wallace, 1997) has been developed by extending the two-component approach to a multi-component model. The resistance network for the ERIN model, with some greater detail of the aerodynamic resistances involved, is illustrated in Figure 8.4D. In the ERIN model, total evaporation from the $(N+1)$ components in the plant/soil community is given by:

$$\lambda E = \sum_{i=1}^{N+1} C_i PM_i \tag{16}$$

where,

$$PM_i = \frac{\Delta A + \{\rho_a c_p D_r - \Delta r_a^i (A - A_i)\}/(r_a^a + r_a^i)}{\Delta + \gamma\{1 + r_s^i/(r_a^a + r_a^i)\}} \tag{17}$$

and

$$C_i = \left(1 + \frac{\sum_{j_{j \neq i}}^{N+1} 1/R_j}{1/R_i + 1/R_a}\right)^{-1} \tag{18}$$

where, R_a is given by:

$$R_a = (\Delta + \gamma) r_a^a \tag{19}$$

and

$$R_i = (\Delta + \gamma) r_a^i + \gamma r_s^i \tag{20}$$

In these formulae, $r_a^i = r_b^i + r_w^i$ and note that $r_w^1 = 0$, since the in-canopy mixing point (D_0) is assumed to be immediately outside the boundary layer of the uppermost canopy (designated 1).

Using the light partitioning part of the ERIN model described in Section 8.2, the energy available for each species is given by:

$$A_i = f_i R_n \tag{21}$$

Hence, the remaining energy that is available for the soil component is:

$$A_{N+1} = \left(1 - \sum_{i=1}^{N} f_i\right) R_n - G \tag{22}$$

The total evaporation from a multi-species system can, therefore, be calculated from Equations (16) to (22), when only the above-canopy climatic conditions are known. It should be noted that the individual component species and soil evaporation rates λE_i are not given by the product $C_i PM_i$. To calculate each λE_i, it is necessary to use the in-canopy vapour pressure deficit, D_0, in the following form of the Penman–Monteith formula:

$$\lambda E_i = \frac{\Delta f_i R_n + \rho_a c_p D_0 / r_a^i}{\Delta + \gamma(1 + r_s^i / r_a^i)} \tag{23}$$

(Wallace, 1997), with D_0 found from:

$$D_0 = D_r + \{\Delta A - (\Delta + \gamma)\lambda E\} r_a^a / \rho_a c_p \tag{24}$$

At this same level, the in-canopy air temperature is found from:

$$T_0 = \{(A - \lambda E) r_a^a / \rho_a c_p\} + T_r \tag{25}$$

Sensible heat fluxes for each component are given by:

$$H_i = A_i - \lambda E_i \tag{26}$$

Note that in the ERIN model, if $(N + 1) = 2$, we obtain the original two-source model described by Shuttleworth and Wallace (1985); and if $(N + 1) = 1$, we obtain the well-known Penman–Monteith equation. It is also interesting to note that in the ERIN model, if we let $N \to \infty$ and make all the surface and aerodynamic resistances the same, we also obtain the Penman–Monteith equation. This is what would be expected for an infinite number of identical components.

Verhoef and Allen (1998, 1999) have used a multi-component model approach to describe the energy balance and CO_2 exchange of a dryland savannah consisting of four components (shrubs, forbs, grasses and bare soil). The main difference in their approach is that it was assumed that all four components receive the full radiation load, R_n (as opposed to $f_i R_n$ in the ERIN model). In the Verhoef and Allen model, the component contributions to the total evaporation are, therefore, dependent on their aerial cover, so λE becomes:

$$\lambda E = \sum_{i=1}^{N+1} \alpha_i PM_i C_i' \tag{27}$$

where, α_i is the fractional coverage for component i. In this formulation, the coefficients, C_i' are different from those specified in Equation (18), since they are affected by ground cover, α_i. The formulae for C_i' in a two-component and a four-component system are given by Huntingford and co-workers (1995) and Verhoef and Allen (1998, 1999), respectively.

This type of model, where each component receives full radiation, works well for relatively sparsely vegetated surfaces (e.g. where $\alpha_i \leqslant 0.3$). A two-component version of the above approach, but without CO_2 exchange, has also been used by a number of authors to describe the energy balance of sparse Sahelian vegetation (Blyth and Harding, 1995; Huntingford *et al.*, 1995; Verhoef, 1995).

The ERIN model also allows for fractional coverage by the various component species, but this is achieved through partitioning the radiation, so that each component receives less than the full radiation load. There is then no subsequent need to introduce partial cover factors (α_i) in the flux equations. In some cases, a mixture of these approaches has been used (Allen *et al.*, 1997; Hall *et al.*, 1998), but care is needed to avoid the mistake of accounting for partial ground cover more than once.

8.3.2 Surface resistances

The surface resistance of a plant canopy is related to the stomatal resistance of the individual leaves, r_l, within it; the reciprocal of r_l is known as the leaf conductance, g_l. When the entire canopy is made up of a single species, its surface resistance is given by the leaf area weighted sum of the resistances (r_l^j) of the individual (m) layers in the canopy:

$$r_s = \sum_{j=1}^{m} r_l^j / L_j \tag{28}$$

When there are multiple species in a canopy, the surface resistance of each component, r_s^i, can be similarly calculated if the leaf resistances and leaf areas of each component are known. This procedure for calculating r_s has been used in single species canopies by a number of authors (e.g. Szeicz and Long, 1969; Wallace *et al.*, 1990; Baldocchi *et al.*, 1991; Rochette *et al.*, 1991).

Leaf resistances can be measured directly using porometry (Parkinson, 1985), or estimated from the known response of r_l to environmental variables, such as radiation, air temperature and humidity, and CO_2 concentration. The latter entails fitting a suitable model to the measured leaf resistances, thus yielding a set of parameters representative of a specific canopy. A commonly used approach is the empirical model of Jarvis (1976), later modified by Stewart (1988), and thereafter referred to as the Jarvis–Stewart model. Recent developments, however, use a physiological approach based on the concept of stomatal resistance controlling both the CO_2 assimilation rate and the transpiration rate (Jacobs, 1994; Friend and Cox, 1995; Cox *et al.*; 1998; Verhoef and Allen, 1999).

Sparse canopy and multi-component models also require a knowledge of the soil surface resistance, which describes the resistance to water vapour diffusion from the evaporation 'front' within the soil, where the phase change from liquid to vapour takes place, to the soil surface. The evaporation front will be at or close to the surface for a wet soil. For a dry soil, evaporation will occur from wet soil below a progressively deepening dry layer (Monteith, 1981; Van de Griend and Owe, 1994). With most of the world's vegetation having only partial ground cover, the formulation of surface resistance to soil evaporation, therefore, plays an important role in present-day SVATs. Soil surface resistance can be measured directly with a so-called fast air circulation chamber (Kohsiek, 1981), or by inverse calculation if the evaporative flux is measured directly (Massman, 1992; Verhoef, 1995). When used in a predictive mode, soil surface resistances are usually related to the moisture content of the surface soil (Katerji and Pereira, 1985; Camillo and Gurney, 1986; Van de Griend and Owe, 1994), or sometimes to wind speed (Kondo *et al.*, 1990).

Often, the understorey of a sparse canopy is not just bare soil but an intimate mixture of soil and several undergrowth species. For example, weed species in an agricultural system, grasses and/or forbs in a silvopastoral system. When using a two-component model to describe the evaporation of these surfaces, an effective surface resistance has to be assigned to the complete understorey (e.g. Dolman, 1993; Huntingford *et al.*, 1995; Verhoef, 1995). However, these effective understorey resistances can be misleading, since they are a complex combination of the physical resistance of the soil surface and the physiological resistance of the understorey vegetation. For example, Huntingford and co-workers (1995) found some unrealistic model predictions when they tried to find the stomatal response of the understorey herb layer to solar radiation by inverse modelling using a two-component model. A multi-component model, such as that used in the ERIN model or the one described by Verhoef and Allen (1998, 1999), avoids this problem since it is not necessary to lump the various understorey components together.

It should be noted that under normal conditions surface resistances are usually of the order of 50–300 $\mathrm{s\,m^{-1}}$ and are generally much larger than the aerodynamic resistances (see later for typical values). This means that under dry canopy conditions the net fluxes of water vapour are often dominated by the physiological controls of the leaf stomata.

8.3.3 Aerodynamic resistances

In most SVAT models, fluxes are transferred across a series of aerodynamic resistances, such as those shown in Figure 8.4. This section describes the derivation of the aerodynamic resistance associated with the simplest single source model, i.e. r_a in Figure 8.4A. It then describes how the three most commonly used resistances in multi-source models (r_b^i, r_w^i, r_a^a, Figure 8.4D) are calculated. Finally, it shows how r_a^a can be represented by three components (r_a^I, r_a^R, r_a^c, Figure 8.5) to allow for the enhancement of turbulence in the roughness sublayer. Furthermore, introduction of the roughness sublayer affects the value of r_w^i, therefore a method of altering the formula to account for this is also described.

In the single-component description of latent heat flux (see Figure 8.4A), the aerodynamic transfer resistance, r_a, can be considered as having two components. The first describes the resistance of the air between the mean

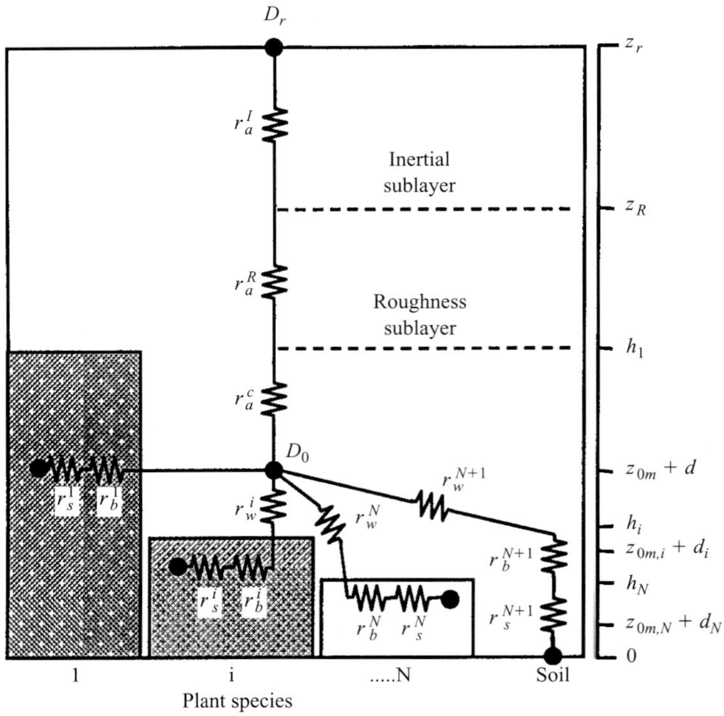

Figure 8.5 Schematic representation of an interactive multi-source soil-vegetation-atmosphere-transfer (SVAT) model, which includes the roughness and inertial sublayers above the canopy. The symbols are explained in the text.

canopy source height, located at $z_{0m} + d$, and the reference (measurement) height, z_r. Estimates of this resistance require values of the momentum roughness length, z_{0m}, and displacement height, d. The second component is referred to as the 'excess resistance'. This is the extra resistance experienced by the latent (and sensible) heat fluxes, as compared to transport of momentum, due to the difference in the location of the sink for momentum ($z_{0m} + d$) and the source of heat and water vapour ($z_{0v} + d$) (Figure 8.4A). It is usually assumed that the roughness length for heat and water vapour are the same (i.e. $z_{0v} = z_{0h}$), in which case the total aerodynamic resistance, r_a, is given by:

$$r_a = \frac{1}{ku_*}\left[\ln\left(\frac{z_r - d}{z_{0m}}\right) - \Psi_h\left(\frac{z_r - d}{\Lambda}\right)\right] + \frac{1}{ku_*}\ln\left(\frac{z_{0m}}{z_{0v}}\right) \qquad (29)$$

In this equation, u_* is the friction velocity, k the Von Karman constant

(= 0.40), Ψ_h the integrated stability function for heat (Paulson, 1970), and Λ the Monin–Obukhov length (m).

The friction velocity, u_*, can be found from (e.g. Tennekes, 1973):

$$u_* = ku_r\left[\ln\left(\frac{z_r - d}{z_{0m}}\right) - \Psi_m\left(\frac{z_r - d}{\Lambda}\right)\right]^{-1} \tag{30}$$

where, u_r is the wind speed at reference height, and Ψ_m the integrated stability function for momentum. Many authors ignore the stability correction, in which case they are assuming neutral stability, with $\Psi_h = \Psi_m = 0$.

In models describing two or more components (Figures 8.4B, C, D and 8.5), the first resistance outside the stomata is the leaf boundary layer resistance, r_b. Transfer across leaf boundary layers is generally described by non-dimensional expressions which reflect largely empirical adaptations of engineering models derived for flat plates (Schuepp, 1993). The leaf boundary layer resistance is usually dependent on leaf dimensions, most commonly leaf width (here denoted as w_i), and wind speed, u_i, at the source level height of the vegetation layer (i) in question:

$$r_b^i = a\sqrt{\frac{w_i}{u_i}} \tag{31}$$

where $i = 1, N$, and a is a constant having a value of around 100 $s^{-0.5}$ m^{-1} (Choudhury and Monteith, 1988; Huntingford *et al.*, 1995). To use Equation (31), it is necessary to calculate the in-canopy wind speed, u_i. To do this, most SVAT models use K-theory in combination with the mixing length approach (see Raupach and Thom, 1981). In this case, the mean in-canopy wind speed decays exponentially from its value u_c ($m\,s^{-1}$) at the top of the canopy, giving:

$$u_i = u_c e^{n(z_i/h_1 - 1)} \tag{32}$$

where, $z_i = z_{0m,i} + d_i$ ($i = 1, N$). In Equation (32), n is the dimensionless extinction coefficient, and a value of 2.5 is often adopted since most values for closed, dense canopies lie between 2 and 3 (Shuttleworth and Wallace, 1985; Choudhury and Monteith, 1988). However, in practice, n is a function of the plant area index, P^*, the projected area of total plant material per unit ground area (Pereira and Shaw, 1980):

$$n = \frac{c_d P^*}{2k^2} \tag{33}$$

where, c_d (= 0.2) is a drag coefficient. This means that the value of n decreases as a canopy becomes sparser. In practice, if P^* is not known it can often be approximated using the canopy leaf area index.

The boundary layer resistances given above (r_b^i) are for single leaves and need to be divided by the appropriate leaf area index, L_i, to obtain the bulk boundary layer resistance of the entire canopy. However, it is important to note that in models using Equation (27) (e.g. Verhoef and Allen, 1998, 1999) it is necessary to use the 'local' leaf area index, L_i^P (i.e. L_i/α_i) to scale up from leaf to canopy level. The same is true for scaling up from r_l^i to r_s^i. The soil boundary layer resistance (r_b^{N+1}) has been given a value of 50 s m^{-1}, following Hall *et al.* (1998).

After traversing the leaf boundary layer, understorey fluxes encounter the within-canopy aerodynamic resistances, r_w^i (Figures 8.4D and 8.5). These are calculated by integrating the reciprocal of the appropriate eddy diffusivity, K_d, between the two heights over which the resistance applies. For example, for r_w^i in Figure 8.4D this means integration between levels $z_{0m,i} + d_i$ and z_1. To specify K_d within the canopy, it is first necessary to know its value at the top of the tallest canopy (Species 1). Assuming K-theory applies directly above the canopy, the diffusivities for momentum, heat and water vapour transport are given by:

$$K_{d_{m,h,v}}(z) = ku_*(z - d)\Phi_{m,h,v}^{-1} \tag{34}$$

Therefore, Equation (34) can be used to specify K_d at the top of the tallest canopy, i.e. $K_d(h_1)$. Now, within the canopy a good approximation to $K_d(z)$ in the top 2/3 of the canopy is given by:

$$K_d(z) = K_d(h_1)e^{n(z/h_1 - 1)} \tag{35}$$

(e.g. Huntingford *et al.*, 1995). This equation is used to calculate the within-canopy resistances, r_w^i, i.e. integrating $1/K_d(z)$ over the height ranges $z_{0m,i} + d_i$ to $z_{0m,1} + d_1$ gives:

$$r_w^i = \int_{z_i}^{z_1} \frac{1}{K_d(z)} dz = \frac{e^n h_1}{nK_d(h_1)}[e^{-nz_i/h_i} - e^{-nz_1/h_1}] \tag{36}$$

with $i = 2, N + 1$.

The final aerodynamic resistance between the in-canopy mixing point (at D_0) and the reference level, r_a^a, is obtained by a further integration of $1/K_d(z)$ over the height ranges $z_{0m,1} + d_1$ to z_r, (see Figure 8.4B and D):

$$r_a^a = \frac{1}{ku_*}\left[\ln\frac{z_r - d}{h_1 - d} - \Psi_h(z_r) + \Psi_h(h_1)\right] + \frac{h_1}{nK_d(h_1)}[e^{n(1-z_1/h_1)} - 1] \tag{37}$$

(e.g. Huntingford *et al.*, 1995; Verhoef and Allen, 1998, 1999), with d the zero plane displacement for the total surface (discussed in Section 8.3.5) and $\Psi_h(z)$ the integrated stability correction to eddy diffusivity for heat and water vapour at height z. Equation (37) shows that the resistance, r_a^a, consists of two terms. The first term is obtained by integrating the reciprocal of Equation (34) above the top of the canopy (i.e. $h_1 \leqslant z \leqslant z_r$) and the second term is found from integrating the reciprocal of Equation (35) within the canopy (i.e. between $z_1 \leqslant z \leqslant h_1$).

8.3.4 *Roughness sublayer effects*

In theory, conventional K-theory is only valid in the inertial sublayer above the roughness sublayer (Figure 8.5). The roughness sublayer is in contact at its lower boundary with the vegetation and it can extend upwards to between two and three times the canopy height (Raupach *et al.*, 1980). Within the roughness sublayer, the form of the diffusivities is different. There is significant experimental evidence (e.g. Raupach, 1979; Raupach *et al.*, 1980; Cellier and Brunet, 1992) that within this sublayer momentum transfer is slightly enhanced but that heat and water vapour transfer may be enhanced by a factor of 2 or 3 over that predicted by conventional K-theory, i.e. Equation (34). Raupach and co-workers (1996) argue that this can be explained as the result of coherent eddies associated with the inflection in the wind profile that occurs at the top of the canopy.

Hall and co-workers (1998) have taken the existence of the roughness sublayer into account by splitting the part of the resistance, r_a^a, between h_1 and z_r, as given in the first term of Equation (37), into two parts. This is illustrated in Figure 8.5, where there is a roughness sublayer aerodynamic resistance between h_1 and z_R, i.e. r_a^R, and an inertial sublayer aerodynamic resistance between z_R and z_r, i.e. r_a^I. Hall and co-workers (1998) used the K-theory approach of Cellier and Brunet (1992), who proposed the simple pragmatic alternative of keeping heat and water vapour diffusivities within the roughness sublayer constant and equal to their value at z_R:

$$K'_{d\,h,v}(z) = K_{d\,h,v}(z_R) = ku_*(z_R - d)\Phi_{h,v}^{-1}(z) \tag{38}$$

where, the prime denotes the roughness sublayer. Now, integrating $1/K'_{d\,h,v}$ between h_1 and z_R gives:

$$r_a^R = \frac{z_R - h_1}{ku_*(z_R - d)} - \Psi_{h,v}^*(z_R) + \Psi_{h,v}^*(h_1) \tag{39}$$

where, $\psi_{h,v}^*(z_R)$ and $\psi_{h,v}^*(h_1)$ are the integrated stability corrections for the roughness sublayer, at z_R and h_1, respectively, given by Cellier and Brunet (1992) as:

$$\Psi_{h,v}(z) = \left(\frac{z - d}{z_R - d}\right)\left(\frac{\Phi_{h,v}^{-1} - 1}{\Phi_{h,v}^{-1} + 1}\right) \tag{40}$$

Integrating the reciprocal of Equation (34) between z_R and z_r gives the aerodynamic resistance within the inertial sublayer, r_a^I, as:

$$r_a^I = \frac{1}{ku_*}\left[\ln\frac{z_r - d}{z_R - d} - \Psi_{h,v}(z_r) + \Psi_{h,v}(z_R)\right] \quad (41)$$

Finally, in this approach, there is the resistance to transfer between the in-canopy source height and the bottom of the roughness sublayer (= top of the main canopy, Species 1), designated r_a^c in Figure 8.5. Note that following Hall and co-workers (1998), it is assumed that the in-canopy mixing-height is not located at $z_1 = z_{0m,1} + d_1$, as in the approach described by Huntingford and co-workers (1995), but rather at $z_{0m} + d$, which varies with the element density of the dominant species and the roughness of the understorey (see below). Again using Equation (38) and integrating between the appropriate heights gives r_a^c as:

$$r_a^c = \frac{h_1 \sinh(n)}{nK'_{h,v}(h_1)}\left\{\ln\left(\frac{e^n - 1}{e^n + 1}\right) - \ln\left(\frac{e^{n(z_{0m}+d)/h_1} - 1}{e^{n(z_{0m}+d)/h_1} + 1}\right)\right\} \quad (42)$$

Note that the hyperbolic sine in Equation (42) is the result of using Equation (33) to calculate n, rather than assuming a value $n = 2.5$ (see Hall *et al.*, 1998).

When the existence of the roughness sublayer and a variable value of n are taken into account, the values of the within-canopy resistances (r_w^i) need to be recalculated. Again, this is done by integration of $1/K'_{d_{h,v}}$, as specified in Equation (38), but this time between ranges $z_{0m,i} + d_i$ and $z_{0m} + d$ (instead of $z_{0m,1} + d_1$). This gives the expression:

$$r_w^i = \frac{h_1 \sinh(n)}{nK'_{h,v}(h_1)}\left\{\ln\left(\frac{e^{n(z_{0m}+d)/h_1} - 1}{e^{n(z_{0m}+d)/h_1} + 1}\right) - \ln\left(\frac{e^{n(z_{0m,i}+d_i)/h_1} - 1}{e^{n(z_{0m,i}+d_i)/h_1} + 1}\right)\right\} \quad (43)$$

for $i = 2, N + 1$ (Hall *et al.*, 1998).

8.3.5 Estimation of aerodynamic properties

The equations described above require values for the momentum roughness parameters, z_{0m} and d, for the total surface and for the individual components, $z_{0m,i}$ and d_i. When dealing with sparse canopies or canopy mixtures where the component species may be dispersed, it is important to allow for variations in z_{0m} and d with canopy density. To account for this, the drag partition model described by Raupach (1994), as modified by Verhoef and co-workers (1997), has been used. In this specification, d is given by:

$$d = h_1\left(-\left(\frac{1 - e^{-\sqrt{2c_{d1}\delta_r}}}{\sqrt{2c_{d1}\delta_r}}\right) + 1\right) \quad (44)$$

where, the roughness density parameter $\delta_r = 0.5L_1$. The value of c_{d1}, which is a free parameter, is given below. The roughness length for the total surface is found from:

$$z_{0m} = h_1\left(\left(1 - \frac{d}{h_1}\right)\mathrm{e}^{-k\Gamma + Y}\right) \tag{45}$$

where, $\Gamma = (C_S + \lambda C_R)$. C_S is the drag coefficient for the substrate free of obstructions (when $\delta_r = 0$), and C_R is the drag coefficient of an isolated, surface-mounted roughness element. Y is the vegetation influence function, which is given by:

$$Y = \ln(c_w) - 1 + c_w^{-1} \tag{46}$$

with $c_w = 2.0$. Verhoef and co-workers (1997) found that the best estimates for c_{d1} and C_R, which apply to a wide range of canopy densities, are 20.6 and 0.35, respectively. When the substrate beneath the main canopy is bare soil, C_S has a value of 0.003 (Raupach, 1992). When the substrate is a complete cover of vegetation, the value of C_S has arbitrarily been set to 0.015 (Verhoef *et al.*, 1997). When the substrate is a mixture of N–1 species each with a fractional ground cover of α_i and soil with a cover of α_{N+1}, the average value of C_S for the complete substrate has been calculated from:

$$C_S = \alpha_{N+1}0.003 + \sum_{i=2}^{N}\alpha_i 0.015 \tag{47}$$

The roughness length and displacement height of the individual components of the mixed canopy, i.e. $z_{0m,1}$ and d_i, have been estimated as 13 and 66% of the species heights (h_i), respectively. This simple formula has been suggested by Monteith (1973) and by Brutsaert (1982), for mature agricultural crops and other dense canopies. For the bare soil, $z_{0m,N+1} = 0.01$ m and $d_{N+1} = 0$.

To calculate r_a^R, Equation (39), the height of the roughness sublayer, z_R, is needed, and can be estimated with several formulae (see Verhoef *et al.*, 1997). Some formulae specify z_R as a function of vegetation height, z_{0m} and d, or spacing between the canopy elements. Here, the more common, simple approach of setting $z_R = 2.5h_1$ is used. This constant value for z_R is based on a wind-tunnel study by Raupach and co-workers (1980), who observed that z_R was approximately 2–2.5 times the obstacle height, independent of the obstacle density.

The relative sizes of the different aerodynamic resistances in the network shown in Figure 8.5 are given in Figure 8.6 for the canopy and climatic conditions specified in Table 8.1 and Section 8.4. The largest resistances are found closest to the canopy, within or close to the leaf boundary layer. Transfer resistances to the in-canopy mixing point are smaller but still significant. This means that these resistances are important in determining the microclimate at the in-canopy mixing point. However, it should be

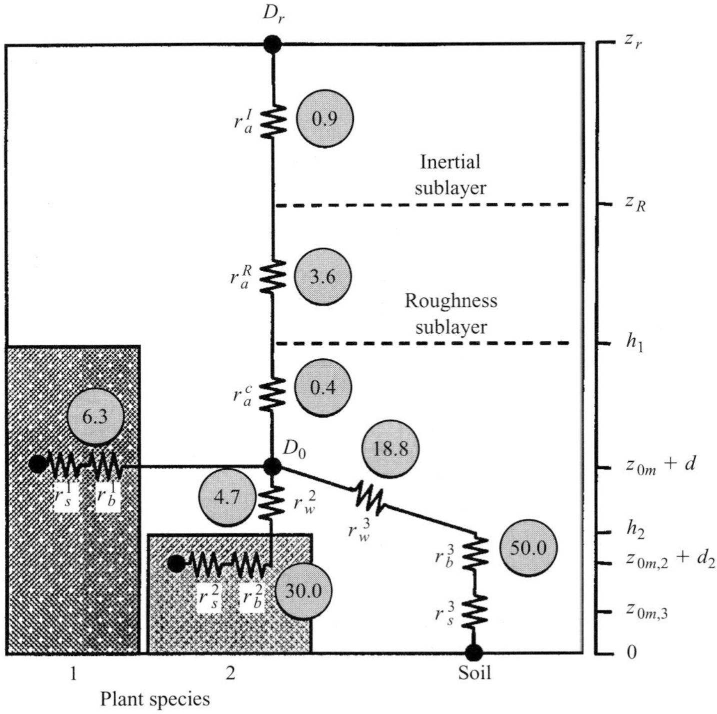

Figure 8.6 Typical values of the aerodynamic resistances ($\mathrm{s\,m^{-1}}$) associated with the resistance network shown in Figure 8.5. The canopy and climatic conditions used to derive these resistances are given in Section 8.4.

Table 8.1 Upperstorey and understorey parameter values used in Figures 8.6 and 8.8.

Parameter	Symbol	Upperstorey	Understorey
Canopy height (m)	h_i	4	1
Canopy thickness (m)	t_i	3	1
Local leaf area index	L_i^P	3	1
Ground cover	α_i	0–1	0.5
Extinction coefficient	K_i	0.5	0.5
Leaf dimension (m)	w_i	0.02	0.03
Species type		C_3	C_3

recognised that the specification of above and within canopy aerodynamic transfer is currently highly uncertain. This can be seen in Figure 8.7, which reproduces the very different values of r_a^a, r_w^2 and r_b^1 reported by Verhoef (1995) from a range of different two-component SVAT models. In some cases the differences can be greater than a factor of four.

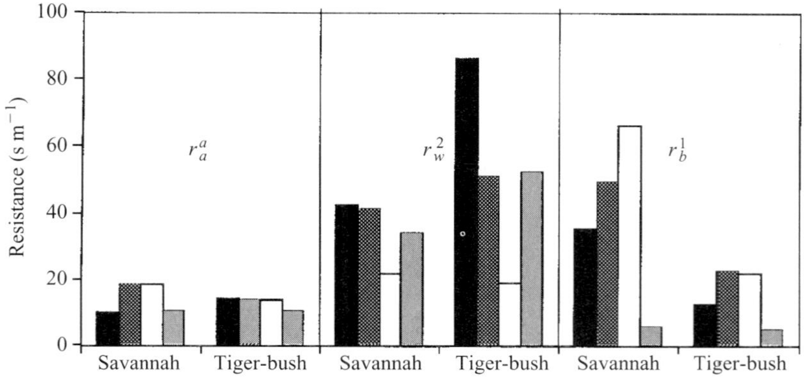

Figure 8.7 The different values of the aerodynamic (r_a^a and r_w^2) and boundary layer (r_b^1) resistances for savannah and tiger bush vegetation as given by four dual source modelling schemes: ■, Shuttleworth and Wallace (1985); ▒, Choudhury and Monteith, (1988); □, Deardorff (1978); and ▒, Huntingford *et al.* (1995) (from Verhoef, 1995).

8.3.6 Photosynthesis and respiration

Only relatively recently have SVAT models estimating latent (and sensible) heat fluxes been extended to include CO_2 exchange between the surface and the atmosphere. Various examples of single-source models of this type have been described (e.g. Woodward *et al.*, 1995; Calvet *et al.*, 1998; Cox *et al.*, 1998). In more complex models, vertical or horizontal heterogeneity in the canopy structure has been explicitly represented. Of these, the first group consists of models featuring many canopy layers, which differ mainly in the amount of radiation they receive, allowing the behaviour of a relatively closed, tall canopy, such as a forest, to be described. Examples of SVATs describing such one-dimensional, multi-layer systems can be found in reports by Sellers and co-workers (1992) and Baldocchi and Harley (1995). In a second group, where the canopy to be modelled is more spatially heterogeneous, a multi-source description is used, for example, a number of two-component models can be found in the literature (e.g. Chen *et al.*, 1996; Olioso *et al.*, 1996). Multi-component descriptions are given by Mobbs and co-workers (1998) for an agroforestry system and by Verhoef and Allen (1998, 1999) for savannah. The latter authors describe an approach which simultaneously links the energy balance, net CO_2 exchange and surface temperature, using a physiological description of leaf stomatal conductance (Jacobs *et al.*, 1996). This is the approach used here and described briefly in the following section.

To calculate leaf stomatal resistance, r_l^i, Jacobs and co-workers (1996) use the diffusion equation relating the net photosynthetic rate of plants, P_n^i (mg m^{-2} leaf s^{-1}) to their leaf stomatal resistance, r_l^i:

$$r_l^i = (C_s^i - C_i^i)/1.6P_n^i \tag{48}$$

Here, C_s^i and C_i^i (mg m^{-3}) are the leaf surface and internal CO_2 concentration for vegetation component, *i*, respectively. The net photosynthetic rate of each species, P_n^i, is given by:

$$P_n^i = (f_\theta^i P_{max}^i + R_D^i)(1 - \mathrm{e}^{-\varepsilon^i I_a^i/(f_\theta^i P_{max}^i + R_D^i)}) - R_D^i \tag{49}$$

where, P_{max}^i (mg m^{-2} leaf s^{-1}) is the photosynthetic rate at saturation light intensity, R_d^i is the dark respiration, I_a^i (W m^{-2}) is the absorbed PAR, and ε^i is the initial quantum use efficiency (mg J^{-1} PAR^{-1}). Jacobs and co-workers (1996) did not account for the effects of soil moisture status on leaf stomatal resistance, r_l^i. To make some allowance for this in the simplest way possible, a linear multiplication factor, f_θ^i, has been added to P_{max}^i; this has the effect of reducing both P_n^i and r_l^i as the supply of water in the soil is reduced. The rate of decline of P_n^i and r_l^i with root zone soil moisture will be determined by the form of the relationship between f_θ^i and soil moisture content.

The component species leaf stomatal resistances (r_l^i) are also dependent on surface temperature, T_s^i (°C), and surface vapour pressure deficit, D_s^i (kPa), as described previously (Jacobs *et al.*, 1996; Verhoef and Allen, 1999). Eight species-dependent parameters are needed to calculate P_n^i and hence r_l^i. For the calculations described later in Section 8.4, the parameters given by Jacobs (1994), which are representative of a typical C_3 species, have been used.

The values of CO_2 concentration, temperature and vapour pressure deficit at the surface of each component of a multi-species mixture are given by:

$$C_s^i = C_0 - 1.4(r_b^i + r_w^i)L_i P_n^i \tag{50}$$

$$T_s^i = T_0 + ((A_i - \lambda E_i)(r_b^i + r_w^i))/\rho c_p \tag{51}$$

and

$$D_s^i = (e_s(T_s^i) - e_0)\Big/\left(1 + \frac{(r_b^i + r_w^i)}{r_s^i}\right) \tag{52}$$

where, $e_s(T_s^i)$ is the saturated vapour pressure (kPa) at surface temperature and $e_0 = e_s(T_0) - D_0$. The total net CO_2 flux is then calculated using:

$$F_c = \sum_{i=1}^{N} L_i P_n^i - R_{soil} \tag{53}$$

In Equation (53), R_{soil} is the total respiration of soil and roots (mg m^{-2} s^{-1}), which is given by:

$$R_{soil} = a' L_t \mathrm{e}^{(b' T_s^{N+1})} \tag{54}$$

where, T_s^{N+1} is the soil surface temperature (°C), L_t the total leaf area index of all species (i.e. $L_t = \Sigma L_i$, m^2 leaf m^{-2} ground) and a' and b' are constants. With F_c obtained from Equation (53), the CO_2 concentration at canopy source height, C_0 (mg m^{-3}) can be calculated using:

$$C_0 = C_r - 1.4(r_a^c + r_a^R + r_a^I)F_c \tag{55}$$

where the factor 1.4 accounts for the difference between boundary layer resistances for CO_2 and water vapour. Here, C_r is the atmospheric CO_2 concentration (mg m^{-3}) at reference level.

The above description of leaf photosynthesis and leaf stomatal resistance is coupled to the ERIN model via the specification of absorbed PAR per unit leaf area, I_a^i, which is given by:

$$I_a^i = f_i PAR / L_i \tag{56}$$

with the values of f_i given by the formulae in Section 8.2. PAR is taken as half the incident solar radiation. The CO_2 fluxes are also linked to the ERIN model, which defines the component surface temperatures. Equations (48) and (49) allow the individual component photosynthetic rates and leaf stomatal conductances (P_n^i and r_l^i) to be calculated. The influence of these on the CO_2 concentration around the neighbouring species can be calculated using Equation (55). The effect of soil CO_2 fluxes on the in-canopy CO_2 concentration can also be calculated. Some of these effects are illustrated in the following section.

8.4 Microclimatic interactions

Microclimatic interactions affecting light, temperature, humidity, wind speed and CO_2 concentration are the key processes by which individual plants within a mixture can affect their neighbours. The original ERIN model, along with the modifications to the aerodynamic transfer and the inclusion of CO_2 fluxes described here, can be used to calculate these microclimate interactions in mixed plant communities containing any number (N) of species. The model can be used for intimate mixtures of plant species where the heights of the plants are similar, e.g. in a grass sward. However, to demonstrate some of the key features of the model, interactions in a two-species mixture where one canopy tends to dominate the other will be described. In this arrangement, there are two primary forms of interaction: firstly, the effects of the upperstorey on the understorey; and, secondly, the effects of the understorey on the upperstorey. These are dealt with separately below.

8.4.1 Upperstorey effects on an understorey canopy

The most obvious type of microclimatic interaction is where a dominant upperstorey vegetation shades an understorey canopy. This is the case in natural vegetation systems, such as savannahs, and in man-made systems, such as agroforestry, some intercropping mixtures (e.g. millet and cowpea), or where weeds grow under a sole crop. The light interactions in this type of mixed canopy have been described in Section 8.2; however, there are also other important microclimatic implications, which are described in more detail here. For example, the upperstorey can change the temperature and humidity around the understorey. It can also alter the understorey temperature and the temperature of the soil. These changes in understorey and soil temperature can affect temperature-dependent processes, such as plant development and soil respiration, as well as the transfer of heat and water vapour back to the upperstorey.

Figure 8.8 shows some calculations made using the multi-species model described here to simulate a mixture of two species where the upperstorey canopy foliage is entirely above that of the understorey vegetation. The parameters used for the upperstorey and understorey are given in Table 8.1 and were chosen to represent a mixture of woody trees or bushes with a herbaceous understorey. This situation is common in natural vegetation, such as savannah, and in man-made arrangements, such as agroforestry. It should be noted that, in the following example when the cover of the upperstorey (α_1) is varied the local leaf area index, L_l^P, is kept constant. This means that the leaf area index per unit ground area, L_l, will vary from 0 (when $\alpha_1 = 0$) to 3 (when $\alpha_1 = 1$), see Equation (9). This is different to the light interception example shown earlier in Figure 8.2, where the amount of upperstorey canopy was kept constant as ground cover was varied from 0 to 1. In this case L_1 remained constant at a value of 3 and L_1^P varied with cover.

The meteorological conditions at the reference height (12 m) were taken as representative of midday, i.e. net radiation = 500 W m^{-2}, solar radiation = 835 W m^{-2}, temperature = 25°C, vapour pressure deficit = 1.5 kPa, wind speed = 4 m s^{-1}, and CO_2 concentration = 655 mg m^{-3}. The soil surface was chosen to be relatively dry with a surface resistance of 2000 s m^{-1}. Parameter f_θ^i was kept constant at 1.0 for both the upper and understorey, i.e. there was no soil water stress in this example.

Figure 8.8A shows how the in-canopy microclimate changes as the cover of the upperstorey increases. The in-canopy air temperature, to which the understorey is exposed, decreases by just over 1 K as the upperstorey cover changes from 0 to 1. At the same time, vapour pressure deficit around the understorey leaves also decreases by approximately 0.4 kPa or 25%. Figure 8.8A also shows that the in-canopy CO_2 concentration decreases as cover increases, but only by approximately 20 mg m^{-3} or 3% in this example. This relatively small change in in-canopy CO_2 concentration concurs with the calculations made by Huntingford and co-workers (1998), who concluded that CO_2 exchange between the soil and a canopy had a small effect on canopy photosynthesis. The changes in temperature and vapour pressure

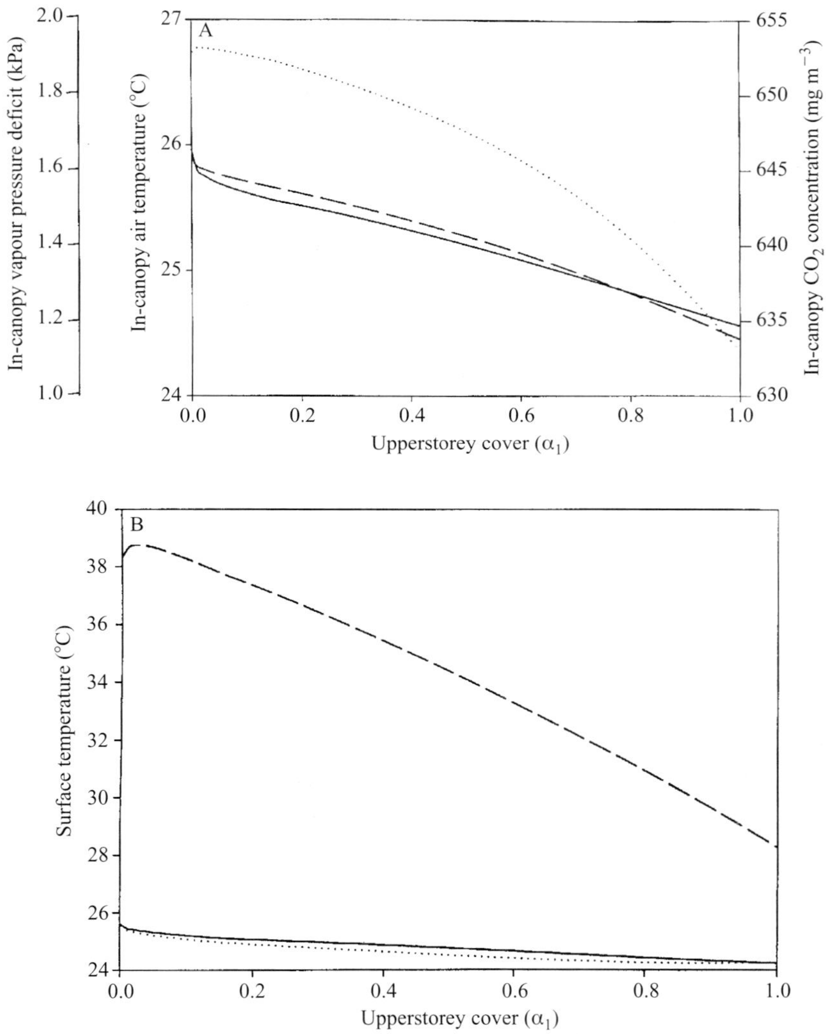

Figure 8.8 Variation of: A) in-canopy microclimate, i.e. canopy temperature (T_0; ——), vapour pressure deficit (D_0; — — —) and CO_2 concentration (C_0; · · · · · ·); and B) the temperatures of the soil (– – –), understorey (· · · · ·) and upperstorey (——) with upperstorey ground cover (α_1) in a two-species mixture with a dominant upperstorey. The meteorological and canopy conditions used for these calculations are described in Section 8.4.

deficit predicted here are similar to the microclimate modifications close to windbreaks reported by Brenner (1991). He quotes data from his own field study in Niger and from studies by other authors showing temperature decreases near to trees of 1–2 K and vapour pressure deficit decreases of 0.1–0.2 kPa. These changes in the temperature, humidity and CO_2 concentration around the understorey will affect its physiological behaviour. These are discussed in more detail in Section 8.5.

The effect of increasing the upperstorey cover on the temperature of the vegetation below and of the soil is shown in Figure 8.8B. The most marked effect of increased upperstorey cover is the decrease in the temperature of the soil surface of approximately 10 K in this simulation. The temperature of the leaves of both the understorey and upperstorey vegetation also decreases, but only by around 1.5 K. In a dryland savannah in Niger, West Africa, Verhoef (1995) found that the upperstorey bushes lowered the understorey herb leaf temperature by a similar amount, approximately 1.9 K. A similar decrease in foliage temperature, i.e. 1.5 K, was found by Corlett and co-workers (1992) in millet shaded by leucaena trees in India. In contrast, in an agroforestry system in Kenya, Allen and co-workers (1997) also used the ERIN model to simulate the effect of tree canopy shade on an understorey maize crop, but they found a higher decrease in leaf temperature of approximately 3.5 K. This decrease was similar to measurements made around midday in the same agroforestry system. Although the canopy architecture and leaf conditions are different in the current simulations, some of the difference between the degree of understorey foliage cooling found here and that found by Allen and co-workers (1997) is likely to be due to the different aerodynamic specification used in the two simulations. Allen and co-workers (1997) did not include the effect of enhanced flux transfer due to the existence of the roughness sublayer. Consequently, their aerodynamic resistances were higher than in the model presented here. This illustrates that the degree of within canopy microclimate modification is highly dependent on the specification of the within and above canopy aerodynamic transfer. However, even a 1–2 K decrease in plant temperature could have a significant effect on plant development. This is illustrated in Section 8.5.

8.4.2 Substrate effects on an upperstorey canopy

The effect of the exchange of heat, water vapour and CO_2 between the soil substrate and/or an understorey crop and the upperstorey canopy is illustrated in Figure 8.9. The example shown uses the same meteorological conditions and the same two-species mixture described above (Table 8.1), but with the upperstorey cover fixed at 0.25. To produce a range of understorey conditions, the value of f_θ^2 has been varied between 1 and 0. This is equivalent to changing from a wet soil to a completely dry soil and, consequently, the

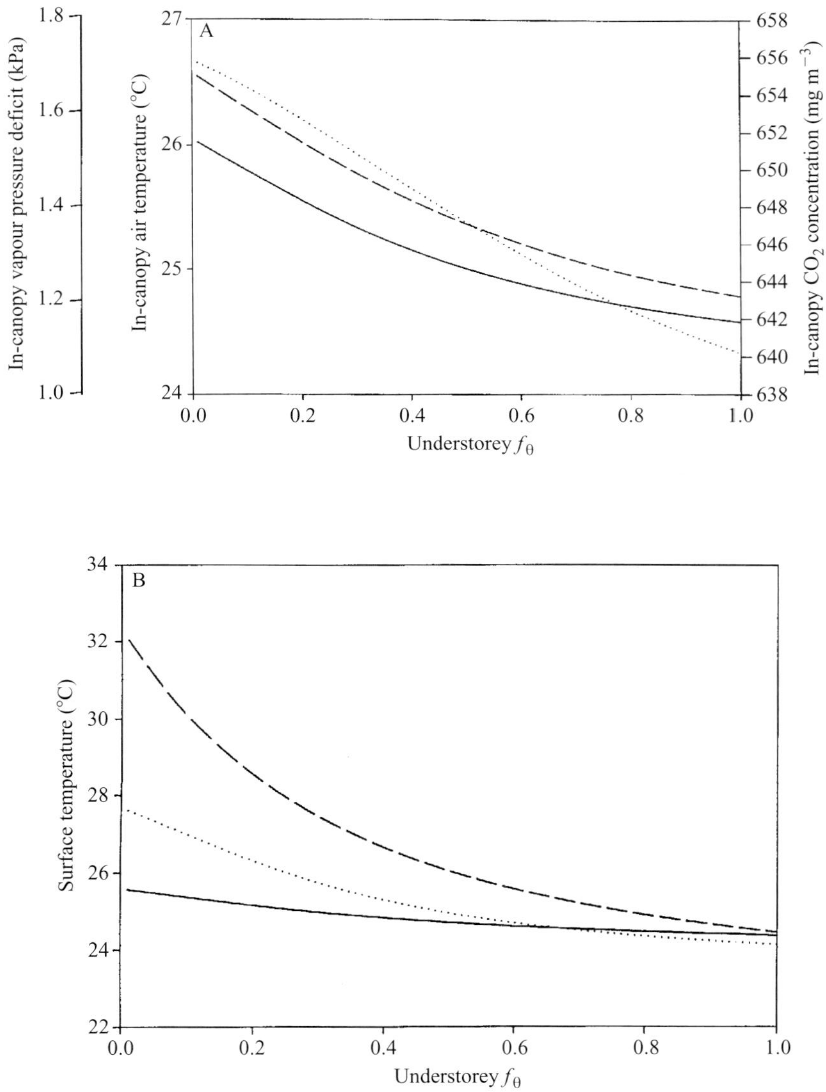

Figure 8.9 Variation of: A) in-canopy microclimate i.e. canopy temperature (T_0; ——), vapour pressure deficit (D_0; — — —) and CO_2 concentration (C_0; · · · · ·); and B) the temperatures of the soil (– – –), understorey (· · · · ·) and upperstorey (——) with understorey soil wetness (f_θ) in a two-species mixture with a dominant upperstorey. The meteorological and canopy conditions used for these calculations are described in Section 8.4.

understorey photosynthetic rates and leaf stomatal conductances (P_n^2 and r_l^2) will be reduced accordingly. To allow for the change in soil surface resistance (r_s^s), which would also occur as the soil dries, r_s^s varies from 100 s m^{-1} when $f_\theta^2 = 1$ to 10,000 s m^{-1} when $f_\theta^2 = 0$. In this simulation, it has been specified that the drying of the topsoil will not affect the upperstorey canopy (i.e. $f_\theta^1 = 1$). This would be the case if the upperstorey vegetation was deep-rooted and abstracted water from beneath the root zone of the understorey.

Figure 8.9A shows that the vapour pressure deficit and temperature around the upperstorey canopy decrease as the soil changes from dry to wet. The drop in in-canopy microclimate temperature is approximately 1.5 K and the drop in vapour pressure deficit is approximately 0.5 kPa. Figure 8.9A also shows that the model predicts a small decline in in-canopy CO_2 concentration of approximately 15 mg m^{-3}. Figure 8.9B shows how the surface temperatures of the soil and the two component species varies as the soil gets wetter. Clearly, soil temperature will be expected to decrease, but the temperature of the understorey also decreases, by nearly 2 K, due mainly to the reduced flux of heat from the soil surface. This decline in heat flux from the soil has a much smaller effect on the upperstorey than it does on the understorey, due to the effect of the resistances, r_b^3 and r_w^3. The microclimate temperature and vapour pressure deficit changes predicted here are similar to those reported by Wallace (1997) using the original ERIN model. Again, when comparing these two studies, it must be recognised that different mixed canopy, climate and aerodynamic specifications have been used.

8.5 Consequences of microclimatic interactions for plant water use, growth and development

Modifications to the microclimate due to the interactions between the soil and different species in a mixed canopy affect water use, growth and development of the component species. Figure 8.10A shows how the evaporation from the components in the mixed canopy described in Table 8.1 would vary as the cover of the upperstorey (α_1) increases from 0 to 1. Transpiration from the upperstorey increases, but this is primarily a consequence of the increase in the amount of upperstorey canopy (i.e. as L_1^P was kept constant, then as α_1 increases so will the leaf area index per unit area of ground, L_1). However, as the upperstorey cover increases, the transpiration from the understorey is reduced because of the reduction in radiant energy reaching the understorey and the reduced in-canopy vapour pressure deficit (D_0) shown in Figure 8.8A. Soil evaporation also decreases with upperstorey cover, but to a lesser extent than the reduction in understorey transpiration.

The effect of varying upperstorey cover on CO_2 fluxes is shown in Figure 8.10B. Again, the photosynthetic rate of the upperstorey increases with cover

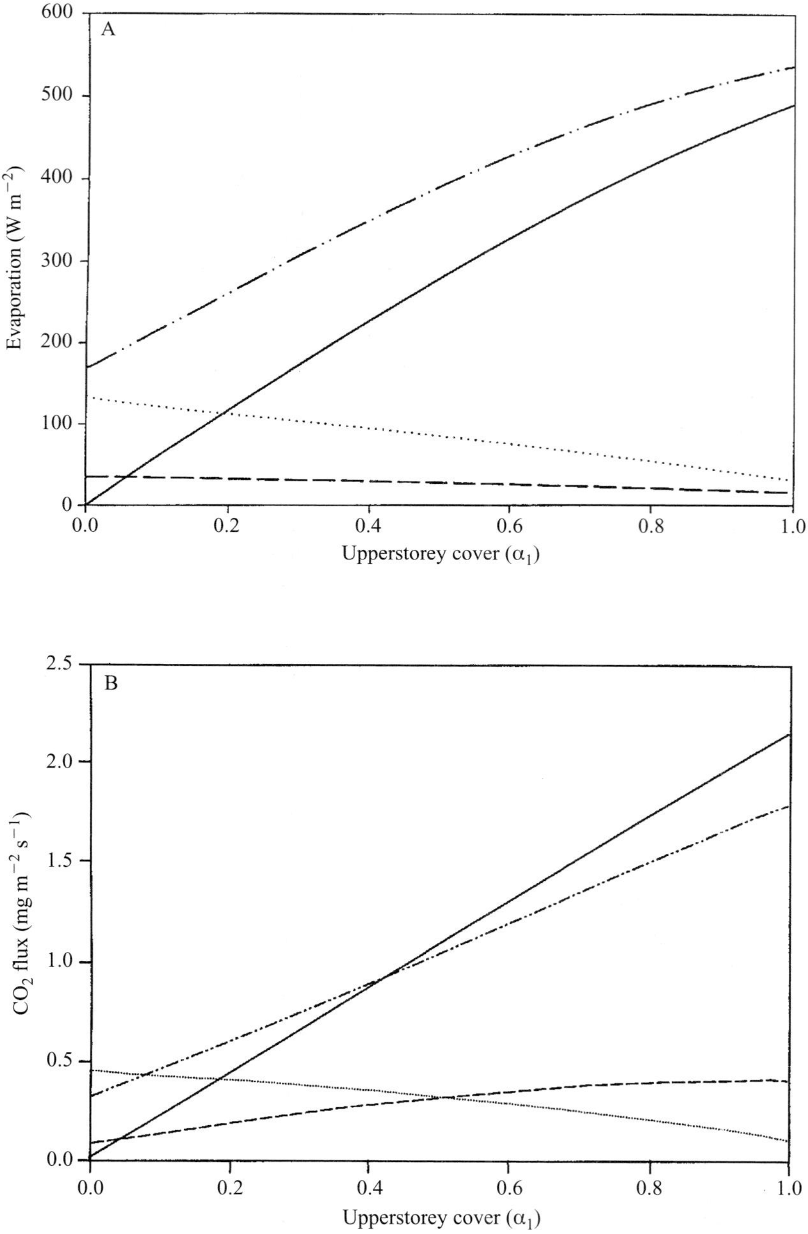

Figure 8.10 Variation of: A) evaporation; and B) CO_2 flux with upperstorey ground cover (α_1) in a two-species mixture with a dominant upperstorey. The meteorological and canopy conditions used for these calculations are the same as those used in Figure 8.8 (see Section 8.4). Key: —— upperstorey; ·····, understorey; – – – –, soil; – ·· – ··, total.

as a consequence of the increased amount of canopy per unit area of ground. In contrast, the rate of photosynthesis of the understorey decreases with upperstorey cover, due to the combined effects of the reduced level of incident PAR and to a lesser extent the decrease in within-canopy CO_2 concentration shown in Figure 8.8A. The net effect of increasing the upperstorey cover is to increase soil respiration rates. This is because the decrease in soil respiration due to the cooler soil (Figure 8.8B) is outweighed by the increase in soil respiration due to the increased below-ground biomass, which is assumed here to be proportional to the total leaf area (see Equation 54).

Canopy fluxes can also be affected by the condition of the understorey and/or soil. Figure 8.11A shows how the evaporation from the components in the mixed canopy described in Table 8.1 varies with the dryness of the soil. Understorey and soil evaporation increase as the soil gets wetter, whilst at the same time transpiration from the upperstorey decreases. CO_2 fluxes from the understorey increase markedly as the soil becomes wetter, reflecting the control that soil water supply has on canopy photosynthesis and leaf conductance (Figure 8.11B). Soil respiration decreases as f_θ^2 increases from 0 to 1 as a consequence of the decrease in soil temperature (Figure 8.9B). This is the opposite to the response of soil respiration shown in Figure 8.10B, since in the current simulation there is no increase in above ground LAI. The CO_2 flux of the upperstorey is virtually unaffected by soil wetness, since in this simulation it was decided that photosynthesis and leaf conductance of the upperstorey would be unaffected by f_θ.

It is well known that plant development rates are strongly influenced by temperature and that air humidity can affect the efficiency with which plants fix carbon per unit of water transpired (e.g. France and Thornley, 1984; Monteith 1986). The magnitude of the effect of canopy shade generated modifications of the microclimate around an understorey crop are shown below. Firstly, the effect of temperature modification on plant development is illustrated using thermal time data for sorghum given by Bradley and Crout (1993). They report that a common Indian sorghum variety will take around 370 K days to develop from emergence to panicle initiation (growth stage 1), followed by 650 K days to reach anthesis (growth stage 2), and a further 640 K days to reach maturity (growth stage 3). Assuming a base temperature for growth of 7°C and an average daily temperature of 20°C, Figure 8.12 shows the time the crop would spend in each developmental phase. The crop takes a significantly greater amount of time to develop in each stage due to the reduction in the temperature around it produced by shade. Even a 2 K decrease in temperature could lengthen developmental stages by approximately 18%. The consequences of this for overall crop growth and yield will depend on the conditions which prevail during the prolonged crop season. If conditions are favourable at key stages, they could be positive, if they are not, growth and yield could decrease.

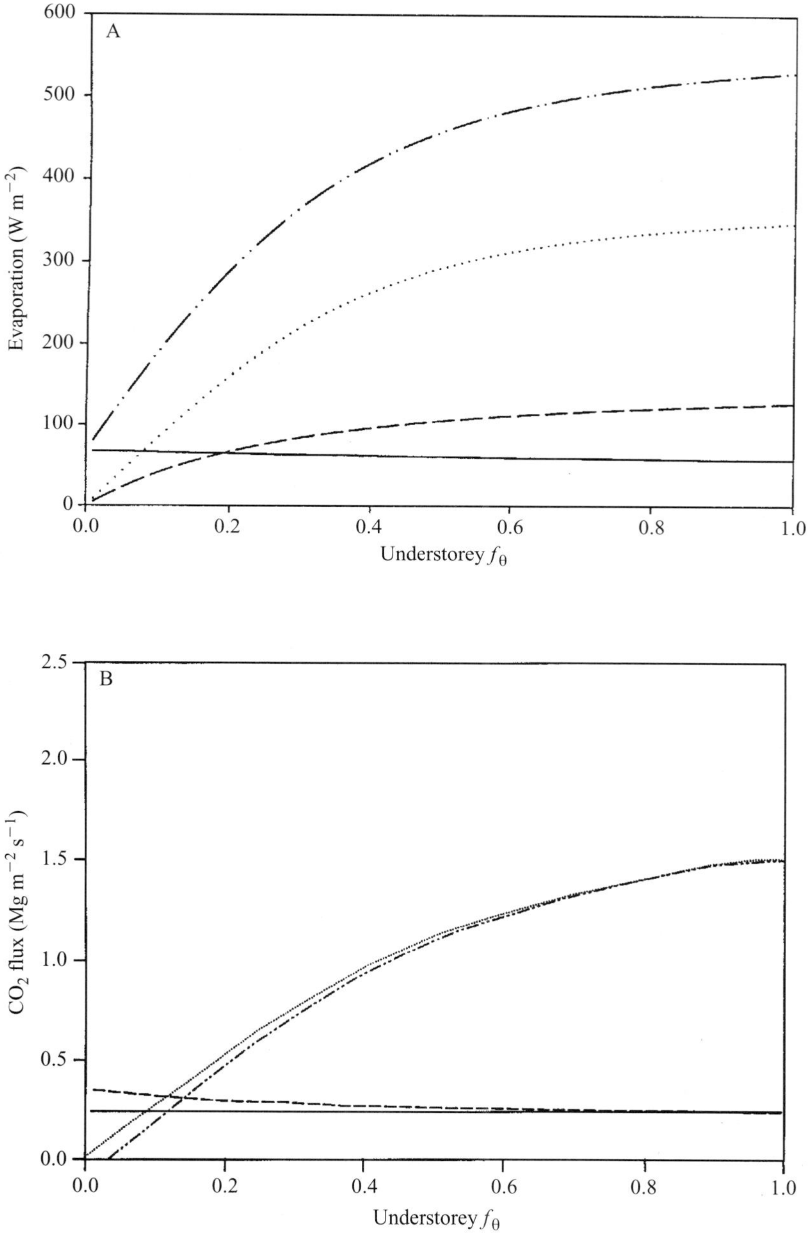

Figure 8.11 Variation of: A) component evaporation; and B) component CO_2 flux with understorey soil wetness (f_θ) in a two-species mixture with a dominant upperstorey. The meteorological and canopy conditions used for these calculations are the same as those used in Figure 8.9 (see Section 8.4). Key: —— upperstorey; · · · · ·, understorey; – – – –, soil; – ·· – ··, total.

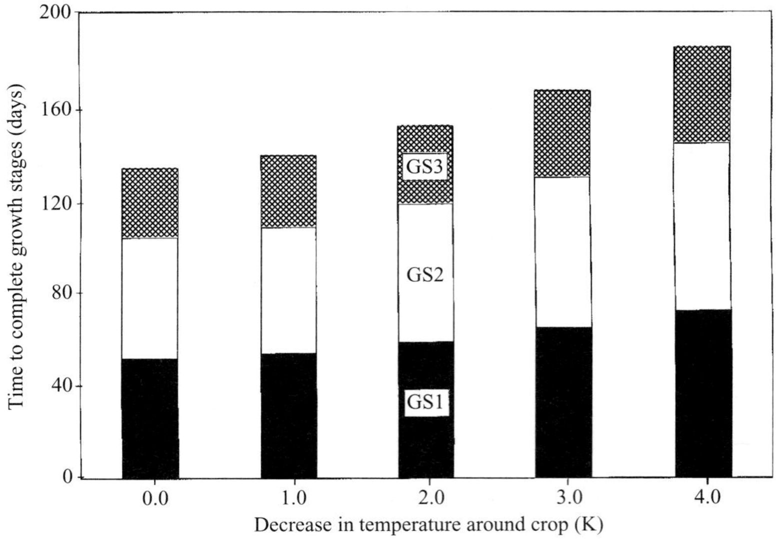

Figure 8.12 The effect on plant development of a decrease in the temperature around a sorghum crop, produced by shade from an upperstorey tree canopy. Three growth stages (GS) are shown: emergence to panicle initiation (GS1); panicle initiation to anthesis (GS2); and anthesis to maturity (GS3). Growth stage and plant development data were taken from Bradley and Crout (1993).

Any decrease in the vapour pressure deficit around a crop produced by shade could also improve crop growth. This is because the water-use ratio $(P^i_n/\lambda E_i)$ is proportional to the reciprocal of the vapour pressure deficit of the air (e.g. Ong *et al.*, 1996). The model described here can be used to calculate the effect of different mixed canopy combinations on both component and total system water-use ratios. This is illustrated in Figure 8.13A and B, which shows the change in water-use ratio with upperstorey cover and understorey condition, respectively. The canopy arrangements and climate are again the same as those used above in Figures 8.8 and 8.9 (see Table 8.1). The current model predicts that both component water-use ratios will increase by approximately 25% as the cover of the upperstorey increases from 0 to 1. The total system water-use ratio includes the evaporation from the soil, so it is lower than the water-use ratios of the component species. However, this also increases by approximately 25% in this simulation.

Figure 8.13B shows the simulation of the effects of soil moisture on water-use ratio. Again, both plant components have an improved water-use ratio when the soil water supply is increased. This is a direct effect of f_θ in the

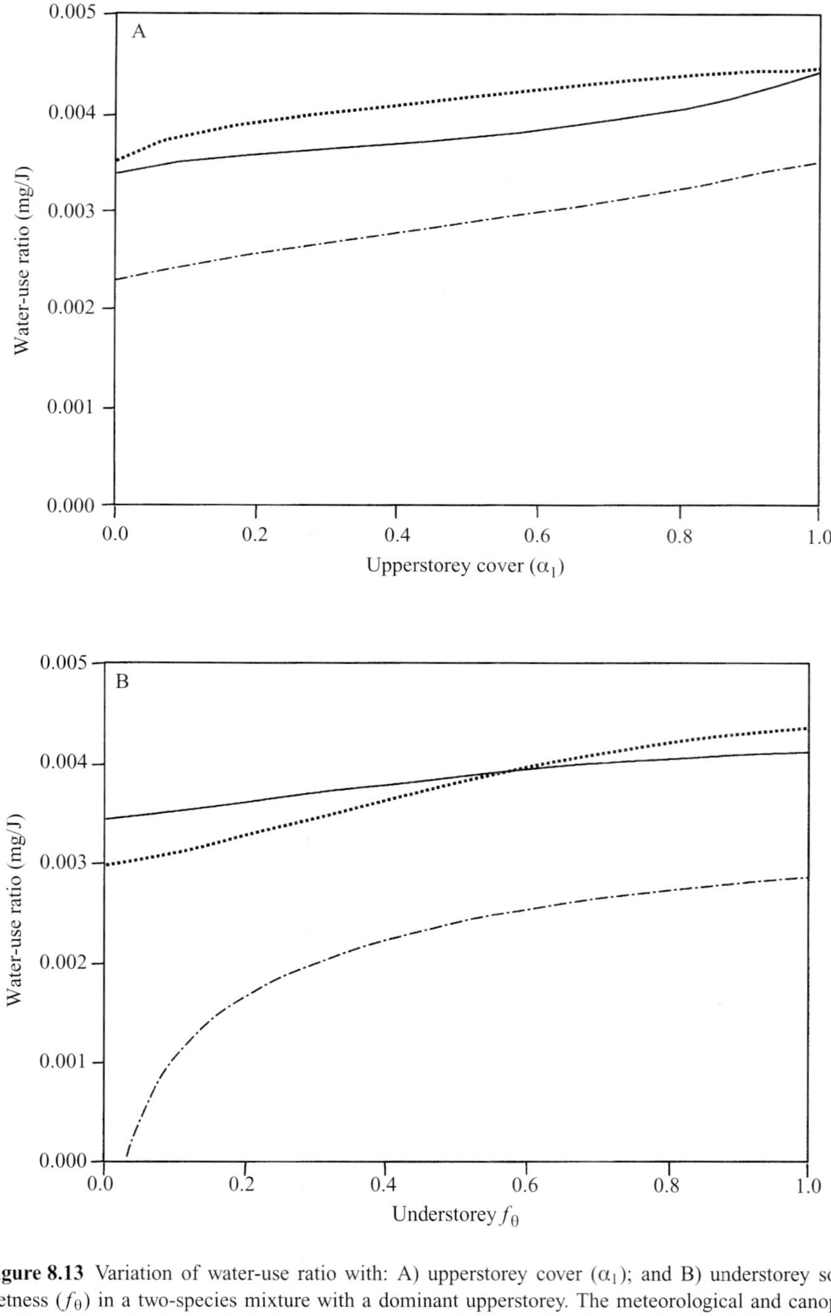

Figure 8.13 Variation of water-use ratio with: A) upperstorey cover (α_1); and B) understorey soil wetness (f_θ) in a two-species mixture with a dominant upperstorey. The meteorological and canopy conditions used for these calculations are the same as those used in Figures 8.8 and 8.9 (see Section 8.4). Key: —— upperstorey; · · · · ·, understorey; – · – · –, total.

case of the understorey, but an indirect effect of the canopy interactions in the case of the upperstorey, since f_θ is not affecting this canopy directly. The total system water-use ratio decreases very sharply as f_θ approaches 0, since the total system photosynthesis rate declines rapidly under these conditions (see Figure 8.11B). It should be noted that many other factors can affect the water-use ratio, such as species type, water stress and disease (see Ong *et al.*, 1996). However, the key point demonstrated here is that the magnitude of the changes in temperature, humidity and CO_2 concentration caused by the microclimatic interactions between the canopies can be significant enough to affect development, growth and water-use efficiency.

8.6 Concluding remarks

This chapter shows how important, above-ground interactions in mixed plant communities can be modelled by combining recent descriptions of light partitioning and using this to drive the energy balance and photosynthetic behaviour of the system. The treatment of radiation partitioning in models such as those described by Dolman (1993), Huntingford and co-workers (1995) and Verhoef and Allen (1998, 1999), is only appropriate in canopies where the upperstorey has a ground cover below approximately 0.3. At or above this point the canopy will start shading the understorey considerably, thus influencing the understorey's available energy and hence its heat and CO_2 exchange. On the other hand, the original two-source model described by Shuttleworth and Wallace (1985) assumed homogeneous ground cover, where the understorey always received less than full radiation. This is clearly a limitation of this model when it is applied to vegetation surfaces where the upperstorey canopy is sparse and/or spatially clumped. The radiation partitioning approach used in the ERIN model (Wallace, 1997) is, therefore, more appropriate for dealing with the radiative interactions in mixed-plant communities.

The light partitioning in the ERIN model has the advantage that it only requires a knowledge of plant height, leaf area and extinction coefficient of each species in order to calculate the individual amounts of radiation intercepted by the components of a multi-species mixture. This is much simpler than the alternative of using a complex multi-layer radiation balance model and, in two-species mixtures, gives very similar results. The ERIN model also has a simple way of taking spatial heterogeneity into account by allowing for partial ground cover and foliage clumping, the latter reducing the effective extinction coefficient of a species. The light sub-model, therefore, has the advantage that it requires no explicit information on the three-dimensional geometry of a mixed plant community or details of the amounts of foliage each species has in different canopy layers. However, this has the disadvantage that, although it allows the area average light interception of the component species to be calculated, it provides no information on the spatial distribution of light intensity within or between the species. If this is important in any particular application, then an explicit three-dimensional model may have to be used, such as that described by Mobbs and co-workers (1999). It should be noted that the only really reliable way to test these different light interception models will be to obtain appropriate data from multi-species mixtures in the field.

Once the incident radiation has been partitioned, there are several approaches to converting this into fluxes of latent and sensible heat. Some of these do not allow for the individual components to interact and these models are not suitable for calculating the micro-climatic interactions in mixed plant communities. The models described here do allow for species interaction by specifying an in-canopy mixing point (D_0, T_0, C_0, etc.) at which fluxes from the individual components interact. This chapter has demonstrated that microclimatic temperature modification may be as much as 1–2 K, humidity modification may be up to 25% and that CO_2 is only altered by a few per cent. However, the combination of these effects is large enough to significantly affect plant growth and development, and could also have a marked effect on both individual and community water use efficiency.

The microclimate at the in-canopy mixing point is highly dependent on the aerodynamic resistances within and above the canopy. The description of aerodynamic transfer used in the original ERIN model is quite simplistic, so in the present chapter the more detailed aerodynamic descriptions used by Verhoef and Allen (1998, 1999) and Hall and co-workers (1998) have been incorporated. However, these theoretical values of aerodynamic resistance within canopies are highly uncertain and will remain so until more direct observations of their values are made in the field.

A further novel feature of the theory described here is the incorporation of the exchange of CO_2 into the ERIN model. This has been achieved using the approach described by Verhoef and Allen (1998, 1999) and allows the photosynthetic rates of the individual components and the entire plant community to be calculated. A model like this could be used to examine the productivity and water-use efficiency of a range of mixed plant systems, such as those used in intercropping and agroforestry. The model could also be used to estimate the possible effects of CO_2 enrichment and climate change on the competitive balance between different species in a plant community (e.g. Hall *et al.*, 1998).

The way in which the current model deals with below-ground influences is very simple, i.e. via the influence of the factor, f_θ, on photosynthesis and leaf stomatal resistance. This could be improved by incorporating a more realistic water supply model, where both the soil moisture content and the amount and activity of the roots of each species directly controlled the water potential of the leaves. Then leaf water potential could be used to determine photosynthesis and leaf stomatal resistance. The ability to specify the rooting patterns of individual species would allow more realistic simulations to be made of plant mixtures where a key factor in the success of an individual is its ability to abstract water more effectively than its neighbours. A species could do this either by being more effective at abstracting water from within a region of the soil where there is competition from other species or by abstracting water from regions of the soil where its competitors do not have roots (e.g. deep-rooted trees mixed with shallow-rooted herbs). A future focus should, therefore, be to link the current above-ground, mixed-species model with an appropriate below-ground, multi-species, soil-root water abstraction model.

Appendix: Extension of light interception model to many species

In the case of two species, the limiting values of light interception were calculated for each species when either Species 1 or Species 2 dominated (Equations 2, 3, 4 and 5). The actual values of f_1 and f_2 were then calculated by weighting these limiting values by an empirical factor (F) depending on species height. When the number of species is greater than 2, all possible combinations of species dominance need to be taken into account. For example, Figure 8.14 shows the six possible arrangements when there are three species. Firstly, consider when Species 1 is above Species 2 and 3, arrangement (a) in Figure 8.14. In this case, the proportion of the total interception of Species 1 which occurs under this arrangement is $h_1^2/(h_1^2 + h_2^2 + h_3^2)$. Below Species 1, there are two further permutations, depending on whether Species 2 or 3 dominates, and each has a dominance factor of $h_2^2/(h_2^2 + h_3^2)$ and $h_3^2/(h_2^2 + h_3^2)$, respectively. When Species 1

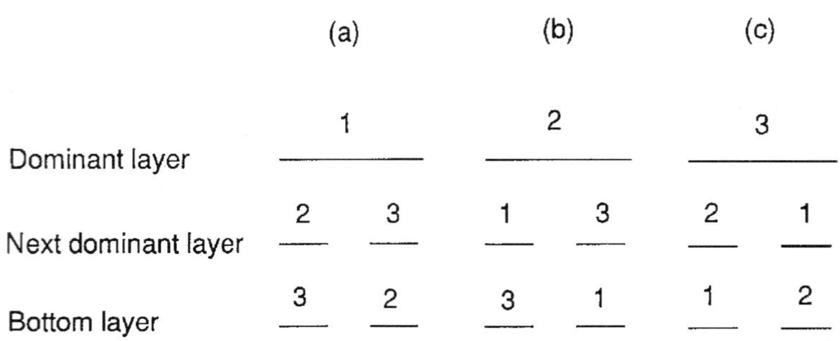

Figure 8.14 Schematic representation of the six different permutations of dominance in a mixture of three species (from Wallace, 1997).

dominates in a three-species mixture, the contributions to the light interception of each species are, therefore, given as follows. For Species 1:

$$f_1^1 = \frac{h_1{}^2}{h_1{}^2 + h_2{}^2 + h_3{}^2}(1 - e^{-K_1 L_1}) \tag{A.1}$$

and, for Species 2:

$$f_2^1 = \frac{h_1{}^2}{h_1{}^2 + h_2{}^2 + h_3{}^2}\frac{h_2{}^2}{h_2{}^2 + h_3{}^2} e^{-K_1 L_1}(1 - e^{-K_2 L_2}) + \frac{h_1{}^2}{h_1{}^2 + h_2{}^2 + h_3{}^2}\frac{h_3{}^2}{h_2{}^2 + h_3{}^2} e^{-K_1 L_1} e^{-K_3 L_3}(1 - e^{-K_2 L_2}) \tag{A.2}$$

and, for Species 3:

$$f_3^1 = \frac{h_1{}^2}{h_1{}^2 + h_2{}^2 + h_3{}^2}\frac{h_3{}^2}{h_2{}^2 + h_3{}^2} e^{-K_1 L_1}(1 - e^{-K_3 L_3}) + \frac{h_1{}^2}{h_1{}^2 + h_2{}^2 + h_3{}^2}\frac{h_2{}^2}{h_2{}^2 + h_3{}^2} e^{-K_1 L_1} e^{-K_2 L_2}(1 - e^{-K_3 L_3}) \tag{A.3}$$

A similar set of equations is obtained when Species 2 dominates, arrangement (b) in Figure 8.14, i.e. the contribution to Species 1 is:

$$f_1^2 = \frac{h_2{}^2}{h_1{}^2 + h_2{}^2 + h_3{}^2}\frac{h_1{}^2}{h_1{}^2 + h_3{}^2} e^{-K_2 L_2}(1 - e^{-K_1 L_1}) + \frac{h_2{}^2}{h_1{}^2 + h_2{}^2 + h_3{}^2}\frac{h_3{}^2}{h_1{}^2 + h_3{}^2} e^{-K_2 L_2} e^{-K_3 L_3}(1 - e^{-K_1 L_1}) \tag{A.4}$$

and, to Species 2 is:

$$f_2^2 = \frac{h_2{}^2}{h_1{}^2 + h_2{}^2 + h_3{}^2}(1 - e^{-K_2 L_2}) \tag{A.5}$$

and, to Species 3 is:

$$f_3^2 = \frac{h_2{}^2}{h_1{}^2 + h_2{}^2 + h_3{}^2}\frac{h_3{}^2}{h_1{}^2 + h_3{}^2} e^{-K_2 L_2}(1 - e^{-K_3 L_3}) + \frac{h_2{}^2}{h_1{}^2 + h_2{}^2 + h_3{}^2}\frac{h_1{}^2}{h_1{}^2 + h_3{}^2} e^{-K_2 L_2} e^{-K_1 L_1}(1 - e^{-K_3 L_3}) \tag{A.6}$$

Finally, when Species 3 dominates, arrangement (c) in Figure 8.14, a third set of equations is obtained, where the contribution to Species 1 is:

$$f_1^3 = \frac{h_3{}^2}{h_1{}^2 + h_2{}^2 + h_3{}^2}\frac{h_1{}^2}{h_2{}^2 + h_1{}^2}\,\mathrm{e}^{-K_3 L_3}(1 - \mathrm{e}^{-K_1 L_1}) + \frac{h_3{}^2}{h_1{}^2 + h_2{}^2 + h_3{}^2}\frac{h_2{}^2}{h_2{}^2 + h_1{}^2}\,\mathrm{e}^{-K_3 L_3}\,\mathrm{e}^{-K_2 L_2}(1 - \mathrm{e}^{-K_1 L_1}) \quad \text{(A.7)}$$

and, to Species 2 is:

$$f_2^3 = \frac{h_3{}^2}{h_1{}^2 + h_2{}^2 + h_3{}^2}\frac{h_2{}^2}{h_2{}^2 + h_1{}^2}\,\mathrm{e}^{-K_3 L_3}(1 - \mathrm{e}^{-K_2 L_2}) + \frac{h_3{}^2}{h_1{}^2 + h_2{}^2 + h_3{}^2}\frac{h_1{}^2}{h_2{}^2 + h_1{}^2}\,\mathrm{e}^{-K_3 L_3}\,\mathrm{e}^{-K_1 L_1}(1 - \mathrm{e}^{-K_2 L_2}) \quad \text{(A.8)}$$

and, to Species 3 is:

$$f_3^3 = \frac{h_3{}^2}{h_1{}^2 + h_2{}^2 + h_3{}^2}(1 - \mathrm{e}^{-K_3 L_3}) \quad \text{(A.9)}$$

Now, the total light interception by any species is given by the sum of the appropriate terms under each arrangement. For example, total light interception by Species 1, f_1, is given by the sum of Equations (A.1), (A.4) and (A.7). Similarly, f_2 is the sum of Equations (A.2), (A.5) and (A.8), and f_3 is the sum of Equations (A.3), (A.6) and (A.9). By the nature of the permutations, when the number of species (N) is greater than 2, there are no simple algebraic shorthand formulae which can be written to give a general form for f_i. However, there is an obvious pattern in the formulae for $N = 2$ and $N = 3$, which can readily be extended to values of $N > 3$. The light interception of Species i in a mixture of N species is comparatively easily calculated using a short computer program.

References

Allen, S., Roberts, J., Smith, M., Jackson, N. and Lawson, G. (1997) Simulating the interaction between tree cover and crop temperature in integrated agroforestry models. *Agroforestry Forum*, **8** 20–23.

Baldocchi, D.D. and Harley, P.C. (1995) Scaling carbon dioxide and water vapour exchange from leaf to canopy in a deciduous forest. II. Model testing and application. *Plant, Cell and Environment*, **18** 1157–1173.

Baldocchi, D.D. and Meyers, T.P. (1988) Turbulence structure in a deciduous forest. *Boundary-Layer Meteorology*, **43** 345–364.

Baldocchi, D.D., Luxmoore, R.J. and Hatfield, J.L. (1991) Discerning the forest from the trees: an essay on scaling canopy stomatal conductance. *Agricultural and Forest Meteorology*, **54** 197–226.

Blyth, E.M. (1995) Using a simple SVAT scheme to describe the effect of scale on aggregation. *Boundary-Layer Meteorology*, **72** 267–285.

Blyth, E.M. and Harding, R.J. (1995) Application of aggregation models to surface heat flux from the Sahelian tiger bush. *Agricultural and Forest Meteorology,* **72** 213–235.

Bradley, R.G. and Crout, N.M.J. (1993) *The PARCH Model for Predicting Arable Resource Capture in Hostile Environments, PARCH-User Guide,* V2.02a. Tropical Crops Research Unit, University of Nottingham, Nottingham, UK.

Brenner, A.J. (1991) *Tree–Crop Interactions within a Sahelian Windbreak System*, Ph.D. Thesis, University of Edinburgh, Scotland, UK.

Brutsaert, W.H. (1982) *Evaporation into the Atmosphere.* D. Reidel, Norwell, MA, USA.

Calvet, J.C., Noilhan, J., Roujean, J.L., Bessemoulin, P., Cabelguenne, M., Olioso, A. and Wigneron, J.P. (1998) An interactive vegetation SVAT model tested against data from six contrasting sites. *Agricultural and Forest Meteorology,* **92** 73–95.

Camillo, P.J. and Gurney, R.J. (1986) A resistance parameter for bare-soil evaporation models. *Soil Science,* **141,** 95–105.

Cellier, P. and Brunet, Y. (1992) Flux-gradient relationships above tall plant canopies. *Agricultural and Forest Meteorology,* **58** 93–117.

Chen, D., Hunt, H.W. and Morgan, J.A. (1996) Responses of a C_3 and C_4 perennial grass to CO_2 enrichment and climate change: comparison between model predictions and experimental data. *Ecological Modelling,* **87** 11–27.

Choudhury, B.J. and Monteith, J.L. (1988) A four-layer model for the heat budget of homogeneous land surfaces. *Quarterly Journal of the Royal Meteorological Society,* **114** 373–398.

Cohen, S., Mosoni, P. and Meron, M. (1995) Canopy clumpiness and radiation penetration in a young hedgerow apple orchard. *Agricultural and Forest Meteorology,* **76** 185–200.

Corlett, J.E., Ong, C.K., Black, C.R. and Monteith, J.L. (1992) Above-ground and below-ground interactions in a leucaena/millet alley cropping system. I. Experimental design, instrumentation and diurnal trends. *Agricultural and Forest Meteorology,* **60** 53–72.

Cox, P.M., Huntingford, C. and Harding, R.J. (1998) A canopy conductance and photosynthesis model for use in a GCM land surface scheme. *Journal of Hydrology,* **212–213** 79–94.

Deardorff, J.W. (1978) Efficient prediction of ground surface temperature and moisture, with inclusion of a layer of vegetation. *Journal of Geophysical Research,* **83** 1889–1903.

Denmead, O.T. and Bradley, E.F. (1985) Flux-gradient relationships in a forest canopy, in *The Forest–Atmosphere Interaction* (eds. B.A. Hutchison and B.B. Hicks), D. Reidel, Dordrecht, The Netherlands, pp. 421–441.

Dolman, A.J. (1993) A multiple-source land surface energy balance model for use in general circulation models. *Agricultural and Forest Meteorology,* **65** 21–45.

Dolman, A.J. and Wallace, J.S. (1991) Lagrangian and K-theory approaches in modelling evaporation from sparse canopies. *Quarterly Journal of the Royal Meteorological Society,* **117** 1325–1340.

France, J. and Thornley, J.H.M. (1984) *Mathematical Models in Agriculture: A Quantitative Approach to Agriculture and Related Sources,* Butterworths, London, UK.

Friend, A.D. and Cox, P.M. (1995) Modelling the effects of atmospheric CO_2 on vegetation–atmosphere interactions. *Agricultural and Forest Meteorology,* **73** 285–295.

Gijzen, H. and Goudriaan, J. (1989) A flexible and explanatory model of light distribution and photosynthesis in row crops. *Agricultural and Forest Meteorology,* **48** 1–20.

Goudriaan, J. (1977) *Crop Micrometeorology: A Simulation Study,* PUDOC, Wageningen, The Netherlands.

Hall, R.L., Milne, R., Lawson, G.J., Verhoef, A., Mobbs, D.C., Huntingford, C., Brown, T., Allen, S.J. and Wallace, J.S. (1998) *Combined Growth and Water Use Modelling of Mixed Vegetation,* CEH Integrating Fund Project T06050P2, CSP 9.2.4, Institute of Hydrology, Wallingford, UK.

Huntingford, C., Allen, S.J. and Harding, R.J. (1995) An intercomparison of single- and dual-source vegetation–atmosphere transfer models applied to transpiration from Sahelian savannah. *Boundary-Layer Meteorology,* **74** 397–418.

Huntingford, C., Hall, R.L. and Verhoef, A. (1998) An argument for the use of two-layer SVAT schemes to simulate terrestrial carbon dioxide fluxes. *Hydrology and Earth System Sciences*, **2** 299–302.

Jackson, J.E. and Palmer, J.W. (1979) A simple model of light transmission and interception by discontinuous canopies. *Annals of Botany*, **44** 381–383.

Jacobs, C.M.J. (1994) *Direct Impact of Atmospheric CO_2 Enrichment on Regional Transpiration*, PhD thesis, Agricultural University, Wageningen, The Netherlands, ISBN, 90-5485-250-X.

Jacobs, C.M.J., van den Hurk, B.J.J.M. and de Bruin, H.A.R. (1996) Stomatal behaviour and photosynthetic rate of unstressed grapevines in semi-arid conditions. *Agricultural and Forest Meteorology*, **80** 111–134.

Jarvis, P.G., (1976). The interpretation of the variations in leaf water potential and stomatal conductance found in canopies in the field. *Philosophical Transactions of the Royal Society, London, Series B*, **273** 593–610.

Katerji, N. and Pereira, A.R. (1985) Determination of canopy resistance to water vapour and its various components: theoretical approach and experimental verification. *Agricultural and Forest Meteorology*, **34** 105–120.

Kiniry, J.R., Williams, J.R., Gassman, P.W. and Debaeke, P. (1992) A general process orientated model for two competing plant species. *Transactions of the American Society of Agricultural Engineers*, **35** 801–810.

Kohsiek, W. (1981) A rapid-circulation evaporation chamber for measuring bulk stomatal resistance. *Journal of Applied Meterology*, **20** 42–52.

Kondo, J., Saigusa, N. and Sato, T. (1990) A parameterization of evaporation from bare soil surfaces. *Journal of Climate and Applied Meteorology*, **29** 385–389.

Kropff, M.J. and van Laar, H.H. (1992) *Simulation of Interplant Competition*. Draft Manuscript for a Simulation Monograph (5 August 1992), International Rice Research Institute, The Philippines.

Kustas, W.P. (1990) Estimates of evapotranspiration with a one- and two-dimensional model of heat transfer over partial canopy cover. *Journal of Applied Meteorology*, **29** 704–715.

Lafleur, P.M. and Rouse, W.R. (1990) Application of an energy combination model for evaporation from sparse canopies. *Agricultural and Forest Meteorology*, **49** 135–153.

Massman, W.J. (1992) A surface energy balance method for partitioning evapotranspiration data into plant and soil components for a surface with partial canopy cover. *Water Resources Research*, **28** 1723–1732.

McNaughton, K.G. (1989) Regional interactions between canopies and the atmosphere, in *Plant Canopies, Their Growth, Form and Function* (eds. G. Russell, B. Marshall and P.G. Jarvis), Society for Experimental Biology Seminar Series **31**, Cambridge University Press, Cambridge, UK, pp. 63–81.

McNaughton, K.G. and Van den Hurk, B.J.J.M. (1995) A 'Lagrangian' revision of the resistors in the two-layer model for calculating the energy budget of a plant canopy. *Boundary-Layer Meteorology*, **74** 261–288.

Mobbs, D.C., Cannell, M.G.R., Crout, N.M.J., Lawson, G.J., Friend, A.D., Arah, J. (1998) Complementarity of light and water use in tropical agroforests – I. Theoretical model outline, performance and sensitivity. *Forest Ecology and Management*, **102 (No. 2–3)** pp. 259–274.

Monteith, J.L. (1965) Evaporation and environment, in *State and Movement of Water in Living Organisms*, 19th Symposium of the Society for Experimental Biology, Swansea 1964, Cambridge University Press, Cambridge, UK, pp. 205–234.

Monteith, J.L. (1973) *Principles of Environmental Physics*, Elsevier, New York, USA.

Monteith, J.L. (1981) Evaporation and surface temperature. *Quarterly Journal of the Royal Meteorological Society*, **107** 1–27.

Monteith, J.L. (1986) How do crops manipulate water supply and demand? *Philosophical Transactions of the Royal Society of London, Series A*, **316** 245–289.

Monteith, J.L. and Unsworth, M.H. (1990) *Principles of Environmental Physics*, 2nd Edn, Edward Arnold, London, UK.

Nichols, W.D. (1992) Energy budgets and resistances to energy transport in sparsely vegetated rangeland. *Agricultural and Forest Meteorology*, **60** 221–247.

Noilhan, J. and Planton, S. (1989) A simple parameterization of land surface processes for meteorological models. *Monthly Weather Review of the USA*, **117** 536–549.

Norman, J.J. (1979) Modelling the complete crop canopy, in *Modification of the Areal Enviromnent of Crops* (eds. B. Barfield and J. Gerber), American Society of Agriculture Engineering, St. Joseph, MI, USA, pp. 249–277.

Norman, J.M. and Welles, J.M. (1983) Radiative transfer in an array of canopies. *Agronomy Journal*, **75** 481–488.

Olioso, A., Carlson, T.N. and Brisson, N. (1996) Simulations of diurnal transpiration and photosynthesis of a water-stressed soybean crop. *Agricultural and Forest Meteorology*, **81** 41–59.

Ong, C.K., Black, C.R., Marshall, F.M. and Corlett, J.E. (1996) Principles of resource capture and utilisation of light and water, in *Tree-Crop Interactions: A Physiological Approach* (eds. C.K. Ong and P. Huxley), CAB International, Wallingford, UK, pp. 73–158.

Parkinson, K.J. (1985) Porometry, in *Instrumentation for Environmental Physiology* (eds. B. Marshall and F.I. Woodward), Cambridge University Press, Cambridge, UK, pp. 171–191.

Paulson, C.A. (1970) The mathematical representation of wind speed and temperature in the unstable atmospheric surface layer. *Journal of Applied Meteorology*, **9** 857–861.

Pereira, A.R and Shaw, R.H. (1980) A numerical experiment on the mean wind structure inside canopies of vegetation. *Agricultural Meteorology*, **22** 303–318.

Raupach, M.R. (1979) Anomalies in flux-gradient relationships over forest. *Boundary-Layer Meteorology*, **16** 467–486.

Raupach, M.R. (1987) A Lagrangian analysis of scalar transfer in vegetation canopies. *Quarterly Journal of the Royal Meteorological Society*, **113** 107–120.

Raupach, M.R. (1992) Drag and drag partition on rough surfaces. *Boundary-Layer Meteorology*, **60** 375–395.

Raupach, M.R. (1994) Simplified expressions for vegetation roughness length and zero-plane displacement height as functions of canopy height and area index (Research Note). *Boundary-Layer Meteorology*, **71** 211–216.

Raupach, M.R. and Thom, A.S. (1981) Turbulence in and above plant canopies. *Annual Review of Fluid Mechanics*, **13** 97–129.

Raupach, M.R., Finnigan, J.J. and Brunet, Y. (1996) Coherent eddies and turbulence in vegetation canopies, the mixing-layer analogy. *Boundary-Layer Meteorology*, **78** 351–382.

Raupach, M.R., Thom, A.S. and Edwards, I. (1980) A wind tunnel study of turbulent flow close to regularly arrayed rough surfaces. *Boundary-Layer Meteorology*, **18** 373–397.

Rimmington, G.M. (1984) A model of the effect of interspecies competition for light on dry matter production. *Australian Journal of Plant Physiology*, **11** 277–286.

Rochette, P., Pattey, E., Desjardins, R.L., Dwyer, L.M., Stewart, D.W. and Dubé, P.A., (1991) Estimation of maize (*Zea mays* L.) canopy conductance by scaling up leaf stomatal conductance. *Agricultural and Forest Meteorology*, **54** 241–261.

Russell, G., Jarvis, P.G. and Monteith, J.L. (1989) Absorption of radiation by canopies and stand growth, in *Plant Canopies, Their Growth, Form and Function* (eds. G. Russell, B. Marshall and P.G. Jarvis), Society for Experimental Biology Seminar Series **31**, Cambridge University Press, Cambridge, UK, pp. 21–39.

Schuepp, P.H. (1993) Tansley Review No. 59, Leaf Boundary layers. *New Phytologist*, **125** 477–507.

Sellers, P.J., Berry, J.A., Collatz, G.J., Field, C.B. and Hall, F.G. (1992) Canopy reflectance, photosynthesis and transpiration. III. A reanalysis using improved leaf models and a new canopy integration scheme. *Remote Sensing of Environment*, **42** 187–216.

Shuttleworth, W.J. and Gurney, J.R. (1990) The theoretical relationship between foliage temperature and canopy resistance in sparse crops. *Quarterly Journal of the Royal Meteorological Society*, **116** 497–519.

Shuttleworth, W.J. and Wallace, J.S. (1985) Evaporation from sparse crops: an energy combination theory. *Quarterly Journal of the Royal Meteorological Society*, **111** 839–855.

Sinoquet, H. and Bonhomme, R. (1991) A theoretical analysis of radiation interception in a two-species plant canopy. *Mathematical Biosciences*, **105** 23–45.

Sinoquet, H. and Bonhomme, R. (1992) Modelling radiative transfer in mixed and row intercropping systems. *Agricultural and Forest Meteorology*, **62** 219–240.

Sinoquet, H. and Caldwell, R.M. (1995) Estimation of light capture and partitioning in intercrop systems, in *Ecophysiology of Tropical Intercropping* (eds. H. Sinoquet and P. Cruz), INRA Editions, Paris, France, pp. 79–97.

Sinoquet, H., Mouha, B., Gastal, F., Bonhomme, R. and Varlet-Grancher, C. (1990) Modeling the radiative balance of the components of a well-mixed canopy: application to a white clover tall fescue mixture. *Acta Oecologica*, **11** 469–486.

Smith, R.C.G. and Barrs, H.D. (1988) Inferring stomatal resistance of sparse crops from infrared measurements of foliage temperature. *Agricultural and Forest Meteorology*, **42** 183–198.

Spitters, C.J.T. (1990) On descriptive and mechanistic models for interplant competition with particular reference to crop–weed interaction, in *Theoretical Production Ecology, Reflections and Prospects* (eds. R. Rabbinge, J. Goudriaan, H. van Keulen, F.W.T. Penning de Vries and H.H. van Laar), Simulation Monographs 34, PUDOC, Wageningen, The Netherlands, pp. 217–236.

Stewart, J.S. (1988). Modelling surface conductance of pine forest. *Agricultural and Forest Meteorology*. **43** 19–35.

Szeicz, G. and Long, I.F. (1969) Surface resistance of crop canopies. *Water Resources Research*, **5** 622–633.

Tennekes, H. (1973) The logarithmic wind profile. *Journal of Atmospheric Science,* **30** 234–238.

Tournebize. R., Sinoquet, H. and Brusiére, F. (1996) Modelling evapotranspiration partitioning in a shrub/grass alley crop. *Agricultural and Forest Meteorology*, **51** 255–272

Van de Griend, A.A. and Owe, M. (1994) Bare soil surface resistance to evaporation by vapor diffusion under semiarid conditions. *Water Resources Research*, **30 (2)** 181–188.

Van den Hurk, B.J.J.M., Verhoef, A., de Bruin, H.A.R. and Van den Berg, A. (1995) An intercomparison of three vegetation/soil models for a sparse vineyard canopy. *Quarterly Journal of the Royal Meteorological Society*, **121** 1867–1889.

Verhoef, A. (1995) *Surface Energy Balance of Shrub Vegetation in the Sahel*, PhD thesis, Department of Meteorology, Wageningen Agricultural University, Wageningen, The Netherlands, ISBN 90-5485-458-8.

Verhoef, A. and Allen S.J. (1998) The relative importance of surface and aerodynamic resistances in a multi-source energy–CO_2 model. *Physics and Chemistry of the Earth*, **23** 459–463.

Verhoef A. and Allen, S.J. (1999) A SVAT scheme describing energy and CO_2 fluxes for multi-component vegetation, calibration and test for a Sahelian savannah. *Ecological Modelling*, (*in press*).

Verhoef, A., McNaughton, K.G. and Jacobs A.F.G. (1997) A parameterization of momentum roughness length and displacement height for a wide range of canopy densities. *Hydrology and Earth Systems Sciences*, **1** 81–91.

Wallace, J.S. (1997) Evaporation and radiation interception by neighbouring plants. *Quarterly Journal of the Royal Meteorological Society*, **123** 1885–1905.

Wallace, J.S., Roberts, J.M. and Sivakumar, M.V.K. (1990) The estimation of transpiration from sparse dryland millet using stomatal conductance and vegetation area indices. *Agricultural and Forest Meteorology*, **51** 35–49.

Wang, Y.P. and Jarvis, P.G. (1990) Description and validation of an array model—MAESTRO. *Agricultural and Forest Meteorology*, **51** 257–280.

Willey, R.W. and Rao, M.R. (1980) A competitive ratio for quantifying competition between intercrops. *Experimental Agriculture*, **16** 117–125.

Woodward, F., Smith, T.M. and Emanuel, W.R. (1995) A global land primary productivity and phytogeography model. *Global Biochemical Cycles*, **9** 471–490.

9 Gap Dynamics

Ronald J. Ryel and Wolfram Beyschlag

9.1 Introduction

Disturbance plays a significant role in the formation and persistence of plant communities. This ubiquitous and important process affects community succession (Shugart and West, 1980; Shugart 1984), composition (White, 1979), stability (van der Maarel, 1993; Lertzman, 1992) and biodiversity (Kohyama, 1993; Solé and Manrubia, 1995). Functioning as a process of canopy renewal, disturbance permits regeneration of species (Watt, 1947), shifts in species composition (Veblen, 1985) and spatial and temporal coexistence of species (Fox, 1977; Rebertus and Veblen, 1993).

Within plant communities, disturbance refers to a disruption of portions of the community that changes resource availability (Pickett and White, 1985). Often, disturbance manifests itself in the plant community as a gap or patch, where foliage or even whole plants are removed or die. Typically, the resulting gaps are characterised by a difference in the level of light penetration from that of the surrounding vegetation but other environmental conditions can also be altered. Canopy gap formation and subsequent recovery can be the predominant process in plant community dynamics (van der Maarel, 1996).

This chapter considers canopy turnover in the context of canopy gap dynamics and the response of plants to these transient spatial opportunities. A presentation of dynamic gap models follows the discussion, with reference to their utility in assessing community dynamics and climatic changes in forests.

9.2 Role of gaps in plant community dynamics

Native plant communities are rarely comprised of individuals with similar age and canopy structure. Differences in structure among species and variations in age, growth form and available resources all contribute to differences in the canopy structure. Gaps, resulting from disturbances to the canopy crown, enhance the structural heterogeneity. Variations in response to canopy recovery among plant species in these gaps can also have a profound influence on canopy structure and community composition, and can be responsible for much of the heterogeneity in plant size structure and species diversity (White, 1979; Pickett and White, 1985). These gaps increase resource availability and enhance recruitment or growth by existing plants.

Gaps in canopies are created by the loss of foliage from one or more individual plants, which often results in increased light penetration to the soil surface or to plants within the lower canopy. The increased light permits the regeneration of crown species through seed germination or increased growth rates of small individuals within the canopy (Figure 9.1). Similarly, other species which may be inhibited by light limitations may exploit the gap and join the canopy crown.

Gaps can also affect soil water and nutrients (Bazzaz, 1983; Vitousek and Denslow, 1986). The increased light penetration can warm the soil and increase or decrease the rate of nutrient mineralisation (Stark, 1994), thus affecting availability to plants. Evaporation of water from the soil surface can also be affected by the increased light within the gap. In addition, the death or loss of large canopy plants can result in substantial increases in nutrient and water availability both from reduced uptake and from decomposition of the dead plant. 'Gaps' can also function as the area under sparse vegetation, such as desert shrubs, where lower temperatures and shade enhance seed germination and promote establishment and growth (Smith *et al.*, 1987; Valiente-Banuet *et al.*, 1991).

Gaps provide a different resource availability from the relative homogeneity of the rest of the canopy. The variable size of gaps and the heterogeneity within them produces increased microsite variability and opportunities for niche differentiation, by spatially and temporally benefiting different species or age and size cohorts (Ricklefs, 1977). Strategies for dealing with this environmental heterogeneity differ among plant species. Responses to canopy heterogeneity permit regeneration of shade-intolerant canopies species, coexistence of species with different life-history strategies and, ultimately, increased richness of species in many plant communities (Watt, 1947; White, 1979; Ricklefs, 1977; van der Maarel, 1996). Gap dynamics have also been implicated as an important mechanism in community succession, the transition from one species assemblage to another with different dominant species (Shugart, 1984). Through growth and filling of canopy gaps by species different from those dominating the canopy, the composition of the plant community can change over time.

The dynamics of canopy gap formation and subsequent plant response have been found to be a major process of regeneration, succession and species coexistence in many types of plant communities. These include temperate (Canham, 1998; Runkle, 1990) and tropical forests (Brokaw and Franklin, 1989; Kohyama, 1993), continuous (Sarukhán and Harper, 1973; Glenn and Collins, 1990) and sparse grasslands (Coffin and Urban, 1993; Hook *et al.*, 1994), shrub communities (Valiente-Banuet *et al.*, 1991) and

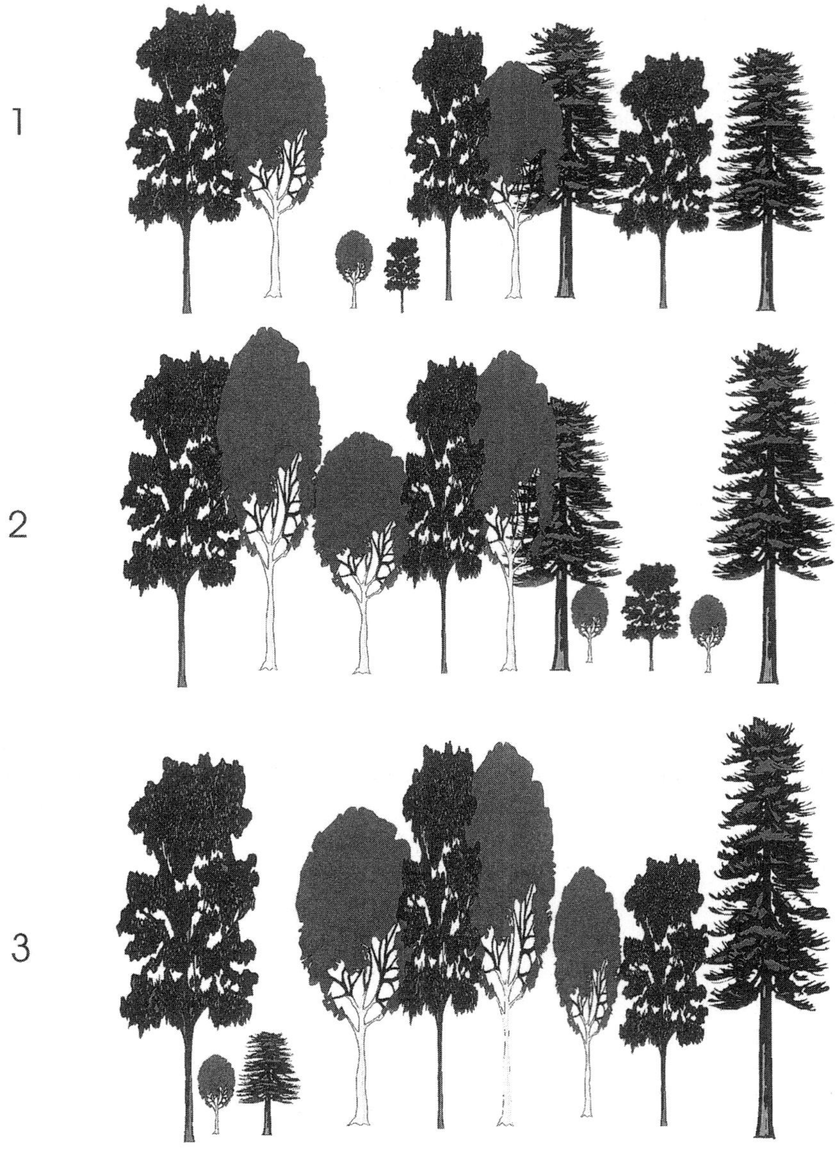

Figure 9.1 Hypothetical cross-section of forest canopy gap formation and regeneration in a mixed-species canopy at three sequential time periods. Tree fall opens the upper canopy and the increased light permits some saplings to eventually grow to fill the gap at crown level. Individually-based patch models would use a patch size equivalent to one to three of the large canopy trees (see text).

heathlands (Watt, 1947; de Hullu and Gimingham, 1984). While disturbances occur naturally in all plant communities, the frequency, duration, magnitude and source vary widely between communities and depend on the canopy structure, growth characteristics of the associated plants and the source of disturbance. And in some plant communities, factors other than gap dynamics may be more important in community dynamics. These factors include clone dynamics (Peterson and Squiers, 1995), dominance by generalist species (Lieberman *et al.*, 1995), and random replacement (Midgley *et al.*, 1995).

9.3 Characteristics of canopy gaps

The area of canopy gaps within plant communities can vary greatly depending on the gap forming process (Figure 9.2). Even within similar forest types, the portion of canopies containing gaps can vary greatly, from as little as 3% to over 70% in forests in eastern North America (Runkle, 1982; Worrall and Harrington, 1988). Canopy gaps can result from many processes, both endogenous and exogenous to the plant community, and vary greatly in their spatial extent and temporal dynamics. Death of individual plants is often cited as a common source of canopy gaps within many communities, but branch loss can be an important source of gaps as well (Lawton, 1990). Individual plants may die for many reasons, including old age, drought, severe herbivory, disease, physical damage from wind or animals, and extreme cold or heat. Leaf loss through herbivory, wind and seasonal phenological changes can also create gaps through which light can penetrate further into the canopy. Faecal deposits from large ruminants are a significant source of canopy gaps in some grasslands (Vázquez-Yanes and Orozco-Segovia, 1994).

Larger scale, but often less frequent, canopy gaps, can occur from fires, winds from severe storms, including hurricanes and tornados, flooding, diseases affecting the plant community, landslides, earthquakes and human activity. Gaps on this scale may range in size from thousands to millions of square metres. In addition, sizable new gaps may form along rivers as a result of alluvial deposits and filling in of small lakes and ponds.

9.3.1 Spatial attributes of canopy gaps

Not all gaps are equal, their size and the frequency of recurrence can vary greatly, even within the same plant community, with significant consequences for their function. Also, the effects of gaps of similar size can differ substantially between plant communities and by latitude. What is similar among all gaps is the change in resources as compared to that found in the adjacent canopy.

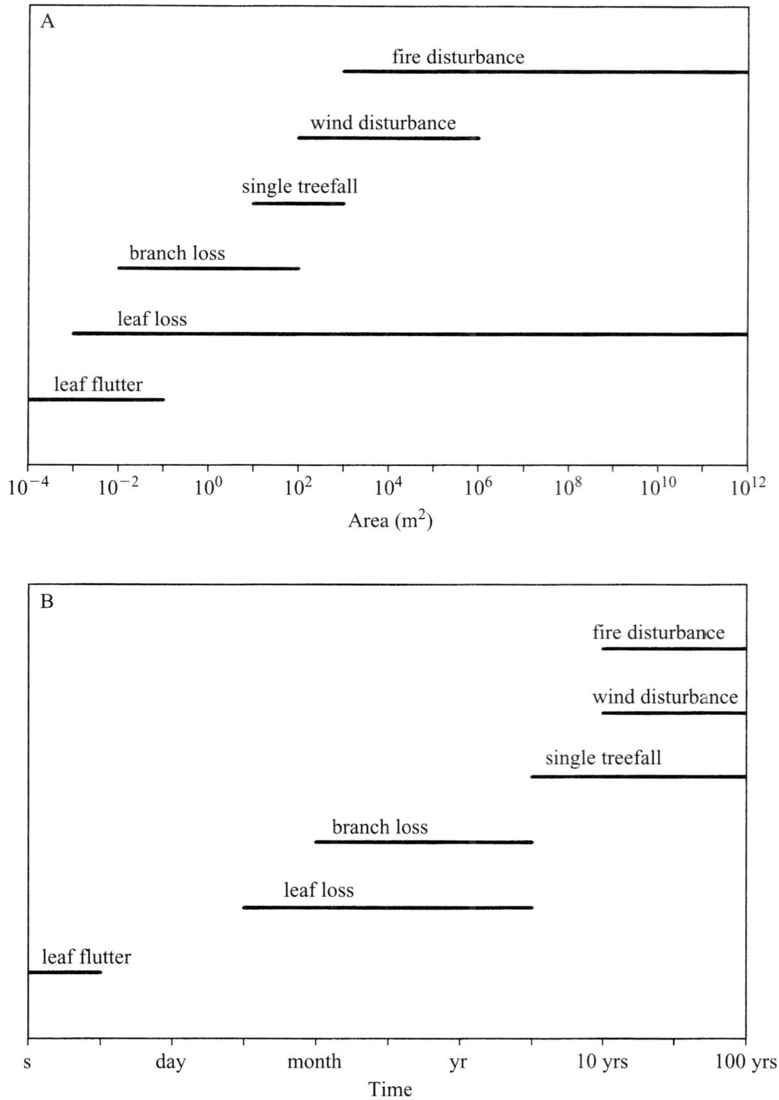

Figure 9.2 A) Spatial and B) temporal characteristics of some gap-forming processes in a forest canopy. 'Area' represents the horizontal extent of the gap as projected to the canopy floor. 'Time' is the period necessary for the canopy to close the gap.

Within canopies, gaps can range in diameter from millimetres to hundreds of metres. On the millimetre to centimetre scale, gaps are primarily sunflecks and can be very short in duration, often in the order of seconds to minutes (Canham *et al.*, 1990; Pearcy *et al.*, 1994). Sunflecks permit short-term increases in photosynthesis in leaves within the canopy (Chazdon and Pearcy, 1986). These small and short-term gaps, however, are not of sufficient duration to permit physiological acclimation by leaves (Chazdon and Field, 1987; Sims and Pearcy, 1993), and have minimal effects on water status and soil temperature (Bazzaz and Wayne, 1994).

Of importance to canopy development, however, are larger canopy gaps which affect the morphological development and regeneration of the canopy. In forest communities, gaps that function for community regeneration have been identified in several cases to be the size of one or more individual trees (e.g. Runkle, 1990; Spies *et al.*, 1990; Dahir and Lorimer, 1996). In addition, the size of tree creating the gap can be an important factor in gap size (Abe *et al.*, 1995). The gap size necessary to affect the understorey can vary depending on canopy height. In 70 m tall, old-growth Douglas fir/hemlock forests, it was found that single-tree gaps were not sufficient to affect understorey light regimes, but single-tree gaps were sufficient in four other shorter (25–35 m) forest types (Canham *et al.*, 1990). Smith and co-workers (1989) suggested that the ratio of gap diameter to canopy height is a useful measure of gap function, although they applied this ratio primarily to penumbral effects. Latitude also affects the distribution of light within canopy gaps, with less light reaching the centre of the gap at higher latitudes and more reaching vegetation bordering the gap edge (Canham *et al.*, 1990). Thus, larger gaps may be necessary at higher latitudes to have the same effect on the understorey as smaller gaps in low latitudes.

Because of their much smaller stature, grasslands communities respond to much smaller gaps. In short-grass steppe, gaps of at least 50 cm diameter were considered necessary for regeneration where rooting density was sufficiently low (Hook *et al.*, 1994). Even smaller gaps, on the scale of 1 cm^2 were sufficient for seasonal species replacement in a species-rich, permanent pasture in North Wales (Thórhallsdóttir, 1990). Rusch and van der Maarel (1992) suggest that regeneration gaps need not be larger than single seeds in some grass communities.

Gap size can affect resource availability and environmental conditions, and, in turn, the species most suited to exploit the gap. Larger gaps permit more direct beam radiation to penetrate to lower canopy plants and to the canopy floor (Smith *et al.*, 1989). In addition, the direct beam radiation can penetrate for longer periods of time in larger gaps (Canham *et al.*, 1990). As a result, smaller gaps often contain more shade-tolerant species in forests, and shade-intolerant species can be more common in larger patches (Forcier, 1975; Barden, 1981; Runkle, 1982; Abe *et al.*, 1995). Nutrient concentrations and soil moisture can also vary with gap size, influenced by differences in temperature, rooting density and microbial activity (Lee, 1978; Phillips and Shure, 1990; Parsons *et al.*, 1994). Increases in soil nutrients and moisture can enhance fine root growth in gaps (Raich 1980; Sanford, 1989, 1990), and thus resource acquisition. Resource levels vary with gap size, so that different plant species can benefit. Species richness can also increase with gap size (Phillips and Shure, 1990), and may be associated with more variability in resources in the large gap (McClure and Lee, 1993).

Canopies often have a range of gap sizes, reflecting the varying sizes or numbers of plants creating the gap. In many plant communities, ranging from forests (Welden *et al.*, 1991; Dahir and Lorimer, 1996) to grasslands (Hook *et al.*, 1994), small gaps are much more numerous than large gaps. The distribution of gap sizes has been described as power-law in a Panamanian tropical forest (Manrubia and Solé, 1997), negative exponential in northern Appalachian spruce-fir forest (Foster and Reiners, 1986), and log-normal in North American old-growth mesic forests (Runkle, 1982). A variety of canopy gap sizes has been linked to higher species diversity in some communities (McClure and Lee, 1993; Abe *et al.*, 1995; Solé and Manrubia, 1995).

9.3.2 Temporal aspects of canopy gaps

Another key attribute of canopy gaps is their temporal dynamics. The length of time gaps are 'open' and the time between recurrence of a gap is important in the response of plants. Recurrence intervals can range from seconds to many years, depending on the source and community type (Figure 9.2).

Canopy gaps caused by foliage movement (sunflecks) are typically too fast for direct structural response. The recurrence intervals of these transient gaps are in the order of seconds or minutes. However, these gaps can be very important for carbon assimilation of understorey plants. In forests, understorey plants may obtain 10–70% of their daily carbon gain from these sunflecks (Pearcy and Calkin, 1983; Weber *et al.*, 1985; Pearcy, 1987; Pfitsch and Pearcy, 1989a,b). Differences in the portion of carbon assimilation from sunflecks reflect variable physiological responses among understorey species and perhaps different diffuse light climates among canopies (Pearcy *et al.*, 1994).

Canopy gaps of longer duration and interval of recurrence are important for structural and community characteristics. In forests, gaps caused by the mortality of a mature tree would recur in the maximum lifetime of a tree, 100–400 yrs for many species (Shugart, 1984). However, given that many mature trees do not reach maximum age before death, the average gap turnover rate is often less than this. Simple exponential mortality rates (e.g. Botkin, 1993) would result in a mean lifespan of 25–100 yrs for trees living maximally 100–400 yrs. Such differences in mean gap formation rates and maximum lifespan have been measured in hardwood and conifer stands (Dahir and Lorimer, 1996), and in tropical forests (Lieberman *et al.*, 1985). Forest turnover from gaps created by treefall can be lengthy, with estimates of 300–500 yrs for southern beech forests (Rebertus and Veblen, 1993). Lieberman and co-workers (1985), however, point out that there may be no close correspondence between gap formation rates and tree mortality because trees can die without creating a significant gap and several trees may die in a single gap.

In many forest communities, gap formation occurs more frequently than would result from old-age mortality of mature trees. Instead, factors such as fire, wind, flooding or human disturbance may form gaps in shorter time intervals. These disturbances are temporally more variable and can result in gaps of much greater extent. Also, these disturbances can produce synchronized (as opposed to temporally random) gap formation (Iwasa and Kubo, 1995), and large scale disturbance may result in periods of extensive even-aged forests (Sakio, 1997). In addition, trees removed by these exogenous events may be of all ages, instead of just older, mature trees (Greenberg and McNab, 1998). If gap formation occurs frequently enough, many trees may never reach maturity (O'Brien *et al.*, 1992).

Periodicity of gaps can be less than a year. Shedding of leaves in deciduous forests creates gaps, which may be exploited by understorey species. Seasonal variability in growth responses can occur among species and affect gap dynamics (Cornelissen, 1996). In some grassland communities, short life-spans can result in the rapid recurrence of gaps, which can be filled by a variety of species (Walker and Peet, 1983; Willems *et al.*, 1993; Sykes *et al.*, 1994).

The rate at which gaps form is another aspect of gap dynamics. Gaps may form quickly, through treefall, or more gradually, with loss of branches over a period of several years as a result of insect infestation, chronic wind stress or disease (Worrall and Harrington, 1988; Krasny and Whitmore, 1992). Also, gaps created by a treefall may enlarge over time with additional loss of trees along the edges due to the increased vulnerability of the bordering trees (Kubo *et al.*, 1996). This has been observed in a variety of forest types, including spruce-fir (Foster and Reiners, 1986; Perkins *et al.*, 1992), deciduous old-growth forest (Runkle, 1984), tropical rainforest (Hubbell and Foster, 1986; Kanzaki *et al.*, 1994) and tropical cloud forest (Lawton and Putz, 1988). Temporal availability of resources may differ between these slowly forming gaps and those formed quickly (Krasny and Whitmore, 1992), with consequences for the species which can best exploit these gaps.

9.4 Response of individual plants to gaps

Plants respond to changes in the resource availability and microclimatic conditions of canopy gaps. These responses can occur in one of three ways: physiological, structural and seed germination. The response of a plant to a canopy gap depends on the phenological stage and physiology of the species involved, and the characteristics of the gap. Physiological responses can affect carbon assimilation performance and water-use efficiency, while seed germination and structural responses affect filling and canopy recovery in gaps.

9.4.1 Physiological response to gaps

Plants respond physiologically to gaps through changes in photosynthesis, respiration and stomatal conductance. The physiological response depends on the exposure time to increased light in the gap. Sunflecks are the shortest duration canopy gaps eliciting physiological responses. These short-term gaps can expose leaves to greatly increased levels of sunlight for periods of a few seconds to minutes. The response of a leaf to this brief increase in light intensity depends on its state of light induction, stomatal behaviour and physiological adaptation. Lightflecks can be responsible for much of the photosynthesis of an understorey plant.

Leaves that have been shaded or in darkness for extended periods show a slow induction response when suddenly exposed to a much brighter light (Schreiber *et al.*, 1994). When a sunfleck strikes a shaded leaf, the leaf exhibits a 10–30 min rise in the rate of photosynthesis to a maximum, light-saturated rate. However, for a leaf recently exposed to high light, reaching full induction and then shaded for a minute or two prior to the sunfleck, the transition to the maximum rate of photosynthesis is much quicker, in the order of a minute (Pearcy *et al.*, 1994). On repeated exposure to sunflecks, an initially shaded leaf exhibits increasingly rapid transition to the maximum rate (Pearcy *et al.*, 1985). At the end of sunfleck exposure, the leaf continues with a higher rate of photosynthesis for a short period, upwards of a minute for an uninduced leaf but only a few seconds for an induced leaf (Chazdon and Pearcy, 1991). As a result of the post-illumination period increase in photosynthesis, the light-use efficiency of the leaf can actually be greater than 100% for light flecks of 1–10 s duration (Chazdon and Pearcy, 1986).

The stomatal aperture also affects the response of leaves to sunflecks. Low stomatal conductance, resulting from low light or low humidity, can limit the rate of photosynthesis during induction (Pearcy *et al.*, 1994). In addition, it may take 15–20 min for stomatal conductance to fully respond to a sunfleck (Kirschbaum *et al.*, 1988). The rate of stomatal opening with exposure to light can be faster in shade-tolerant tree species than shade-intolerant species (Woods and Turner, 1971), and faster in monocotyledons than in dicotyledons (Johnsson *et al.*, 1976).

The response of leaves to newly formed canopy gaps that increase light levels over several hours each day is rather different from the response to transient sunflecks. Changes to leaves exposed to long-term gaps involve acclimation, a process which is too slow to happen during sunflecks (Pearcy and Sims, 1994). Moreover, these responses are often affected by concurrent changes in temperature, and nutrient and water availability within the gap.

Leaves of the same plant often differ physiologically and structurally depending on whether they develop in shade or sun environments. Physiological differences relate primarily to higher photosynthetic capacity in sun leaves (Björkman, 1981). This difference seems to result from different investments in electron transport and carboxylation capacities (Pearcy and Sims, 1994). Structural differences also relate to higher photosynthetic capacity in sun leaves and include higher mass per leaf area (Sims and Pearcy, 1993), higher chlorophyll content (Björkman, 1981), and higher chlorophyll *a*/*b* ratios (Anderson, 1986). Leaves developing in sun environments also experience less photoinhibition when exposed to sustained high irradiances (Björkman, 1981; Anderson and Osmond, 1987). When shade leaves are suddenly exposed to high irradiances, as in a new gap, some but not all species are able to increase their photosynthetic capacity. This was found to take as long as 4–14 days in herbs (Pearcy and Sims, 1994) and up to 45 days in the evergreen, *Hedera helix* (Bauer and Thöni, 1988). Acclimation appears to involve coordinated changes in Rubisco activity, electron transport capacity, cytochrome *f* content and chlorophyll *a*/*b* ratio (Chow and Anderson, 1987a,b). When exposed to canopy gaps, acclimation in shade leaves can result in not only higher photosynthesis but also reduced photoinhibition. However, because increases in carboxylation and electron transport rates require a sizable investment in nitrogen (Evans, 1989), limited nitrogen supply may delay or limit acclimation (Ferrar and Osmond, 1986; see also Chapter 7 in the present book). Similarly, water stress may limit acclimation (Anderson and Osmond, 1987). Leaf replacement is another method of acclimation for plants exposed to a canopy gap. At present, no clear patterns emerge relating leaf senescence and production with the dynamics of photosynthetic capacity and photoinhibition (Pearcy and Sims, 1994).

9.4.2 Structural responses to gaps

In addition to leaf replacement in response to gap formation, plants can exhibit considerable structural changes to the altered environment. These structural changes are primarily directed toward capturing space within the gap. Structural responses can involve major canopy plants or understorey individuals, and form much of the basis of gap dynamics in plant communities. Gap filling can occur from existing canopy trees or from growth of understorey seedlings (Runkle, 1990). There are strong indications that morphological responses of plants are more important than physiological characteristics or responses in filling gaps (Küppers, 1985, 1994; Field, 1988;

Beyschlag *et al.*, 1992). The foliage allocation patterns, as they relate to successful growth into the open space of the gap, strongly reflect the competitive ability of individual plants.

A common response to newly formed gaps is for large canopy plants to grow new branches into the gap. This response is most prominent in small canopy gaps where existing, dominant canopy plants can fill the newly formed space leaving little opportunity for plants in the understorey. While often growing quite symmetrically without competition, plants show a great deal of plasticity in morphology when competing for resources (Hallé *et al.*, 1978). When canopy gaps open, changes in red/far-red (R/FR) ratios of light reflected from neighbouring plants can be detected by plants through phytochrome receptors (Ballaré *et al.*, 1992; see Chapter 5). Changes in the R/FR ratio and, in some cases, blue/UV-A light levels due to gap formation can increase stem elongation properties (Smith, 1982), affect branch orientation (Novoplansky *et al.*, 1990) and increase branching as a result of reduction in apical dominance (Tucker and Mansfield, 1972), all promoting the filling of light gaps in the canopy (Ballaré, 1994). Total irradiance may also affect branch orientation and the ability to gain light higher in the canopy (Grime *et al.*, 1986).

Branching patterns are very important to plants in exploiting the space formed by a new canopy gap, and a variety of branching patterns have been identified (Küppers, 1989; Porter, 1989). Lateral spread of canopy elements, including stoloniferous species in grasslands (Thórhallsdóttir, 1990), can be most effective in filling gaps by harvesting more light while shading competitors. Multiple layers of branches filling a space, however, may be less effective, since some layers are at least partially shaded (Küppers, 1994). Established individuals of plant species that have effective branching patterns and rapid elongation rates may be the most competitive in filling gaps. However, growth in a particular direction or genetic limitation on branching and branch elongation may limit the effective capture of space in canopy gaps (Küppers, 1994).

When existing large canopy plants cannot fill a canopy gap because of branching limitation or because the gap is too large, the opening often provides the opportunity for seedlings and saplings to grow rapidly into the gap because resources are released. Some of the smaller plants, responding to the opportunity created by the larger gap, may eventually become dominant individuals in the canopy. This regeneration can be by the same or different species from the one or more removed to create the gap initially.

In many plant communities, seedlings and saplings may respond primarily to increased irradiances within the gap i.e. through considerably enhanced rates of photosynthesis leading to higher growth rates. The growth response is often similar to that of large canopy plants, with foliage increases primarily filling the open portions of the gap. Similarly, changes in R/FR ratios and blue/UV-A light levels can alter growth patterns of the seedlings to more effectively fill space both horizontally and vertically.

Canopy gaps can provide significant benefits to species which are relatively shade-intolerant. These pioneer species are characterised by low establishment rates and high mortality in shade environments but high growth rates when light is available. With gaps providing sufficient light for establishment and growth, such species can regenerate and survive as part of the mature canopy crown (Akashi, 1996). Less shade-tolerant species are often more abundant in larger gaps and gap centres (Brokaw, 1985; Runkle, 1990; McClure and Lee, 1993). It appears that many of these relatively shade-intolerant species are able to dominate species with greater shade-tolerance when more light is available. Thus, persistence of shade-intolerant species in a community is dependent upon the frequency of occurrence of large gaps.

In contrast to shade-intolerant species, some shade-tolerant species require shade and can grow to nearly full height without sizable gap formation. These shade-obligate species are characterised by high survival in understorey sites and low survival in gaps. However, some gap is eventually necessary if the plant is to become part of the crown, and death or loss of foliage by other individuals is necessary for full crown development in this case. The growth of these species in the understorey can differ with the shading canopy species, and in some cases even exhibit negative associations with mature conspecific individuals (Akashi, 1996). Such negative association may result from differences in shade tolerance and canopy density below overstorey trees, significant competition resulting from overlap in root distribution between saplings and canopy trees of the same species, or ease of pathogen movement between overstorey trees and saplings of conspecifics (Woods, 1979,1984).

Many species, however, are more generalist and, after becoming established in the understorey, have low growth rates. These species are shade-tolerant and have relatively high survival rates both in the understorey and in gaps. They remain relatively small in the understorey for many years (Silvertown and Doust, 1993), even for periods of 150 yrs (Rebertus and Veblen, 1993). When a gap opens, they can respond quickly with increased growth rates to fill the gap. These growth rates can exceed the rate of branch elongation by large canopy trees (Runkle, 1990). Within some plant communities, this can be the predominant growth response of understorey saplings exposed to a gap (Welden *et al.*, 1991; Midgley *et al.*, 1995; Lieberman *et al.*, 1995). Because small individuals of these generalist species are already established in the understorey, establishment by shade-intolerant species when a gap opens can be greatly reduced and limited to the largest gaps (Runkle, 1990; Abe *et al.*, 1995).

Other resources also affect seedling and sapling growth in gaps. Soil moisture and nutrient availability can be greater in gaps leading to increased growth rates. They can also become limiting, reducing the growth potential of gap filling individuals (van der Maarel, 1996).

The density of seedlings and saplings established in the canopy understorey are often much higher than the density of canopy plants. High densities of individuals can persist in the understorey due to their relatively small size. However, when canopy gaps open and rapid growth occurs, the number of individuals declines due to limited resources and space. Individuals that are most able to sequester resources are most likely to survive the filling of gaps. Mortality can be density-dependent within a species or group of species (Kohyama, 1993; Alvarez-Buylla, 1994), or can be differentially focused on individuals of species which are less able to sequester resources (Shugart, 1984). Density-independent mortality of seedlings and saplings can also occur, resulting from such factors as drought, closing of the gap by lateral growth of dominant canopy plants and fall of individual trees along the edges of gaps.

9.4.3 Seed germination in gaps

Seed germination is another response of plants to the altered environment in canopy gaps. This response is typical of pioneering species that have little chance of survival in an understorey setting. However, seed dispersal patterns and microsite niche differences can have a profound influence on the spatial distribution of seedlings and subsequent response to gap formation.

The altered environment in gaps provides cues that can be detected by some seeds, thus promoting germination. Lack of these cues results in dormancy. The greatest environmental change within gaps is an increase in light intensity. This is also accompanied by an increase in the R/FR ratio of light. Many photoblastic seeds, those requiring light for germination, respond through light-sensitive phytochrome pigments. These pigments respond primarily to increases in photosynthetically active radiation and changes in the R/FR ratio. In canopy gaps, seeds can be directly exposed to changes in the light environment or moved to the surface by uprooting vegetation (Vázquez-Yanes and Orozco-Segovia, 1994). Changes in the light environment must be of sufficient duration to allow for germination; short-term sunflecks are not sufficient (Duke *et al.*, 1977; Vázquez-Yanes and Orozco-Segovia, 1987).

Other environmental changes can also trigger germination of seeds. Diurnal temperature fluctuations, which can increase in gaps, have been shown to enhance germination of seeds with or without exposure to increased light (Thompson *et al.*, 1977; Thompson and Grime, 1983). Increased temperature resulting from exposure to fire or to increased soil temperature can also break dormancy in seeds (Cushwa *et al.*, 1967; Vázquez-Yanes, 1974).

Increased nitrogen availability has been linked to breaking dormancy and promoting seed germination (Pons, 1989).

A litter layer on the soil surface can inhibit seed germination in gaps, by shading photoblastic seeds and reducing temperature fluctuations (Beatty and Sholes, 1988). Chemical inhibitors in the litter can also reduce germination rates (Facelli and Pickett, 1991). As part of gap formation, disturbance of the litter layer can be important in the establishment of pioneer species (Putz, 1983; Carson and Peterson, 1990). Where such disturbance is limited, species with photoblastic seeds may be relatively rare within the community (Vázquez-Yanes and Orozco-Segovia, 1994). Other physical factors within gaps can be important for seed germination e.g. the presence of decaying logs (Botkin, 1993) and optimal soil texture (Coffin and Lauenroth, 1994).

Seed distribution is an important aspect of seedling establishment in gaps. Seed distribution through the canopy floor can vary greatly depending on dispersal mechanisms. Where dispersal is limited, most seeds are found close to the parent plants. This may facilitate establishment of seedlings near gap-forming parents (Itoh *et al.*, 1997). For pioneering species, dispersal mechanisms that ensure extensive coverage are important in successfully establishing in randomly forming gaps (Estrada *et al.*, 1984; Vázquez-Yanes and Orozco-Segovia, 1994). Where seed germination occurs without gap formation, the distribution of seeds is still important, determining the quantity and composition of seedlings that are available to exploit a subsequent gap.

9.5 Canopy dynamics

As mentioned previously, canopy gaps which can elicit structural changes in canopies can have profound influences on community structure and dynamics. Although plants respond to gaps through changes in physiology, structure and germination in all plant communities, there is much variability in how a specific community will respond to gaps. In addition, the responses to gaps are not always consistent across similar communities. However, three general facets of community dynamics are affected by gap dynamics: community regeneration, succession and maintenance of species diversity. Central to these is the concept that plant communities are temporally and spatially dynamic, not static entities (Watt, 1947; Remmert, 1991).

9.5.1 Community regeneration

Mature plant communities can be maintained by the regeneration of dominant canopy individuals through recruitment in gaps. Mature, closed canopies provide little opportunity for young individuals to become part of the canopy crown. However, when large canopy individuals die, gaps are created, allowing new individuals to grow and eventually become part of the canopy crown (Figure 9.1).

By definition, dominant canopy species must be able to successfully reproduce and grow into the canopy gaps if the existing communities are to be regenerated. This may be: from seed germination and seedling establishment in the light-enhanced environment of the gap; from the accelerated growth of seedlings already established in the understorey; or by clonal propagation from the dominant individuals in the canopy neighbouring the gap.

Recruitment rates must match mortality rates for the species composition of mature plant communities to remain constant. If replacement of individuals favours some species over others, the community will change composition. However, mechanisms to maintain similar species composition are not fully understood. Possible mechanisms include: enhanced survival in association with co-occurring species (Runkle, 1990; Akashi, 1996); constancy in the stochasticity of growth response to environment (Bugmann, 1997); and the fractal nature of gap-size distributions (Solé and Manrubia, 1995).

9.5.2 Community succession

When the replacement of large canopy individuals does not match the existing species distribution, the community composition changes. These community changes are predictable, based on the character of the species involved, and are typically from early to late successional species (Figure 9.3). For example, when the canopy is mostly open, pioneering, shade-intolerant species can dominate because they rapidly establish and grow to form the canopy. However, the shade environment induced by these individuals is not conducive to re-establishment, and shade-tolerant species dominate the understorey. When a dominant canopy tree falls, it is more likely to be replaced by an individual from the species already established in the understorey than from germination of seed from the shade-intolerant species. Ultimately, the species composition changes.

The actual evolution of the community composition is dependent on the availability of seeds from appropriate successional species. If seeds of shade-tolerant species are not available, the community may continue to regenerate itself with pioneer species. These seeds may be from adjacent canopy crowns, transported via effective dispersal mechanisms or left over in the soil seed bank. The frequency of recurring large-scale disruptive events is also important, each major event returning the site towards a pioneering state favouring continued re-colonisation by shade-intolerant species.

Species replacement during succession involves four processes: facilitation, interference, tolerance and chronic patchiness (Connell and Slatyer, 1977). Facilitation occurs when one species is necessary for the occurrence

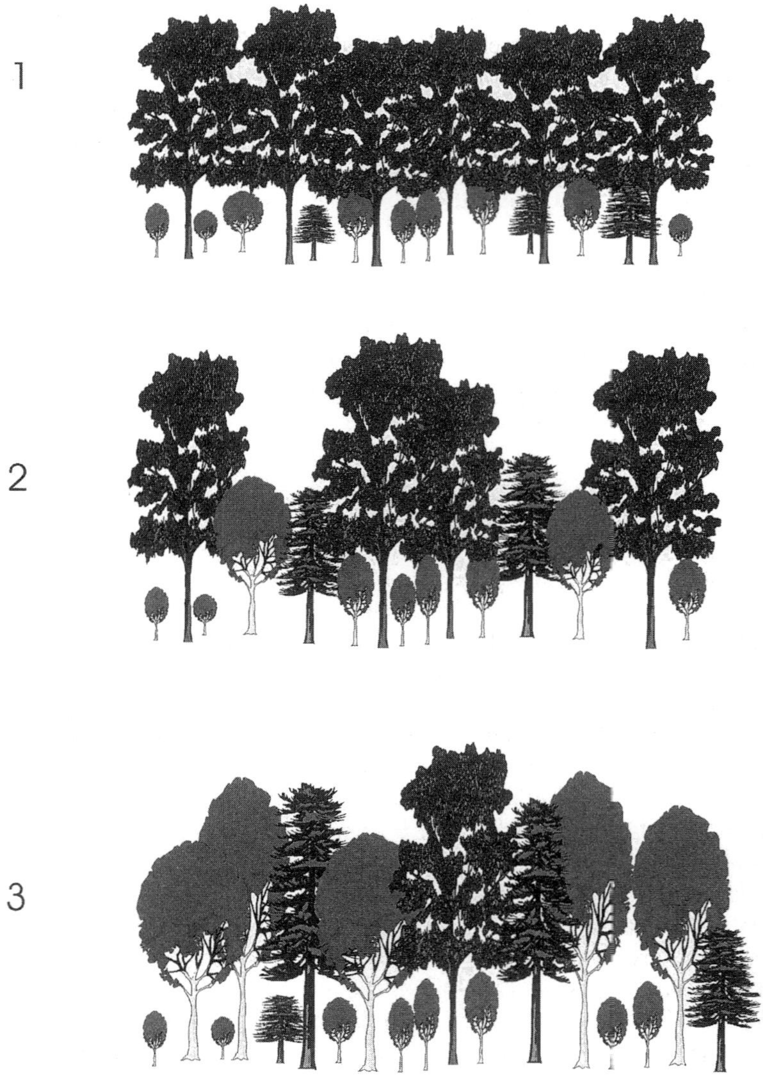

Figure 9.3 Hypothetical cross-section of forest canopy gap formation and filling, resulting in succession from a shade-intolerant single species to two other shade-tolerant species in three sequential time periods. An old individual of the shade-intolerant species persists to the third time period adding to species diversity and providing seeds for potential exploitation of a large (multi-treefall) gap.

of a second species. Interference results when one species restricts the establishment of other individuals, while tolerance means neither species affects the growth, recruitment or survival of the other. Chronic patchiness occurs when disturbances are frequent, and species tend to replace one another at random. The importance of each of these mechanisms varies greatly between communities and among species, resulting in different patterns and rates of succession. Within forests, interference and tolerance alone can account for much of the canopy dynamics during successional changes (Botkin, 1993).

Succession also works on relatively stable communities, but over longer time scales, where environmental conditions are only slowly shifting. Such slow changes in community composition have been documented for forests over the past 10,000 years (Davis, 1981), and are important topics in assessing recent trends in global climatic change (Loehle and LeBlanc, 1996; Shugart and Smith, 1996).

9.5.3 Species diversity

The dynamics of canopy gaps have been implicated in maintaining or increasing species diversity. Probably, the most cited mechanism is that gaps provide possibilities for niche differentiation (Ricklefs, 1977). While the environment under a rather homogeneous canopy tends to be relatively uniform, canopy gaps provide much heterogeneity. Variation in gap size and frequency of recurrence enhances heterogeneity. Species may respond differentially to this heterogeneity, forming associations with specific attributes resulting in niche differentiation.

In a simple plant community, frequent gap formation can help to maintain shade-intolerant species within the community by providing sufficient light levels in gaps. This can allow for coexistence of shade-tolerant and shade-intolerant species. In the extreme, tropical rainforests maintain very high species diversity due to the environmental variability induced by gap dynamics in an otherwise relatively stable environment (Solé and Manrubia, 1995).

Gap dynamics, however, may not be the only major factor affecting species diversity. Previous studies (Welden *et al.*, 1991; Lieberman *et al.*, 1995; Midgley *et al.*, 1995) have indicated that in three different tropical forests with relatively high species diversity, species composition appeared to be established in the understorey prior to gap formation. This suggests that the species diversity paradigm as related to gaps for these forests needs reviewing. In a simulation study, Kohyama (1993) showed that species diversity without gap formation could be maintained with a limited range of species life-history characteristics, but that gap dynamics widened the permissible range of these characteristics that would allow for coexistence. Inclusion of gap dynamics also reduced the convergence time to equilibrium.

9.6 Gap models

Many simulation models have been constructed to investigate the effects of gaps on the structure and dynamics of the community. Model simulation has been very important in assessing complex dynamics over long time periods. These gap models are generally of two types: individual-based patch models and cellular automaton models. While the structure is somewhat different, the processes included in the models are similar.

Different approaches to modelling canopy growth and dynamics have been developed. Models of individual plant growth were developed to follow changes in growth of each plant within a canopy and have been applied to single- or multiple-species canopies. Larger-scale canopy growth models follow changes in total canopy biomass and cover but do not consider plants individually. Models of canopy gap dynamics may be similar in structure to the individual plant growth models, but are characterised by the inclusion of plant regeneration in canopy gaps and generally include two or more species.

Gap models typically include the following dynamics: recruitment, growth, symmetric and asymmetric competition, resource limitations and mortality (Shugart, 1984). The detail of each of these processes varies greatly among different models, depending on the study objectives and information available. For example, recruitment in forest applications often begins with numbers of established seedlings and does not consider the complex and often unknown dynamics of seed viability and germination (Botkin, 1993).

9.6.1 Individual-based patch models

Stand simulation models have been widely used to study the community responses to canopy gaps. Most of these models have been developed for forest canopies (e.g. Botkin *et al.*, 1972; Shugart, 1984), but they have also been applied to quite different communities, such as grasslands (Coffin and Lauenroth, 1990; Coffin and Urban, 1993). These gap models simulate establishment, growth and mortality of individual plants within a small patch or gap (Figure 9.1). The plants within a patch may be of one or more species.

The size of the simulated gap is generally based on the zone of influence of a dominant canopy individual. This is typically 500–1000 m^2 for forests down to 0.1 m^2 in grasslands (Coffin and Lauenroth, 1990; Loehle and LeBlanc, 1996). Resource availability is typically averaged over the patch, with acquisition determined by the size of an individual plant. With larger gap size, the effect of the dominant individual is reduced, and extremely large plots lose the character of gap dynamics (Shugart, 1984). An entire community is simulated by modelling a large number of these small patches; at least 200 is preferable (Bugmann *et al.*, 1996a). Alternately, simulation on one plot can be conducted for extended periods (e.g. 31,000 yrs in Bugmann, 1997) to determine steady-state composition. The time-step of simulation is often a year, but can be shorter when grass or herbaceous canopies are considered.

The growth functions of these gap models (Figure 9.4) are calculated for each species under optimal conditions and they can pertain to incremental growth in height, diameter or volume. Often, relationships are made between these growth parameters (see e.g. Botkin, 1993; Shugart and Smith, 1996). The maximal growth rates are reduced by functions that relate growth response to environmental limitations, such as light, water, nutrients and temperature (Figure 9.4).

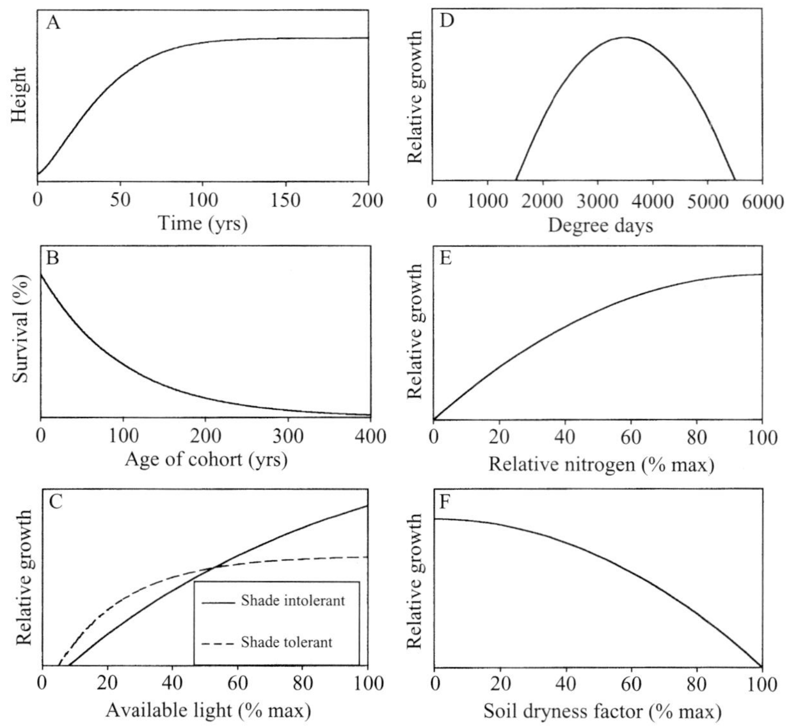

Figure 9.4 Example of growth and survival functions for an individual-based patch model. A) shows the maximum growth without resource limitations; B) survival rate of individual trees of the same age; C) relative growth rate of a shade-tolerant (- - - -) and shade-intolerant (——) species; D) growth response to accumulated temperature; E) relative growth related to nitrogen availability; and F) relative growth rate related to water availability. Relationships modified from Shugart (1984) and Botkin (1993).

Mortality of plants is modelled as a stochastic process and can have three components. Firstly, a random percentage of individuals die in a year (e.g. 1–2% of a tree population). Secondly, reduced growth rate from limitation in resources or environmental conditions is related to an increased probability of mortality, and a minimum threshold of resources can be set below which the plant always dies. Finally, mortality can result from random events not occurring on an annual basis, such as fires or severe windstorms.

When mortality occurs, space is opened up for colonisation by seedlings or existing understorey plants (Figure 9.1). A stochastic process is linked to establishment by all species in the canopy. Multiple individuals may become established, with survival determined by the growth response to environmental conditions and random mortality. Establishment patterns can also be considered as transitions between states based on the dominant species in the gap, and Markov chains have been used in this context to describe transition probabilities (Acevedo *et al.*, 1996).

9.6.2 Cellular automaton models

Cellular automaton models, also known as 'game-of-life models', are designed for complex systems with local interactions and consist of a two-dimensional matrix of points (Wolfram, 1984). The current value at each automaton (lattice point) is dependent on the previous values of a defined neighbourhood of surrounding automata. Each automaton represents a plant species (tree in forest model) or a gap. Since interactions at the automaton are influenced by the local environment, these models can simulate much larger plots than are typical with individual-based patch models (Figure 9.5).

As with individual-based patch models, probabilities are assigned to each plant (automaton) for growth, death and recruitment, and all can be linked to environmental parameters. The number of neighbouring automata that affect an automaton is determined by the range of influence, and may vary by resource. Gap formation occurs when a plant dies and the size of the gap can be related to the size of the dead plant. Mortality in adjacent cells can be increased to simulate the process of widening gaps (Kubo *et al.*, 1996). The dimensions of the matrix need to be sufficient to avoid finite-size effects, with 100×100 cells probably being a suitable minimum (Solé and Manrubia, 1995; Kubo *et al.*, 1996). Multiple simulation runs are necessary to fully characterise the dynamics. The time step is similar to that for the individual-based patch models.

9.6.3 Applications

Gap models have been applied to address a variety of community-based questions; in particular, to demonstrate that gap dynamics can account for the stable species compositions observed in communities (Doyle, 1981; Shugart,

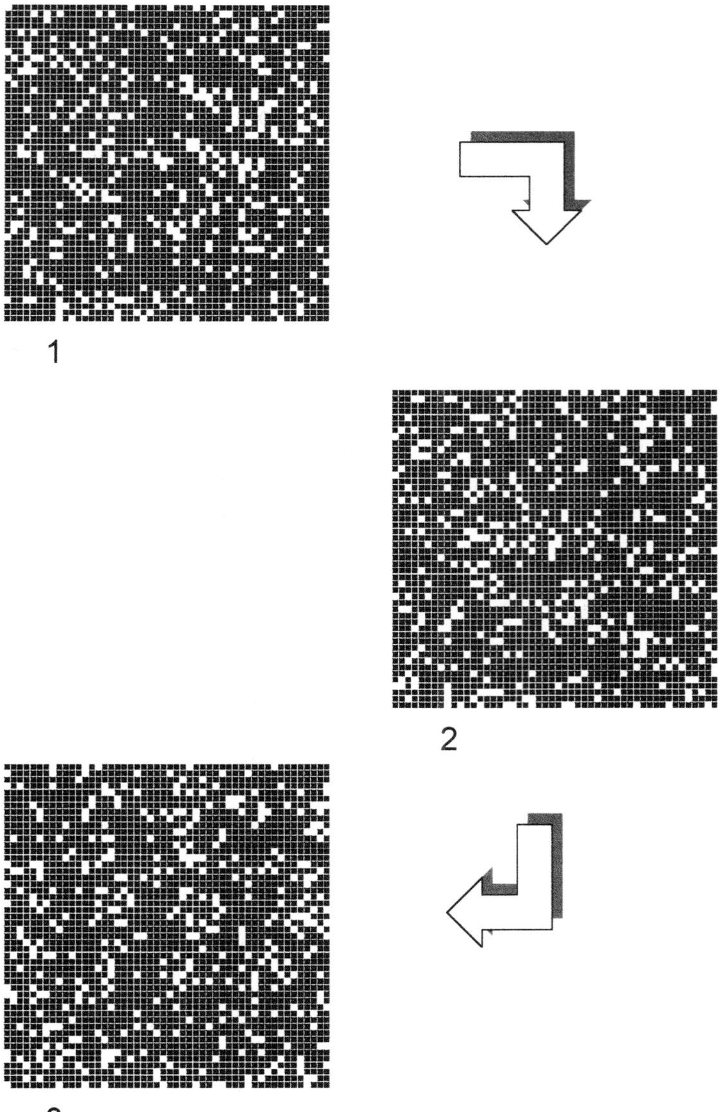

Figure 9.5 Gap dynamics as simulated with cellular automaton model in three sequential time periods. Each cell represents an individual plant with random mortality in any time step. Total gap area is approximately 15% of the total area.

1984; Botkin, 1993), successional changes (Shugart and Noble, 1981; Botkin, 1993), and maintenance of species diversity (Solé and Manrubia, 1995). In addition, they have been used to show that gap dynamics affect community structure and behaviour in contrasting communities: forests with incremental growth form and asymmetric competition and grasslands with clonal growth and more symmetric competition (Coffin and Urban, 1993).

Another area of expanding application is in assessing the response of plant communities to change in global climate. Climate is predicted to change in several characteristics, including increased CO_2 concentration, warming and changes in precipitation patterns (Gates, 1993). Gap models are closely linked to climatic variables in growth and mortality responses, and reach relative equilibrium with climate (Malanson, 1993). With changes in climatic conditions, the models may predict a new equilibrium based on the responses of species to climatic variables. However, the present models are generally lacking in sufficient mechanisms to adequately address this issue (Loehle and LeBlanc, 1996), and respond too sensitively to changes in climate (Bugmann *et al.*, 1996b). More physiologically based models for growth would help reduce the dependence on the temperature response (e.g. Friend and Shugart, 1993). Better, more detailed representations of the changes in light environment as canopy composition changes, should also improve the accuracy of such models (e.g. Wang and Jarvis, 1990; Ryel *et al.*, 1993). However, rates of change in vegetation, compounded by slow migration of species from other regions and possible changes in disturbance regimes will continue to be sources of uncertainty, limiting the precision of model predictions (Loehle and LeBlanc, 1996).

Acknowledgement

This chapter was funded in part by NSF grant DEB-9807097.

References

Abe, S, Masaki, T. and Nakashizuka, T. (1995) Factors influencing sapling composition in canopy gaps of a temperate deciduous forest. *Vegetatio*, **120** 21–31.

Acevedo, M.F., Urban, D.L. and Shugart, H.H. (1996) Models of forest dynamics based on roles of tree species. *Ecological Modelling*, **87** 267–284.

Akashi, N. (1996) The spatial pattern and canopy-understory association of trees in a cool temperate, mixed forest in western Japan. *Ecological Research*, **11** 311–319.

Alvarez-Buylla, E.R. (1994) Density dependence and patch dynamics in tropical rain forests: matrix models and applications to a tree species. *American Naturalist*, **143** 155–191.

Anderson, J.M. (1986) Photoregulation of the composition, function and structure of the thylakoid membranes. *Annual Review of Plant Physiology*, **37** 93–136.

Anderson, J.M. and Osmond, C.B. (1987) Shade-sun responses: compromises between acclimation and photoinhibition, in *Photoinhibition* (eds. D.J. Kyle, C.B. Osmond and C.J. Arntzen), Elsevier, Amsterdam, The Netherlands, pp. 1–38.

Ballaré, C.L. (1994) Light gaps: sensing the light opportunities in highly dynamic canopy environments, in *Exploitation of Environmental Heterogeneity by Plants* (eds. M.M. Caldwell and R.W. Pearcy), Academic Press, New York, NY, USA, pp. 73–110.

Ballaré, C.L., Scopel, A.L., Radosevich, S.R. and Kendrick, R.E. (1992) Phytochrome-mediated phototropism in de-etiolated seedlings: occurrence and ecological significance. *Plant Physiology*, **100** 170–177.

Barden, L.S. (1981) Forest development in canopy gaps of a diverse hardwood forest of the southern Appalachian Mountains. *Oikos*, **37** 205–209.

Bauer, H. and Thöni, W. (1988) Photosynthetic light acclimation in fully developed leaves of the juvenile and adult life cycle phases of *Hedera helix*. *Physiologia Plantarum*, **73** 31–37.

Bazzaz, F.A. (1983) Characteristics of populations in relation to disturbance in natural and man-modified ecosystems, in *Disturbances in Ecosystems* (eds. H.A. Mooney and M. Gordon), Springer-Verlag, New York, USA, pp. 259–275.

Bazzaz, F.A. and Wayne, P.M. (1994) Coping with environmental heterogeneity: the physiological ecology of tree seedling regeneration across the gap-understory continuum, in *Exploitation of Environmental Heterogeneity by Plants* (eds. M.M. Caldwell and R.W. Pearcy), Academic Press, New York, NY, USA, pp. 349–390.

Beatty, S.W. and Sholes, O.D.V. (1988) Leaf litter effect on plant species composition of deciduous forest treefall pits. *Canadian Journal of Forest Research*, **18** 553–559.

Beyschlag, W., Ryel, R.J. and Ullmann, I. (1992) Experimental and modelling studies of competition for light in roadside grasses. *Botanica Acta*, **105** 285–291.

Björkman, O. (1981) Responses to different quantum flux densities, *Encyclopedia of Plant Physiology, New Series, Volume 12A* (eds. O.L. Lange, P.S. Nobel, C.B. Osmond and H. Ziegler), Springer-Verlag, Berlin, Germany, pp. 57–107.

Botkin, D.B. (1993) *Forest Dynamics, An Ecological Model*, Oxford University Press, Oxford, UK.

Botkin, D.B., Janek, J.F. and Wallis, J.R. (1972) Some ecological consequences of a computer model of forest growth. *Journal of Ecology*, **60** 849–873.

Brokaw, N.V.L. (1985) Gap phase regeneration in a tropical forest. *Ecology*, **66** 682–687.

Brokaw, N.V.L. and Franklin, S.M. (1989) Species composition in gaps and structure of a tropical forest. *Ecology*, **70** 538–541.

Bugmann, H. (1997) An efficient method for estimating the steady-state species composition of forest gap models. *Canadian Journal of Forest Research*, **27** 551–556.

Bugmann, H., Fischlin, A. and Kienast, F. (1996a) Model convergence and state variable update in forest gap models. *Ecological Modelling*, **89** 197–208.

Bugmann, H.K.M., Yan, X., Sykes, M.T., Martin, P., Lindner, M., Desanker, P.V. and Cumming, S.C. (1996b) A comparison of forest gap models: model structure and behavior. Climatic Change, **34** 289–313.

Canham, C.D. (1998) Growth and canopy architecture of shade-tolerant trees: the response of *Acer saccharum* and *Fagus grandifolia* to canopy gaps. *Ecology*, **69** 786–795.

Canham, C.D., Denslow, J.S., Platt, W.J., Runkle, J.R., Spies, T.A. and White, P.S. (1990) Light regimes beneath closed canopies and tree-fall gaps in temperate and tropical forests. *Canadian Journal of Forest Research*, **20** 620–631.

Carson, W.P. and Peterson, C.J. (1990) The role of litter in an old-field community: impact of litter quantity in different seasons on plant species richness and abundance. *Oecologia*, **85** 8–13.

Chazdon, R.L. and Field, C.B. (1987) Determinants of photosynthetic capacity in six rainforest *Piper* species. *Oecologia*, **73** 222–230.

Chazdon, R.L. and Pearcy, R.W. (1986) Photosynthetic responses to light variation in rain forest species. II. Carbon gain and light utilization during lightflecks. *Oecologia*, **69** 524–531.

Chazdon, R.L. and Pearcy, R.W. (1991) The importance of sunflecks for forest understory plants. *BioScience*, **41** 760–766.

Chow, W.S. and Anderson, J.M. (1987a) Photosynthetic responses of *Pisum sativum* to an increase in irradiance during growth. I. Photosynthetic activities. *Australian Journal of Plant Physiology*, **14** 1–8.

Chow, W.S. and Anderson, J.M. (1987b) Photosynthetic responses of *Pisum sativum* to an increase in irradiance during growth. II. Thylakoid membrane components. *Australian Journal of Plant Physiology*, **14** 9–19.

Coffin, D.P. and Lauenroth, W.K. (1990) A gap dynamics simulation model of succession in a semiarid grassland. *Ecological Modelling*, **49** 229–266.

Coffin, D.P. and Lauenroth, W.K. (1994) Successional dynamics of a semiarid grassland: effects of soil texture and disturbance size. *Vegetatio*, **110** 67–82.

Coffin, D.P., and Urban, D.L. (1993) Implications of natural history traits to system-level dynamics: comparisons of a grassland and a forest. *Ecological Modelling*, **67** 147–178.

Connell, J.H., and Slatyer, R.O. (1977) Mechanisms of succession in natural communities and their role in community stability and organization. *American Naturalist*, **111** 1119–1144.

Cornelissen, J.H.C. (1996) Interactive effects of season and light environment on growth and leaf dynamics of evergreen tree seedlings in the humid subtropics. *Canadian Journal of Botany*, **74** 589–598.

Cushwa, C.T., Martin, R.E. and Miller, R.L (1967) The effects of fire on seed germination. *Journal of Range Management*, **21** 250–254.

Dahir, S.E. and Lorimer, C.G. (1996). Variations in canopy gap formation among developmental stages of northern hardwood stands. *Canadian Journal of Forest Research*, **26** 1875–1892.

Davis, M.B. (1981) Quaternary history and the stability of forest communities, *Forest Succession: Concepts and Application* (eds. D.C. West, H.H. Shugart and D.B. Botkin), Springer-Verlag, New York, USA, pp. 132–153.

de Hullu, E. and Gimingham, C.H. (1984) Germination and establishment of seedlings in different phases of the *Calluna* life cycle in a Scottish heathland. *Vegetatio*, **58** 115–121.

Doyle, T.W. (1981) The role of disturbance in the gap dynamics of a montane rain forest: an application of a tropical forest succession model, in *Forest Succession: Concepts and Application* (eds. D.C. West, H.H. Shugart and D.B. Botkin), Springer-Verlag, New York, USA, pp. 56–73.

Duke, O.S., Egley, G.H. and Reger, B.J. (1977) Model for variable light sensitivity in imbibed dark-dormant seeds. *Plant Physiology*, **59** 244–249.

Estrada, A., Coates-Estrada, R. and Vázquez-Yanes, C. (1984) Observation on fruiting and dispersers of *Cecropia obtusifolia* at Los Tuxtlas, Mexico. *Biotropica*, **16** 315–318.

Evans, J.R. (1989) Photosynthesis: the dependence on nitrogen partitioning, *Causes and Consequences of Variation in Growth Rate and Productivity of Higher Plants* (eds. H. Lambers, M.L. Cambridge, H. Konings and T.L. Pons), SPB Academic Press, The Hague, The Netherlands, pp. 159–174.

Facelli, J.M. and Pickett, S.T.A. (1991) Plant litter: its dynamics and effects on plant community structure. *Botanical Review*, **57** 1–32.

Ferrar, P.J. and Osmond, C.B. (1986) Nitrogen supply as a factor influencing photoinhibition and photosynthetic acclimation after transfer of shade-grown *Solanum dulcamara* to bright light. *Planta*, **168** 563–570.

Field, C.B (1988) On the role of photosynthetic responses in constraining the habitat distribution of rainforest plants. *Australian Journal of Plant Physiology*, **15** 343–358.

Forcier, L.K. (1975) Reproductive strategies and the co-occurrence of climax tree species. *Science*, **189** 808–811.

Foster, F.R. and Reiners, W.A. (1986) Size distribution and expansion of canopy gaps in a north Appalachian spruce-fir forest. *Vegetatio*, **68** 109–114.

Fox, J.F. (1977) Alternation and coexistence of tree species. *American Naturalist* **111** 69–89.

Friend, A.D. and Shugart, H. H. (1993) A physiology-based gap model of forest dynamics. *Ecology*, **74** 792–797.

Gates, D.M. (1993) *Climate Change and its Biological Consequences*. Sinauer Associates, Sunderland, MA, USA.

Glenn, S.M. and Collins, S.L. (1990) Experimental analysis of patch dynamics in tall grass prairie plant communities. *Journal of Vegetation Science*, **4** 157–162.

Greenberg, C.H. and McNab, W.H. (1998) Forest disturbance in hurricane-related downbursts in the Appalachian mountains of North Carolina. *Forest Ecology and Management*, **104** 179–191.

Grime, J.P., Crick, J.C. and Rincon, J.E. (1986) The ecological significance of plasticity, in *Plasticity in plants, Symposia of the Society for Experimental Biology*, 40 (eds. D.H. Jennings and A.J. Trewavas), Company of Biologists, Cambridge, UK, pp. 5–29.

Hallé, F., Oldeman, R.A.A. and Tomlinson, P.B. (1978) *Tropical Trees and Forests: An Architectural Analysis*. Springer-Verlag, Berlin, Germany.

Hook, P.B., Lauenroth, W.K. and Burke, I.C. (1994) Spatial patterns of roots in a semiarid grassland: abundance of canopy openings and regeneration gaps. *Journal of Ecology*, **82** 485–494.

Hubbell, S.P. and Foster, R.B. (1986) Canopy gaps and the dynamics of a neotropical forest, in *Plant Ecology* (ed. M.J. Crawley), Blackwell, Oxford, UK, pp. 77–96.

Itoh, A., Yamakura, T., Ogino, K., Lee, H.S. and Ashton, P.S. (1997) Spatial distribution of two predominant emergent trees in a tropical rainforest in Sarawak, Malaysia. *Plant Ecology*, **132** 121–136.

Iwasa, Y. and Kubo, T. (1995) Forest gap dynamics with partially synchronized disturbances and patch age distribution. *Ecological Modelling*, **77** 257–271.

Johnsson, M., Issaias, S., Brogardh, T. and Johnsson, A. (1976) Rapid, blue-light induced transpiration response restricted to plants with grass-like stomata. *Physiologia Plantarum*, **36** 229–232.

Kanzaki, M., Yoda, K. and Dhanmaonda, P. (1994) Mosaic structure and tree growth pattern in a monodominant tropical seasonal evergreen forest in Thailand, in *Vegetation Science in Forestry* (eds. R.K. Peet and E.O. Box), Kluwer Academic Publishers, The Netherlands, pp. 499–517.

Kirschbaum, M.U.F., Gross L.J. and Pearcy, R.W. (1988) Observed and modelled stomatal response to dynamic light environments in the shade plant, *Alocasia macrorrhiza*. *Plant, Cell and Environment*, **11** 111–121.

Kohyama, T. (1993) Size-structured tree populations in gap-dynamic forests: the forest architecture hypothesis for the stable coexistence of species. *Journal of Ecology*, **81** 131–143.

Krasny, M.E. and Whitmore, M.C. (1992) Gradual and sudden forest canopy gaps in Allegheny northern hardwood forests. *Canadian Journal of Forest Research*, **22** 139–143.

Kubo, T., Iwasa, Y. and Furumoto, N. (1996) Forest spatial dynamics with gap formation: total gap area and gap size distribution. *Journal of Theoretical Biology*, **180** 229–246.

Küppers, M. (1985) Carbon relations and competition between woody species in a Central European hedgerow. IV. Growth form and partitioning. *Oecologia*, **66** 343–352.

Küppers, M. (1989) Ecological significance of above-ground architectural patterns in woody plants: a question of cost-benefit relationships. *Trends in Ecology and Evolution*, **4** 375–379.

Küppers, M. (1994) Canopy gaps: competitive light interception and economic space filling—a matter of whole-plant allocation, in *Exploitation of Environmental Heterogeneity by Plants* (eds. M.M. Caldwell and R.W. Pearcy), Academic Press, New York, NY, USA, pp. 111–144.

Lawton, R.O. (1990) Canopy gaps and light penetration into a wind-exposed tropical lower montane rain forest. *Canadian Journal of Forest Research*, **20** 659–667.

Lawton, R.O. and Putz, F.E. (1988) Natural disturbance and gap-phase regeneration in a wind-exposed tropical cloud forest. *Ecology*, **69** 764–777.

Lee, R. (1978) *Forest Microclimatology*. Columbia University Press, New York, USA.

Lertzman, K.P. (1992) Patterns of gap-phase replacement in a subalpine, old-growth forest. *Ecology*, **73** 657–669.

Lieberman, D., Lieberman, M., Peralta, R. and Hartshorn, G.S. (1985) Mortality patterns and stand turnover rates in a wet tropical forest in Costa Rica. *Journal of Ecology*, **73** 915–924.

Lieberman, M., Lieberman, D., Peralta, R., and Hartshorn, G.S. (1995) Canopy closure and the distribution of tropical forest tree species at La Selva, Costa Rica. *Journal of Tropical Ecology*, **11** 161–178.

Loehle, C. and LeBlanc, D. (1996) Model-based assessments of climate change effects on forests: a critical review. *Ecological Modelling*, **90** 1–31.

Malanson, G.P. (1993) Comment on modeling ecological response to climatic change. *Climatic Change*, **23** 95–109.

Manrubia, S.C. and Solé, R.V. (1997) On forest spatial dynamics with gap formation. *Journal of Theoretical Biology*, **187** 159–164.

McClure, J.W. and Lee, T.D. (1993) Small scale disturbance in a northern hardwood forest: effects on tree species abundance and distribution. *Canadian Journal of Forest Research*, **23** 1347–1360.

Midgley, J.J., Cameron, M.C. and Bond, W.J. (1995) Gap characteristics and replacement patterns in the Knysna Forest, South Africa. *Journal of Vegetation Science*, **6** 29–36.

Novoplansky, A., Cohen, D. and Sachs, T. (1990) How portulaca seedlings avoid their neighbors. *Oecologia*, **82** 490–493.

O'Brien, S.T., Hayden, B.P. and Shugart, H.H. (1992) Global climate change, hurricanes, and a tropical forest. *Climate Change*, **22** 175–190.

Parsons, W.F.J., Knight, D.H. and Miller, S.L. (1994) Root gap dynamics in lodgepole pine forests: nitrogen transformations in gaps of different size. *Ecological Applications*, **4** 354–362.

Pearcy, R.W. (1987) Photosynthetic gas exchange responses of Australian tropical forest trees in canopy gap and understory microenvironments. *Functional Ecology*, **1** 169–178.

Pearcy, R.W. and Calkin, H. (1983) Carbon dioxide exchange of C_3 and C_4 tree species in the understory of a Hawaiian forest. *Oecologia*, **58** 19–25.

Pearcy, R.W. and Sims, D.A. (1994) Photosynthetic acclimation to changing light environments: scaling from the leaf to the whole plant, in *Exploitation of Environmental Heterogeneity by Plants* (eds. M.M. Caldwell and R.W. Pearcy), Academic Press, New York, NY, USA, pp. 145–174.

Pearcy, R.W., Osteryoung, K. and Calkin, H.W. (1985) Photosynthetic response to dynamic light environments by Hawaiian trees: the time course of CO_2 uptake and carbon gain during sunflecks. *Plant Physiology*, **79** 896–902.

Pearcy, R.W., Chazdon, R.L., Gross, L.J., Mott, K.A. (1994) Photosynthetic utilization of sunflecks: a temporally patchy resource on a time scale of seconds to minutes, in *Exploitation of Environmental Heterogeneity by Plants* (eds. M.M. Caldwell and R.W. Pearcy), Academic Press, New York, NY, USA, pp. 175–208.

Perkins, T.D., Klein, R.M., Badger, G.J. and Easter, M.J. (1992) Spruce-fir decline and gap dynamics on Camels Hump, Vermont. *Canadian Journal of Forest Research*, **22** 413–422.

Peterson, C.J. and Squiers, E.R. (1995) Competition and succession in an aspen white-pine forest. *Journal of Ecology*, **83** 449–457.

Pfitsch, W.A. and Pearcy, R.W. (1989a) Daily carbon gain by *Adenocaulon bicolor*, a redwood forest understory herb, in relation to its light environment. *Oecologia*, **80** 465–470.

Pfitsch, W.A. and Pearcy, R.W. (1989b) Steady-state and dynamic photosynthetic response of *Adenocaulon bicolor* in its redwood forest habitat. *Oecologia*, **80** 471–475.

Phillips, D.L., and Shure, D.J. (1990) Patch size effects on early succession in southern Appalachian forests. *Ecology*, **71** 204–212.

Pickett, S.T.A. and White, P. (1985) *The Ecology of Natural Disturbance and Patch Dynamics*, Academic Press, Orlando, FA, USA.

Pons, T.L. (1989) Breaking of seed dormancy by nitrate as a gap detection mechanism. *Annals of Botany*, **63** 139–143.

Porter, J.R. (1989) Modules, models and meristems in plant architecture, in *Plant Canopies: Their Growth, Form and Function, Society for Experimental Biology, Seminar Series 31* (eds. G. Russell, B. Marshall and P.G. Jarvis), Cambridge University Press, Cambridge, UK, pp. 143–159.

Putz, F.E. (1983) Treefall pits and mounds, buried seeds and the importance of soil disturbance to pioneer tree species in the tropics. *Ecology*, **64** 1069–1074.

Raich, J.W. (1980) Fine roots regrow rapidly after forest felling. *Biotropica*, **12** 231–232.

Rebertus, A. J. and Veblen, T.T. (1993) Structure and tree-fall gap dynamics of old-growth *Nothofagus* forests in Tierra del Fuego, Argentina. *Journal of Vegetation Science*, **4** 641–654.

Remmert, H. (1991) The mosiac-cycle concept of ecosystems: an overview, in *The Mosiac-Cycle Concept of Ecosystems, Ecological Studies* **85** (ed. H. Remmert), Springer-Verlag, Berlin, Germany, pp. 1–21.

Ricklefs, R.E. (1977) Environmental heterogeneity and plant species diversity: a hypothesis. *American Naturalist*, **111** 376–381.

Runkle, J.R. (1982) Patterns of disturbance in some old-growth mesic forests of eastern North America. *Ecology*, **63** 1533–1546.

Runkle, J.R. (1984) Development of wood vegetation in treefall gaps in a beech sugar maple forest. *Holarctic Ecology*, **7** 157–164.

Runkle, J.R. (1990) Gap dynamics in an Ohio *Acer-Fagus* forest and speculations on the geography of disturbance. *Canadian Journal of Forest Research*, **20** 632–641.

Rusch, G. and van der Maarel, E. (1992) Species turnover and seedling recruitment in limestone grassland. *Oikos*, **63** 139–146.

Ryel, R.J., Beyschlag, W. and Caldwell, M.M. (1993) Foliage orientation and carbon gain in two tussock grasses as assessed with a new whole-plant gas-exchange model. *Functional Ecology*, **7** 115–124.

Sakio, H. (1997) Effects of natural disturbance on the regeneration of riparian forests in the Chichibu Mountains, central Japan. *Plant Ecology*, **132** 181–195.

Sanford, R.L. (1989) Fine root biomass under a tropical forest light gap opening in Costa Rica. *Journal of Tropical Ecology*, **5** 251–256.

Sanford, R.L. (1990) Fine root biomass under light gap openings in an Amazon rain forest. *Oecologia*, **83** 541–545.

Sarukhán, J. and Harper, J.L. (1973) Studies on plant demography: *Ranunculus repens* (L), *R. bulbosa* (L) and *R. acris* (L). 1. Population flux and survivorship. *Journal of Ecology*, **61** 675–716.

Schreiber, U., Bilger, W. and Neubauer, C. (1994) Chlorophyll fluorescence as a nonintrusive indicator for rapid assessment of *in vivo* photosynthesis, in *Ecophysiology of Photosynthesis, Ecological Studies* **100** (eds. E.D. Schulze and M.M. Caldwell), Springer-Verlag, Berlin, Germany, pp. 49–70.

Shugart, H.H. (1984) *A Theory of Forest Dynamics*, Springer-Verlag, New York, USA.

Shugart, H.H. and Noble, I.R. (1981) A computer model of succession and fire response of the high-altitude Eucalyptus forest of the Brindabella Range, Australian Capital Territory. *Australian Journal of Ecology*, **6** 149–164.

Shugart, H.H., and Smith, T.M. (1996) A review of forest patch models and their application to global change research. *Climate Change*, **34** 131–153.

Shugart, H.H. and D.C. West (1980) Forest succession models. *BioScience*, **30** 308–313.

Silvertown, J.W. and Doust, J.L. (1993) *Introduction to Plant Population Ecology*, 3rd Edition, Blackwell Scientific Publications, Oxford, UK.

Sims, J.A. and Pearcy, R.W. (1993) Sunfleck frequency and duration affects growth rate of the understory plant, *Alocasia macrorrhiza* (L.) G. Don. *Functional Ecology*, **7** 683–689.

Smith, H. (1982) Light quality, photoreception and plant strategy. *Annual Review of Plant Physiology*, **33** 481–518.

Smith, S.D., Patten, D.T., and Monson, R.K. (1987) Effects of artificially imposed shade on a Sonoran desert ecosystem: microclimate and vegetation. *Journal of Arid Environments*, **13** 65–82.

Smith, W.K., Knapp, A.K. and Reiners, W.A. (1989) Penumbral effects on sunlight penetration in plant communities. *Ecology*, **70** 1603–1609.

Solé, R.V. and Manrubia, S.C. (1995) Are rainforests self-organized in a critical state? *Journal of Theoretical Biology*, **173** 31–40.

Spies, T.A., Franklin, J.F. and Klopsch, M. (1990) Canopy gaps in Douglas-fir forests of the Cascade Mountains. *Canadian Journal of Forest Research*, **20** 649–658.

Stark, J.M. (1994) Causes of soil nutrient heterogeneity at different scales, in *Exploitation of Environmental Heterogeneity by Plants* (eds. M.M. Caldwell and R.W. Pearcy), Academic Press, New York, USA, pp. 255–284.

Sykes, M.T., van der Maarel, E., Peet, R.K. and Willems, J.H. (1994) High species mobility in species-rich communities: intercontinental comparison. *Folia Geobotanica et Phytotaxonomica*, **29** 439–448.

Thompson, K. and Grime, J.P. (1983) A comparative study of germination responses to diurnally-fluctuating temperatures. *Journal of Applied Ecology*, **20** 141–156.

Thompson, K., Grime, J.P. and Mason, G. (1977) Seed germination in response to diurnal fluctuations of temperature. *Nature*, **267** 147–149.

Thórhallsdóttir, T.E. (1990) The dynamics of a grassland community: a simultaneous investigation of spatial and temporal heterogeneity at various scales. *Journal of Ecology*, **78** 884–908.

Tucker, D.J. and Mansfield, T.A. (1972) Effects of light quality on apical dominance on *Xantium strumarium* and the associated changes in endogenous levels of abscisic acid. *Planta*, **102** 140–151.

Valiente-Banuet, A., Bolongaro-Crevenna, A., Briones, O., Ezcurra, E., Rosas, M., Nuñez, H., Barnard, G. and Vázquez, E. (1991) Spatial relationships between cacti and nurse shrubs in a semiarid environment in central Mexico. *Journal of Vegetation Science*, **2** 15–20.

van der Maarel, E. (1993) Some remarks on disturbance and its relations to diversity and stability. *Journal of Vegetation Science*, **4** 733–736.

van der Maarel, E. (1996) Pattern and process in the plant community: fifty years after A.S. Watt. *Journal of Vegetation Science*, **7** 19–28.

Vázquez-Yanes, C. (1974) Studies on the germination of seeds of *Ochroma lagopus* Swartz. *Turrialba*, **24** 176–179.

Vázquez-Yanes, C. and Orozco-Segovia, A. (1987) Light gap detection by the photoblastic seeds of *Cecropia obtusifolia* and *Piper auritum*, two tropical rain forest trees. *Biologia Plantarum*, **29** 234–236.

Vázquez-Yanes, C. and Orozco-Segovia, A. (1994) Signals for seeds to sense and respond to gaps, in *Exploitation of Environmental Heterogeneity by Plants* (eds. M.M. Caldwell and R.W. Pearcy), Academic Press, New York, NY, USA, pp. 209–236.

Veblen, T.T. (1985) Forest development in tree-fall gaps in the temperate rainforests of Chile. *National Geographic Research*, **1** 162–183.

Vitousek, P.M. and Denslow, J.S. (1986) Nitrogen and phosphorus availability in tree-fall gaps of a lowland tropical rainforest. *Journal of Ecology*, **74** 1167–1178.

Walker, J. and Peet, R.K. (1983) Composition and species diversity of pine wiregrass savannas of the Green Swamp, North Carolina. *Vegetatio*, **55** 166–179.

Wang, Y.P. and Jarvis, P.G. (1990) Description and validation of an array model—MAESTRO. *Agricultural and Forest Meteorology*, **51** 257–280.

Watt, A.S. (1947) Pattern and process in the plant community. *Journal of Ecology*, **35** 1–22.

Weber, J.A., Jurik, T.W., Tenhunen, J.D. and Gates, D.M. (1985) Analysis of gas exchange in seedlings of *Acer saccharum*: integration of field and laboratory studies. *Oecologia*, **65** 338–347.

Welden, C.W., Hewett, S.W., Hubbel, S.P. and Foster, R.B. (1991) Sapling survival, growth and recruitment: relationship to canopy height in a neotropical forest. *Ecology*, **72** 35–50.

White, P.S. (1979) Pattern, process, and natural disturbance in vegetation. *Botanical Review*, **45** 230–299.

Willems, J.H., Peet, R.K. and Bik, L. (1993) Changes in chalk-grassland structure and species richness resulting from selective nutrient additions. *Journal of Vegetation Science*, **4** 203–212.

Wolfram, S. (1984) Cellular automata as models of complexity. *Nature*, **311** 419–424.

Woods, K.D. (1979) Reciprocal replacement and the maintenance of co-dominance in a beech maple forest. *Oikos*, **33** 31–39.

Woods, K.D. (1984) Patterns of tree replacement: canopy effects on understory pattern in hemlock northern hardwood forests. *Vegetatio*, **56** 87–107.

Woods, D.B. and Turner, N.C, (1971) Stomatal response to changing light by four tree species of varying shade tolerance. *New Phytologist*, **70** 77–84.

Worrall, J.J. and Harrington, T.C. (1988) Etiology of canopy gaps in spruce-fir forests at Crawford Notch, New Hampshire. *Canadian Journal of Forest Research*, **18** 1463–1469.

10 Plant and canopy diversity

G.R. Squire

10.1 Introduction

In plant canopies, diversity exists in the shape and size of leaf and stem, in the range of plant form and in the combination of forms that give a canopy architecture (Harper, 1989; Porter, 1989; Niklas, 1999). In the opening chapter, Niklas reminds us that the general properties of plant material have hardly changed over evolutionary time. Diversity arises from the way that the material is arranged and organised. Contrast the virtually one-layered, grazed sward of acidic, upland grassland, where four or five species occupy most of the space, with tiered high forest where canopy emergents of great individual mass coexist with smaller trees and shrubs, saplings, herbs, lianas and epiphytes in a highly diverse assemblage.

The changes that occur in a canopy originate in the fecundity and death of individuals. Individuals differ in the extent to which they can change their structure as they grow, and may differ further in being able to exchange genes with other plants and thereby adapt their structures between generations. However, the individuals are not independent of each other. The canopy they become part of acts back on them to alter their structure. So while the physical limitations of plant structure impose a boundary on what the canopy can look like, this boundary in turn has a bearing on the structures that individuals growing as part of the canopy can assume (see Chapter 9).

The central problem, as Corner (1978) stated when describing the swamp forest of the Sedili Rivers, is 'what are all these different plants doing side by side in the same forest?' This link between plant and canopy is the main theme of the present chapter and is developed in three sections. The first considers some of the interdependent functions of plants: capturing resource, staying upright, seeding, and defence. The second explores diversity of form as the interplay between individual and canopy. The third considers the plant and canopy in the broader context of spatial and temporal biodiversity. Although the species is used throughout ecological studies of diversity, the 'growth form' is generally preferred as the instructive unit of 'ecological botany' (Warming, 1909). Among other advantages, forms allow functional comparisons among species and smoother transitions between individuals and canopies.

10.2 Functions and properties

The aerial parts of plants capture sunlight and carbon dioxide, hold reproductive organs to encourage pollination and dispersal, provide defence against being eaten and transport water for temperature regulation. Such processes can be measured at the level of both the plant and the canopy. However, a conscious attempt is made here not to imbue a canopy with functional characteristics as if it were, without question, an entity rather than a collection of interacting parts. Terminology is therefore important: individual plants are described by physiological 'traits' that, acting singly or in tandem, mediate 'functions' for the individual. In contrast, and at the risk of being pedantic, canopies are described by 'properties', which will only have a functional connotation if the canopy is viewed as having a specified purpose.

The dichotomy between leaves and stems, referred to in the opening chapter, has an important bearing. As an individual grows, it can capture more light, but the balance between leaf lamina and the supporting petiole and stem changes. As the individual extends in height, the balance between the height and diameter of the stem also changes. Such give-and-take between leaf and stem or height and girth is commonly called a 'trade off' by ecologists and evolutionists. Especially when energy and resources are limited, plants put more into one function than another: the one is traded off against the other. The perfect plant is unattainable because of biological and physical limitations to the possible combinations of inter alia development, architecture, and resource use.

10.2.1 Investment and return

Such arguments are given some reality by considering the capture of sunlight by archetypes: a small, annual form, represented by the tropical millets, and a large, perennial form represented by the tropical oil and coconut palms. Some typical dimensions and other functions and properties (Table 10.1) show the palms to have a mass and longevity several orders of magnitude greater than those of the millets. However, the forms far more analogous in terms of capturing solar energy. The shapes and orientations of the two types of lamina are similar, so they intercept a comparable amount of sunlight per unit leaf area. (See Chapter 8 for a full description of the physics of light interception by canopies.)

Structure and function can be combined in a useful comparative term if a plant's growth of capturing structures is considered an 'investment', and the sunlight captured as a 'return'. In the following analysis, the symbol, τ, is used to signify the return from investment by a plant or canopy, i.e. solar radiation captured per unit canopy mass, which has units of MJ g^{-1}. For the millets and the palms, τ is far less different than might be intuitively expected

Table 10.1 Comparison of structural trais and radiation capture in tropical palms and millets

Trait	Palms	Millets
Development	Indeterminate	Determinate
Reproduction	Polycarpic, intermittent	Monocarpic, synchronous
Longevity (yrs)	20–100	0.15–0.3
Leaf area at maturity:		
stand (leaf area index)	6	3
single plant (m^2)	500	0.1
Mass at maturity:		
stand ($t\ ha^{-1}$)	50	5
single plant (kg)	400	0.02
Height at maturity (m)	3–15	1–2
Light extinction coefficient (for total solar radiation)	0.3–0.4	0.3–0.4
Annual interception of solar radiation:		
by stand ($GJ\ m^{-2}$)	5	0.5
by individual (GJ)	400	0.02
Investment efficiency ($MJ\ g^{-1}$)	2–5	1–2

Values are generalisations to illustrate differences for uniform stands. Sources for palms: oil palm, Corley (1973), Squire (1990); coconut, Moss (1992). Sources for sorghum and pearl millet detailed in Squire (1990).

from the size of the plants. Differences in the properties of the canopy, therefore, lie primarily in structural mass and longevity.

This investment efficiency, τ, is not a constant or conservative value but changes through time as the canopy develops, and particularly as the balance alters between intercepting and supporting structures. The value of τ is expected to increase with longevity of the lamina and to decrease as a plant gets larger, because higher leaves shade those below and more of the mass is allocated to stem than lamina (Figure 10.1). These opposing effects interact in a complex manner. In the example of the millet (Figure 10.2), the fraction of sunlight captured reached a maximum around 40 days after seeds were sown, vegetative production ceased after 50 days but τ continued increasing. Three phases occurred in the rise of τ. In the first, up to 40 days, it increased more or less constantly in time, since most of the vegetative growth, V, went into leaf, and each increment of leaf brought a high return of captured sunlight, since there was little self-shading. However, once maximum interception was achieved, the rise in τ began to tail-off, since new leaves shaded older leaves and more energy was put into stems. If the stems had continued to increase in mass, then τ might well have levelled and declined. Instead, once the stems stopped growing at the onset of flowering, τ increased again as the canopy brought a substantive return on its investment by continuing to intercept sunlight without further change in leaf area.

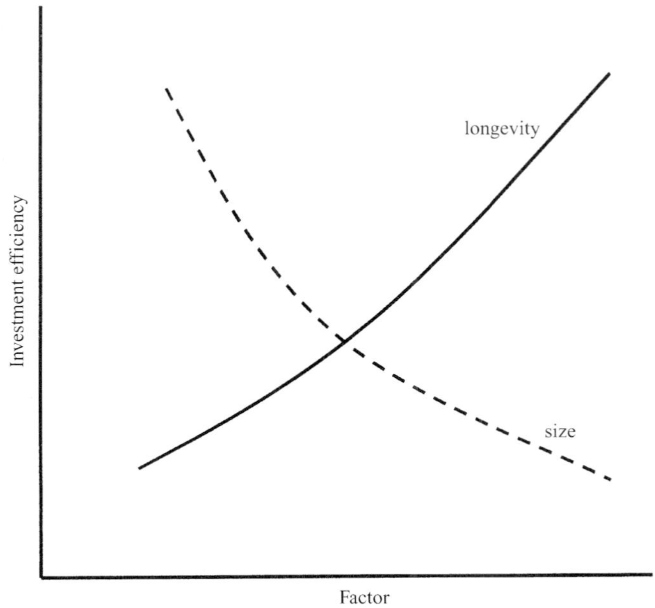

Figure 10.1 Idealised effects on investment efficiency (solar radiation) of size and longevity of foliage.

In the example of the oil palm (Figure 10.3), the time sequence shown starts well after the initial rising phase of canopy expansion. Fractional interception is already 66% and rises to more than 99% over the next 8 yrs. Even though the canopy intercepted most of the light by 10 yrs, it still kept increasing in vegetative mass by producing larger fronds and extending its trunk. The result was that τ declined throughout the period. Note, however, that the absolute values of τ, the return from investment, are greater in the palms. More detailed analysis indicated that this is because a unit of leaf tissue lives for about 2 yrs in the palms compared to about 50 days in the millets, so a unit investment repays in the longer term.

10.2.1.1 Trade-offs through height growth and reproduction

The same botanical traits determine τ in the two forms, but their distinct trade-offs, acting in the local environment, brought about the differences in the value of τ. Both forms underwent growth in height and structural mass, and both were in stands where the foliage of adjacent individuals mingled. However, the structural extension of the millet ceased well before maturity, when the seed heads were 1 m above the ground, thereby allowing continued

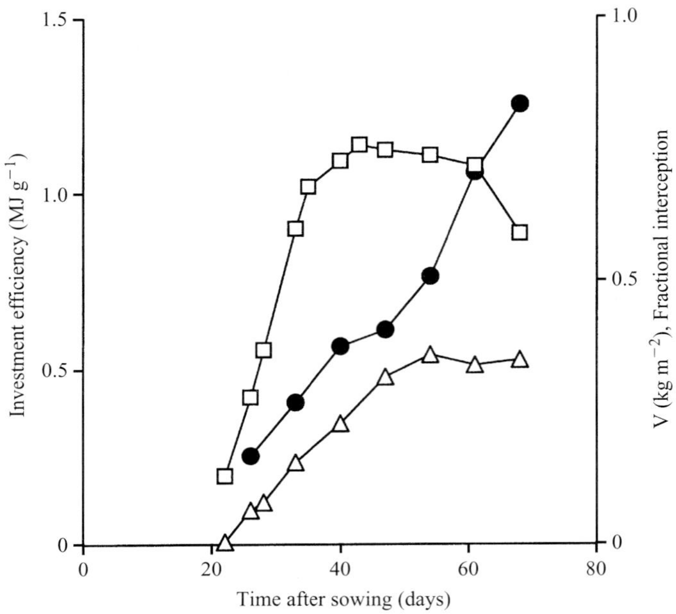

Figure 10.2 Relations between the mass of leaves and stems, V (–△–), fractional interception of total solar radiation (–□–) and the investment efficiency (–●–) for a pearl millet canopy growing for 70 days (sources in Table 10.1).

increasing return from limited investment in leaf. In the palms, the response to being surrounded by other individuals was continuous investment in competitive tissue and supporting structural mass, to the extent that τ suffered in the attempt to outgrow neighbours.

The way these two forms balance the functions of resource capture and growth in height also has implications for fecundity and defence against herbivory. The millets achieve about an equal balance between vegetative and reproductive mass, while the palms allocate material to vegetative mass when competing with other individuals: the decline in τ in Figure 10.3 means less assimilate for fruits and, in extreme instances, fecundity is sacrificed to stay in the race for sunlight. However, the palms are far less prone to being eaten and destroyed by invertebrates and mammals; not only do they grow out of reach of many herbivores, they also have much tougher tissue.

The traits and trade-offs shown by these two forms are strongly influenced by the environments in which they evolved. The millets have to get through their life cycle rapidly to avoid drought or fire, and while growth in height and competitive ability are relevant to survival, they are not the dominant

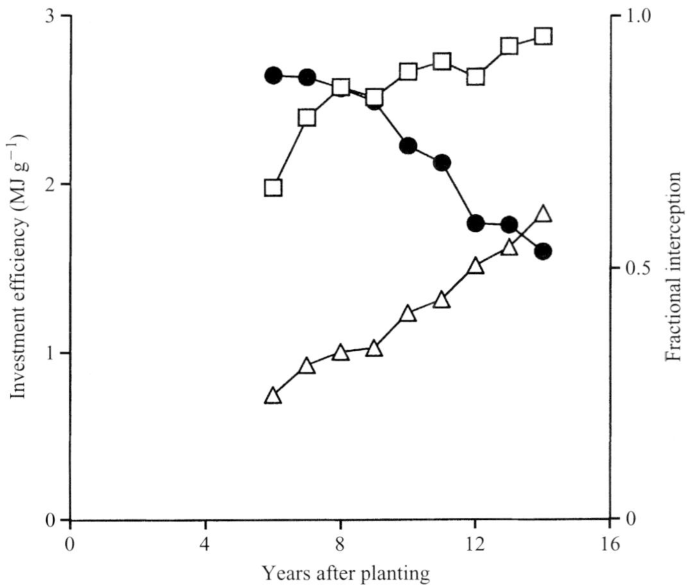

Figure 10.3 Relationships between the mass of leaves and stems, V (–△–), fractional interception of total solar radiation (–□–) and the investment efficiency (–●–) for an oil palm canopy between 6 and 14 yrs after planting at Rengam in Johore, Malaysia (sources in Table 10.1).

factors controlling whether the plant will leave seed. The palms, being indeterminate and polycarpic, have primarily to extend themselves above the highly competitive neighbours of forest margins and clearings, need greater protection against predators (as do all perennial forms) but need only accrue enough excess resource to fruit occasionally.

The effects of continued growth in mass and height can be generalised. The benefits include the ability to capture sunlight in competition with other plants, to rise above the grazing height of herbivores and to increase dispersal of fruits. The costs include less efficient return of sunlight in investment and the concomitant lower allocation of energy to fruits. Neither the millets nor the palms, nor any other plant form, are able to maximise all functions, but each exists in a different optimum state. Associations and trade-offs between resourcing and defence have been discussed and examples given in previous reports (Wardle *et al.*, 1998; Van der Meijden *et al.*, 1988).

10.2.2 Season and space

The relationship between growth form and seasonal capture of sunlight is illustrated for three species which co-habit in field margins and land near agriculture, in parts of the UK. One species, *Thlaspi arvense*, has similar general traits to the millets, in being a rapidly developing, determinate annual. Another species, the grass *Holcus lanatus*, is analogous to the palms, since it is perennial and can grow 100 times larger in mass than the *Thlaspi*, though little taller in height. Between them is the more or less indeterminate, sprawling, bristly annual, *Anchusa arvensis*.

In the experiment from which the example is taken, the species were artificially established as both small patches and various interacting fronts (Wright and Squire, 1998). Figure 10.4A shows the change in the fraction of solar radiation intercepted within 3 × 3 m patches of each of the species grown singly. They were sown at the same time but the canopies expanded at different rates: determinate annual, indeterminate annual, perennial, in that order. Typically, faster development was associated with earlier reproduction and death and, consequently, less energy capture and smaller mass. Therefore, the area under an envelope around the maximum values of fractional interception obtained by any of the species is greater than that by any one of the species. These phenological trade-offs also separated the millets and the palms, and distinguish species more generally (Wardle *et al.*, 1998).

10.2.2.1 The effect of variation in length of season

Which form succeeds in trapping the most resource in a year will, however, depend on local factors, such as soil disturbance, weather, pests and herbivores. The point can be illustrated using curves of cumulative interception (Figure 10.4B). In a growing period lasting, for example, 50 days, form a in Figure 10.4 intercepts most radiation 220 MJ m^{-2}, compared to 130 MJ m^{-2} by form b and only 20 MJ m^{-2} by form c. After a further 50 days, form b has overtaken form a, whose development has ceased, but is still ahead of form c (410, 630, 380 MJ m^{-2}, for forms a, b and c, respectively). By 150 days, form c has now intercepted most radiation: 410, 630 and 680 MJ m^{-2}, respectively. The winter set in progressively after 150 days, but the foliage of form c survived, so that at all times during the next year, even though forms a and b re-emerged, form c captured most sunlight.

The maximum energy input over the whole growing season between April and October might be achieved by some combination of forms a, b and c. Over a restricted shorter period—say, for example, fire or heavy herbivory were to curtail growth in early summer—then form c would be redundant. Over several years—again hypothetically—form c would have captured so much of the radiation that the early contribution of form a would be negligible. If, however, at some time in the future, disturbance such as fire or ploughing destroyed much of form c's canopy, then the rapid leaf expansion of forms a and b would be required to maximise inputs. To do this, they would need to have survived, and they are likely to have done so as buried, viable seed in the seedbank.

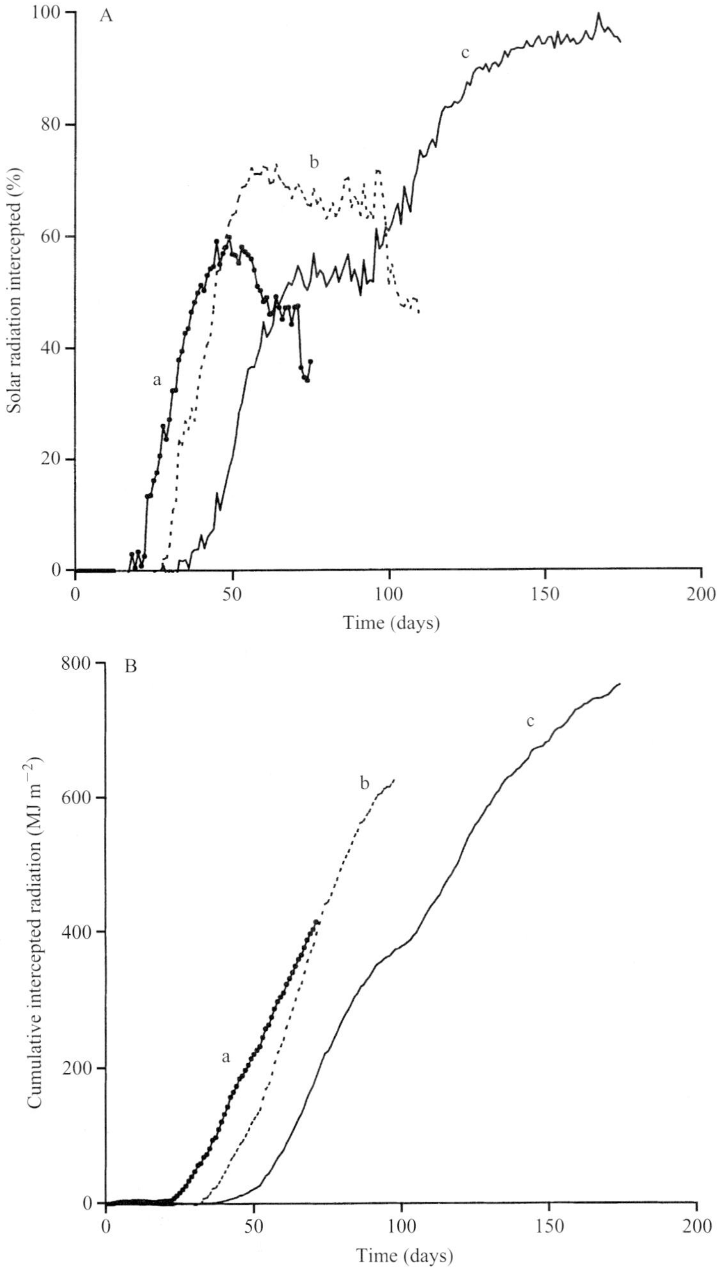

Figure 10.4 A) Seasonal change in the interception of solar radiation by canopies of wayside plants (a) *Thlaspi arvense*, (b) *Anchusa arvensis* and (c) *Holcus lanatus* in Tayside, Scotland. B) Cumulative intercepted radiation for the canopies in panel A.

10.2.2.2 Spatial arrangement and trade-offs

Whether the maximum envelope drawn in Figure 10.4A, would actually be achieved in a hypothetical community of these three species also depends on their spatial arrangement. If each species occurred in large clumps in which the other species were not present, then the maximum radiation interception would not be achieved. Assuming the season progressed to its end, then, when the determinate annual died, radiation would be lost in the time taken for the indeterminate annual or the perennial to cover the space it vacated. Clearly, the spatial configuration of the species interacts with their temporal development to determine the property of the canopy. If the other extreme is considered, where the small determinate annual mingles intimately with the large indeterminate annual, the smaller will intercept less radiation than when in a patch of its own species (assuming all plants at the same density). Therefore, the spatial arrangement that might maximise interception by the canopy would tend to reduce it for the small determinate annual.

10.2.3 The complexity of more species and traits

The analysis summarised in Figure 10.4 dealt only with one function of the plants and one potential property of the canopy. If traits affecting palatability and defence were introduced, then the resulting functions and properties would change at each spatial and temporal level. The bristly annual, *Anchusa*, might well be less prone to being eaten compared with the soft-tissued annual, *Thlaspi*, whilst when tussocks had formed, the perennial, *Holcus*, would have meristems protected by a mass of foliage. Moreover, any plants of the soft-tissued annual growing within the canopy of the bristly annual might well favour the former's ability to survive a passing grazer. In this way, the simple trade-offs affecting development and resource capture become conditional, not just on other traits but on the spatial and temporal arrangements of the different forms.

The argument can, in principle, be extended to more traits and to a canopy of many species. There is a problem of definition, however. While the properties of a complex canopy as a whole can be measured in much the same way as those of a simple monoculture, probing the link between the species and canopy suffers the practical difficulty of measuring the processes and traits of different plant forms growing closely together. Plants can be observed with minimal interference, canopies can be manipulated and the response measured, artificial canopies can be constructed with plants of known traits (as in Figure 10.4) and plants can be removed from site in order to define their physiological traits in controlled environments (Law *et al.*, 1997). The latter course is open to the criticism that the plants are grown out of context and the traits observed might not be relevant to functioning in the real canopy. This notwithstanding, Wardle and co-workers (1998) demonstrated that a range of dicotyledonous herb species taken from pasture ecosystems showed consistent linkages between traits, such as that those with early flowering tended to be smaller at maturity and have a lower leaf area per unit mass (implying lower investment efficiency). These authors proposed that the traits they measured ex situ provided a direct link to properties of the ecosystem from which the plants were removed. Similar experience has confirmed that, where species are sufficiently discrete, their values can be used to examine the links between functions of species and the properties of a system (Chapin *et al.*, 1993; Grime, 1965; Keddy, 1992). Indeed, more generally, such links are the justification for defining functional types and assigning to them a role in ecosystems (Grime *et al.*, 1988).

While much ecological effort has been spent in this direction, traits are largely expressed as mean or bulk values, while variation within a species or functional type is obscured or ignored. This must ultimately be misleading. There are many examples of two or more populations which have a similar mean for a trait but different degrees of variation about the mean (Bradshaw, 1965; Schlichting, 1986; West-Eberhard, 1989). It is not just the range of trait values that might differ between two samples of plants measured at a point in time, but the extent to which the trait values change during the life of the plants in response to external stimuli. The present chapter continues with some observations on variation in traits.

10.3 Individuals in canopies

The variation that exists both intrinsically between individuals and as a result of phenotypic response to environment is now considered. Forms such as the palms and the millets could be grown in a state where development and growth were unconstrained. This state could be defined as a point in the multi-dimensional space created among graphical axes describing the traits. In fact, the individuals would occupy a small cloud rather than a point, because some of the individuals would be genetically different and have different trait values even in their unconstrained state.

This trait space can be easily envisaged within two or three trait-axes, as in Figure 10.5 which defined the palms and millets by (as examples) height, mass and another function. The symbols indicate the unconstrained state at a time when plants have reached reproductive maturity, when the palms have a much greater size and mass than the millets but only a slightly larger τ. Any limitations in the physical or chemical environment or through interactions with other plants will alter the position of an individual in the trait space compared to the point of unconstrained growth. For instance, severe competition for resource will reduce the size of individuals at maturity, but crowding in the stand might increase the height. The directions of the likely change in

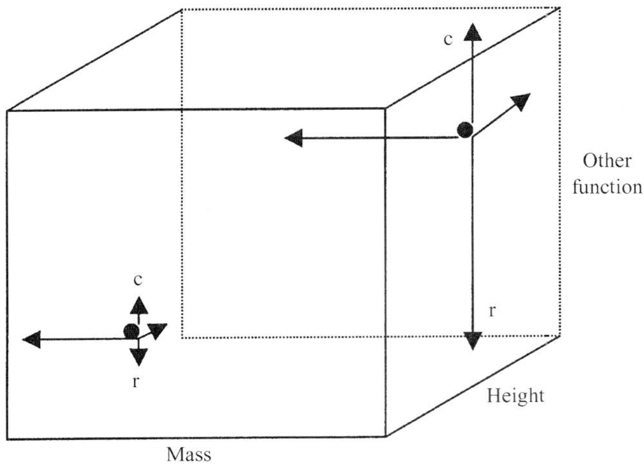

Figure 10.5 Idealised trait space of 'millets' and 'palms' showing points of unconstrained traits (symbols) and effect of increasing population or shading. Arrows show the direction of variation in height and mass and (the vertical axis) the relative concomitant change in two other functions: allocation of resource to reproduction (r) and improved light capturing ability (c).

the three traits as a result of crowding and other influences are shown by the arrows originating from the unconstrained points. If all the possible states were then mapped, each form would occupy a larger cloud than defined by that at unconstrained growth. The shape and density of this cloud describes the potential diversity of the form.

10.3.1 Time-evolution of trait space

The examples compared trait space at one particular instant in the life cycle of the forms, but traits change over time. The phenology of a group of plants can, accordingly, be represented as a distribution of states visited by the plants. A simple example is provided by leaf and flower development in the indeterminate herb, *Vigna subterranea* (Bambara groundnut), a subsistence crop in Africa, whose form and habit are similar to that of the indeterminate temperate annual described previously (Figure 10.4). A plot of the development of leaf number in a sample of around 50 plants (Figure 10.6A) shows the mean stabilising at around 65 days, while the distribution continues to widen. This occurred because some of the larger plants continued producing new leaf as some of the smaller plants lost leaf, presumably as a result of the changing conditions around the plants and the effects of competition between them.

If flower number is now included, the distribution of states can be viewed as a density map, formed by interpolation of a matrix of time–leaf–flower

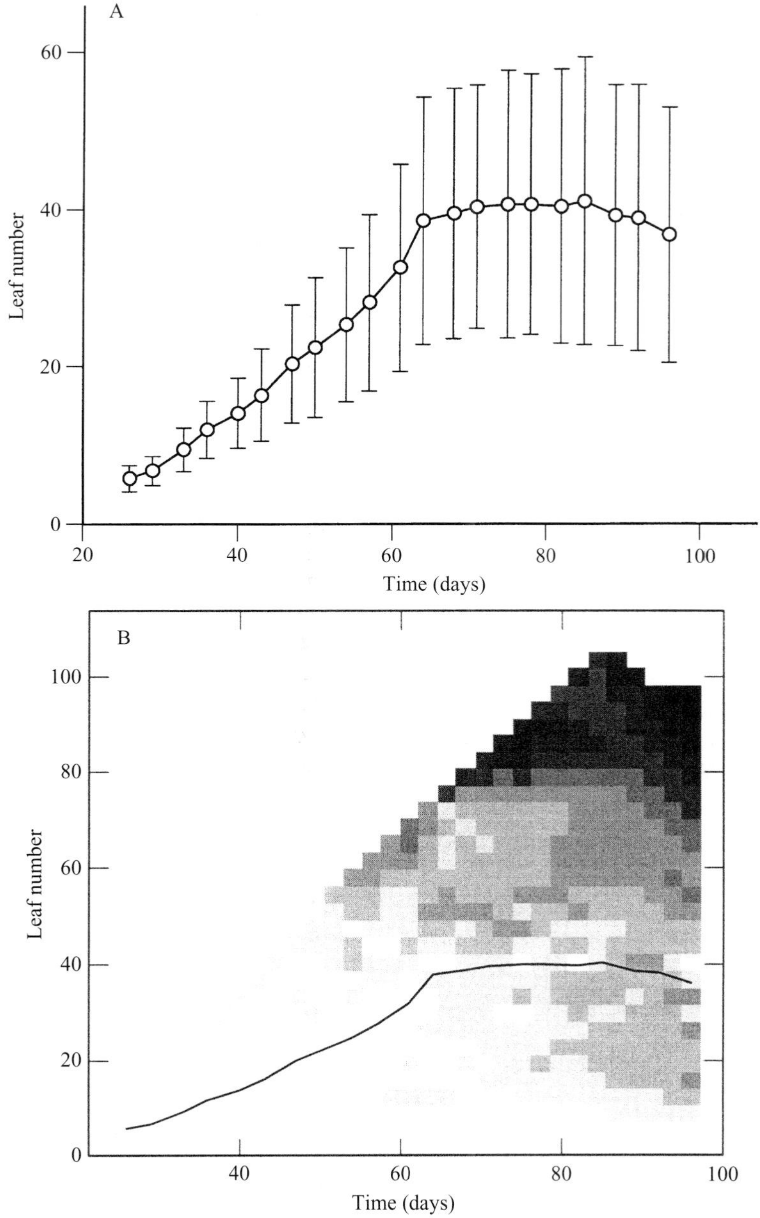

Figure 10.6 Within-population variation of morphological traits in bambara groundnut: A) temporal change of leaf number, shown with standard deviation both sides of mean; B) map of time, leaf number and flower number, values of the latter indicated by shade on a gridded surface (8 divisions; white = 0; darkest shade = 50–55 flowers per plant). For details see text. (Data from the University of Nottingham: Anon, 1996.)

states computed from original data on individual plants. In the map presented (Figure 10.6B), time is along the x-axis, leaf number on the y-axis, while the numbers of flowers at a given leaf–time state are represented as contours on the z-axis. The mean of the leaf number over time is superimposed on the map. The leaf states and flower states do not overlay simply one on the other. In this instance, plants produced more flowers if their leaf number was above the mean, but the map also fragmented at high leaf–time states suggesting that the distribution was not continuous. This association between faster leaf production and faster flowering was not invariably found for this species. In other populations, for example, flowering was more synchronised in time.

This example confirms important features of the phenology of populations. The sample mean is not always a good indicator of the changes that occur in the population, and the trait space, even of one or two traits, is generally skewed or contorted, such that fecundity or yield is unequally distributed within the population.

10.3.1.1 Correlation between states and maintenance of rank

In the trait map (Figure 10.6), individuals were not identified, but an individual's life cycle can be represented as a trajectory through the trait space (Squire and Crawford, 1996; Sumida *et al.*, 1997). Two individuals (by definition) never take the same trajectory. Even if the space is defined by only two or three traits, the development of any pair of individuals can soon be shown to differ. Temporal differences begin at germination (and perhaps earlier on the mother plant). Thereafter, some individuals grow at intrinsically different rates, while others differ because of where they are in the environment. If the ranking of individuals were maintained, then plants producing leaves rapidly would always produce flowers rapidly and vice versa, but this is not necessarily so. Individuals might become more different as environmental heterogeneity continually reinforces past effects. They might become less different if global signals, such as photoperiod, synchronise change between developmental phases (Hay and Kirby, 1991).

Subsequently, a simple step, useful if comparing populations, is to correlate later states with earlier states. For the bambara plants, the correlation coefficient (r) between final leaf–time states of the individuals and previous leaf–time states decreased at states farther back in time. The individuals' trajectories were knocked off course variously by being shaded, having insufficient nutrient or water, or being attacked by disease.

An example of the values of the approach is taken from a study which looked at the architecture, phenology and genetic diversity of wild raspberry in Scotland (Graham *et al.*, 1997; and related unpublished data). This plant is perennial, but each individual stem usually lasts 2 yrs, extending and leafing in the first, flowering and fruiting in the second. Plants from different localities have what appear to be architectural trade-offs between height growth of the main stem and the production of secondary stems. However, when grown in the same environment, node production on the main stem hardly differs; differences in internode expansion are the basis of the height differences. Yet the similarity of mean node number (Figure 10.7A) belies additional diversity in the individuals' trajectories. A simple correlation (r) between final number with the number at earlier time points shows that the states in certain populations decorrelate over time more rapidly than others (Figure 10.7B). They do so because their individuals are more diverse in the rates and duration of node production. Some individuals that produce nodes rapidly cease production earlier and vice versa. Their combined trajectories are more contorted than those of the other populations.

10.3.1.2 Contortion of the trait space and size inequalities

Interactions between individuals within canopies alter such trait distributions and, thereby, the size-inequalities among individuals. The importance of size distribution for both natural populations (Harper, 1977) and crops (Edmeades and Daynard, 1979) has been recognised for many years. A plant's size influences the result of biotic and abiotic threats, such as being eaten, bent over by the wind or out-competed, and will also strongly determine its fecundity. However, an individual's position or rank in a size-distribution will not necessarily be maintained over time, so inferences about its fate based on information at one time point, say late in life, could be incorrect.

An instructive example of the use of time-trajectories is provided by the work of Sumida and co-workers (1997) for the evergreen oak, *Quercus glauca*, growing in a mixed hardwood forest in Japan. They estimated, from destructive sampling, the diameter and height of the stem of a number of individuals that had germinated at different times over a 60 year period of succession. By the 1990s, those early in succession were casting shade on later individuals. The stem diameter/height relationship changed systematically with age and with time in succession. Individuals that developed early in succession had larger diameters relative to height than those that developed later, in their shade (Figure 10.8). The diameters, heights and diameter/height ratio also changed over several years, probably when light penetrating the upper canopy increased or decreased, such that an individual's trajectory 'meandered' about its course (Chapter 9). Their results take the argument back to the dichotomy between leaves and stems, because the authors' interpretation of the trajectories was that late-successional individuals invested more in foliage in order to persist in the shade 'at the cost of gaining height'.

The differences observed between the early and late-successional individuals suggest a trade-off. The total growth of individuals was limited (by carbon or nutrients), the options open to them were curtailed and they took different trade-offs according to the circumstances. The authors recognised

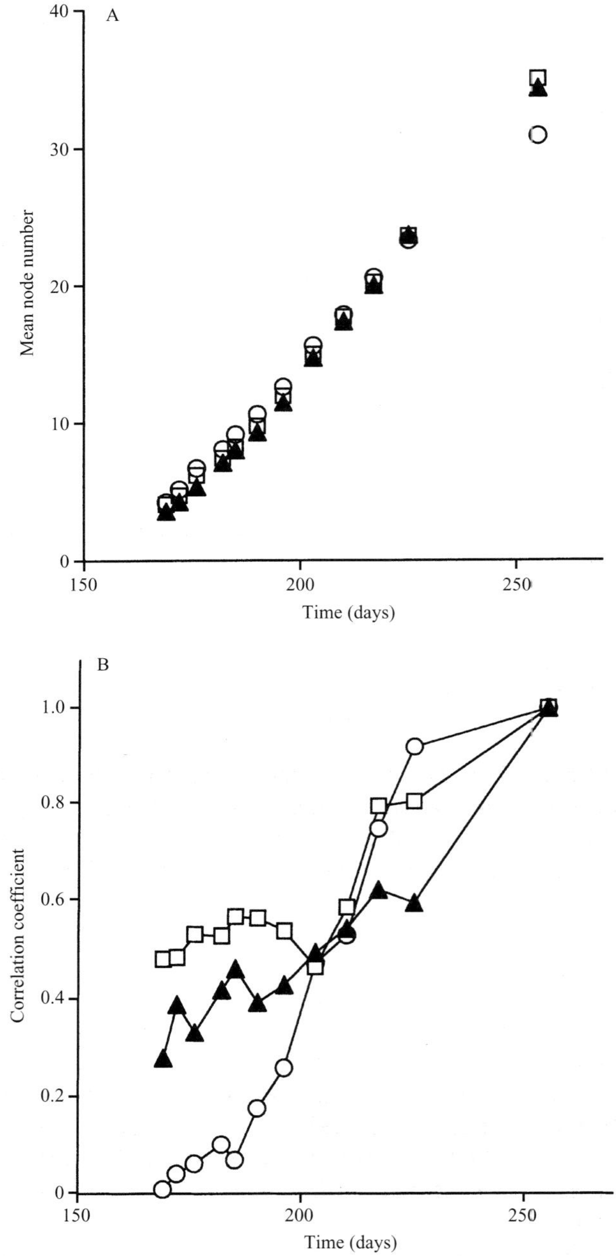

Figure 10.7 A) Growth of leaf node number in three populations of wild raspberry in Scotland; and B) the correlation coefficient (r) for node number at the date of measurement and the final node number.

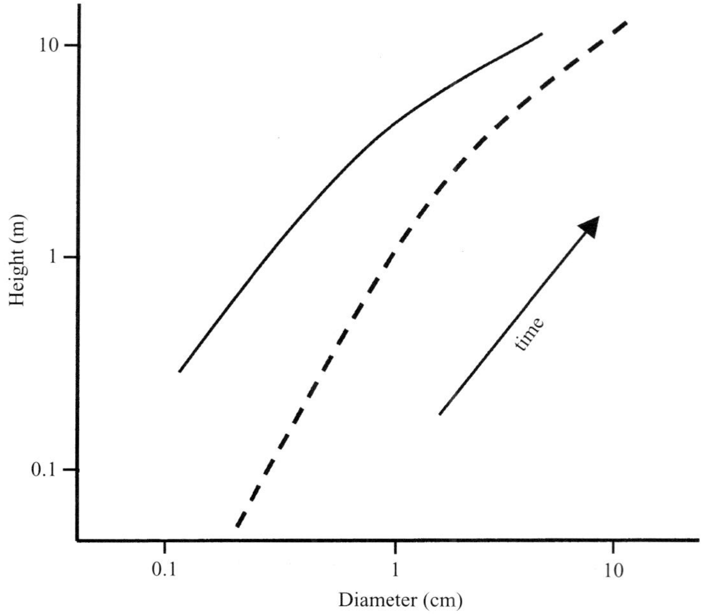

Figure 10.8 Diagrammatic representation of the time trajectories taken by individual trees of *Quercus glauca* through a trait space of stem height and diameter: Key: – – – –, trees in early stages of succession; ——, trees in late stages of succession. (After Sumida *et al.*, 1997.)

that the broad trade-offs caused by long-term position in the canopy, and the meandering of trajectories caused by medium-term changes in light penetration, meant that means or size distributions taken at a point in time or even over successive time intervals were not the whole story and were likely to be misleading.

10.3.2 Trade-offs between vegetative and reproductive states

Harper (1977) and other population biologists have remarked on the great size inequalities that develop in canopies of wild plants and weeds, which give rise to extremely unequal contributions by individuals to the next generation. In more extreme instances, the size inequalities extend to the death of individuals, which is wasteful where structural material or seed is required as the sole product of cropping, and also has implications for the diversity of natural populations. Plants have developed mechanisms that appear to result in a lessening of such inequalities, one such being the sensing of neighbours through light quality (Aphalo and Ballaré, 1995; Ballaré *et al.*, 1994; see also Chapter 5). In a map of the type shown in Figure 10.6, these mechanisms would act to reduce the divergence of leaf–flower–time states.

Indeed, in some data sets examined, flowering began in about half the population on the same day, thereby restricting the occupancy of trait space.

As in previous examples, trajectories of individuals can throw light on the way such inequalities develop. The three arable species in Figure 10.4 were also grown in different combinations with the other species, where individuals were followed from just after emergence to maturity. The determinate, *Thlaspi arvense*, provides an illustration: it can be defined by two traits early in growth before shading between individuals occurred. The first trait indicates progress towards a particular size of the vegetative structure, defined by a number of leaves and leaf nodes as a fraction of 25; this value is arbitrary but indicates a plant which is still small, though large enough not to be uprooted and eaten by a small rodent or bird. The second trait is progress towards the reproductive state, defined as either not reproductive or reproductive.

The traits are compared early in the growing season, day 32 on the time-axis of Figure 10.5. Individuals are ranked by leaf number in Figure 10.9, open symbols indicating the non-reproductive and closed symbols the reproductive state. Four degrees of interaction are presented in panels A, B, C, and D. When the plants were grown spaced well apart (Figure 10.9A), two-thirds had reached the full vegetative state by this time (those at 1 on the y-axis). Figure 10.9B, C and D shows the effect on the size inequalities of the *T. arvense* when grown with increasingly strong interactors. With the strongest, *Anchusa arvensis*, only about 10% of the sample had reached 1 on the vegetative scale. The shape of the distribution of size had also changed. There were fewer individuals in the region 0.5 to 1 compared to the sample from the other interacting populations, therefore indicating that the population was splitting into a fast vegetative group of individuals which were also reproductive, and a much slower and smaller non-reproductive group. The transition to reproduction occurred only on individuals greater than 0.5 on the vegetative scale, but again the strongest interactor had altered the pattern, only one plant being reproductive at vegetative states below 1.

Reference to Figure 10.5 shows that the differences described above occurred while the two other species were still very small, and certainly not mingling with the *T. arvense*. The *Anchusa arvensis* was just registering on the tube solarimeters used to measure light interception, while the *Holcus lanatus* hardly registered at all. This serves to emphasise that interactions through light quality or some other form of signalling, as distinct from competition for resource, were therefore possible.

10.3.3 Trait space and genetic diversity

Developments in genomic technology (Bachmann, 1994; Powell *et al.*, 1996) have prompted questions as to the correspondence between the diversity of

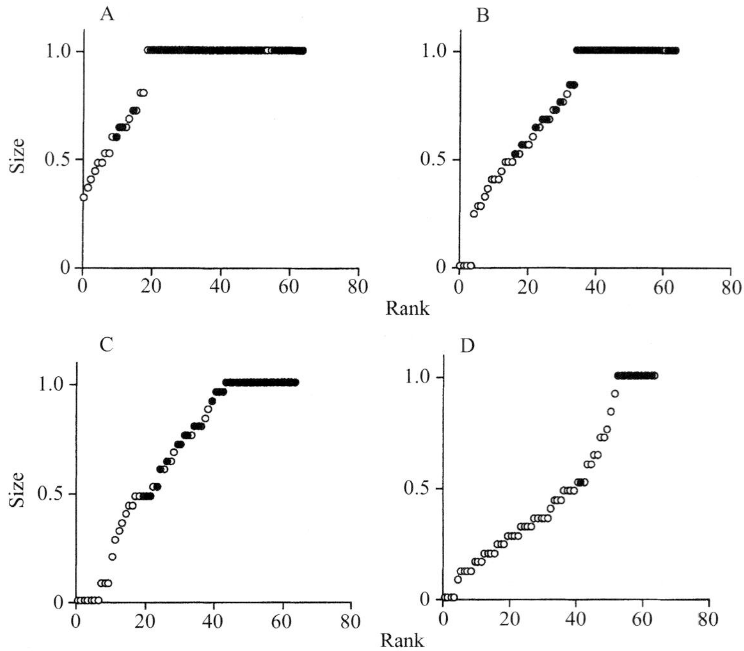

Figure 10.9 Morphological states in the determinate annual, *Thlaspi arvense* grown: A) alone and widely spaced; B) in a monoculture; C) with a weak competitor; and D) with a strong competitor. Individuals ($n = 64$) are ranked by leaf number as a fraction of vegetative state (0 to 1); symbols also show the absence (open symbol) or presence (closed symbols) of the reproductive state.

phenotypic trajectories and the diversity of the genome. DNA-based techniques are proving valuable in two particular instances. Firstly, it may be necessary to know whether two individuals taken from a community, such as a grassland, are different individuals, or are more likely to belong to the same clonal structure. Techniques based on DNA markers can be applied to look for differences; their limitation is that even if two DNA profiles are identical, it does not necessarily mean that the corresponding individuals are genetically the same, since the techniques might not have probed areas of the genome where a difference occurs. Secondly, it is sometimes necessary to prove that gene flow has occurred from one plant to the progeny of another. Even a slight difference between maternal and paternal genomes is usually enough to demonstrate this by one or other of a range of techniques now available (e.g. White and Powell, 1997).

The more general value of broad-scale measures of genetic diversity is uncertain. There are instances where it could be indicative of physiological diversity. For instance, the populations of wild raspberry described earlier were shown by random amplified polymorphic DNA (RAPD) analysis to be genetically distinct (Graham *et al.*, 1997). Moreover, populations whose node numbers decorrelated more rapidly were also more genetically diverse than others. This is useful confirmatory evidence, but the fact that the groups of individuals taken from different locations and grown in the same environment have a consistently different canopy architecture is the primary finding, and could stand without the evidence of the DNA markers.

In order to obtain quantitative relations between markers and traits, two general approaches can be followed (McRoberts *et al.*, 1999): markers can be checked one by one against the distribution of trait values in plants that have and do not have the marker; or the presence and absence of markers among individuals in a population can be converted to, for example, a principal coordinate 'space', which could then be compared with the corresponding 'space' for the traits. However, the knowledge that one population has a different molecular diversity to another has to be very stringently qualified by what the molecular diversity signifies for the functioning of the leaves and stems. As McRoberts and co-workers (1999) have cautioned, the results from any analysis on the same plant populations will vary with the system of DNA primers used, with the particular set of primers chosen out of all potential sets, and with the particular set of physiological traits with which they are compared.

10.4 Coexistence of forms and canopy properties

There are two main sides to the problem (stated in the Introduction) of what all the different plants are doing side by side. One is how the forms coexist within a canopy? The other is, what do they contribute to the properties of the canopy? These questions are to an extent interdependent and are part of a wider and general debate on diversity in ecosystems (Grubb, 1977; Huston, 1979; Rozenzweig, 1995; Givnish, 1999).

10.4.1 Coexistence in canopies

The basis of many interactions leading to increased diversity is that a species might be better than others in one or a few functions but not better at all functions. Looking back to the example given in the opening Chapter of the present volume: the common phyllotaxies tend towards maximising capture of sunlight by a plant, but the function is not perfectly realised and some radiation still hits the ground. Accordingly, the way is open for forms to coexist.

The earlier example of physiological trade-offs in the millet and palm forms provides a further simple illustration. Both forms can exist for a time on stores of water in the soil. In an environment in which there was only 40 days of rain and 40 days worth of store, (it can be hypothesised that) the short season millets would survive and the palms would die. In an environment assured of adequate water and nutrients continuously for several decades, the palms would flourish and shade out the millets. However, many tropical environments are spatially variable, such that areas of deep soil, having a large store of water that would favour the palms, exist close to areas of shallower soil that only the millets could grow in. Neither form alone would occupy all of the area. The trade-offs that characterise each enable both to be grown in a temporally and spatially variable environment. Indeed, palm forms and millet forms are typical of mixed subsistence agriculture in many parts of the tropics.

The environment clearly has a role, therefore. More widely, the observational evidence is that high species number does not occur in environments that support very low productivity (where few species are adapted to live) or very high productivity (where certain species strongly dominate), but occurs in between these states (Tilman and Pacala, 1993; Givnish, 1999). In addition, the more heterogeneous the environment, the less likely are the conditions where one species will be good at everything. The conditions encouraging the coexistence of most forms are likely to be states that optimise the effects of productivity and heterogeneity.

The observational evidence provides an insight into how coexistence might occur. In reality, the processes and mechanisms by which it does occur are very complicated. Those interactions between individuals through competition for resource and through signalling (Chapter 5) or other mechanisms are only part of the story. Individuals also have to be dispersed and find a place to enter a canopy. Since many habitats are highly fragmented, the long-term dynamics of populations, though rooted in the interactions over resource, also depend on a range of factors, such as the mating system, sexual compatibility, pollination vectors and dispersal mechanisms. In his analysis of the origins and maintenance of diversity in tropical forest, Givnish (1999) points to natural enemies, canopy structure and dispersal through ingestion by animals, in addition to competition, as factors influencing diversity.

Fragmentation of habitat or range causes many species or forms to exist as a metapopulation (Husband and Barrett, 1996), in which local turnover and geneflow occurs within interconnected fragments. Here, the individuals influence the dynamics of fragments, which are influenced in turn by the flow of genetic information across the metapopulation as a whole. Such considerations add great complexity to any study of canopies, and though well outside the scope of the present chapter, would be necessary for any comprehensive synthesis of canopy diversity.

10.4.1.1 System descriptors

The processes underlying the evolution of canopy diversity are therefore complex, not fully understood and still the subject of much debate and research. Nevertheless, the outcome—which is the distribution of species or forms across space, or the distribution of individuals among species—can often be represented by simple functions, which are useful for comparing systems. Typically, curves showing the way the number of species changes with the sample area tend to be asymptotic: as area increases from small to large, new species are detected but less often. In one of the swamp forest plots examined by Corner (1978), interpretation of the records (Figure 10.10) shows that the number of tree individuals increased linearly with sample area, while the number of tree species increased asymptotically, reaching one fifth of the number of individuals in the plot as a whole. The author also points out a nearby area of forest in which the canopy emergents had a smaller crown area, allowing more individuals to grow in the spaces between them. This is an example of how a specific canopy structure, brought about by individuals of one form, influence diversity in the canopy as a whole. In other systems, the number of individuals can be many thousands times more than

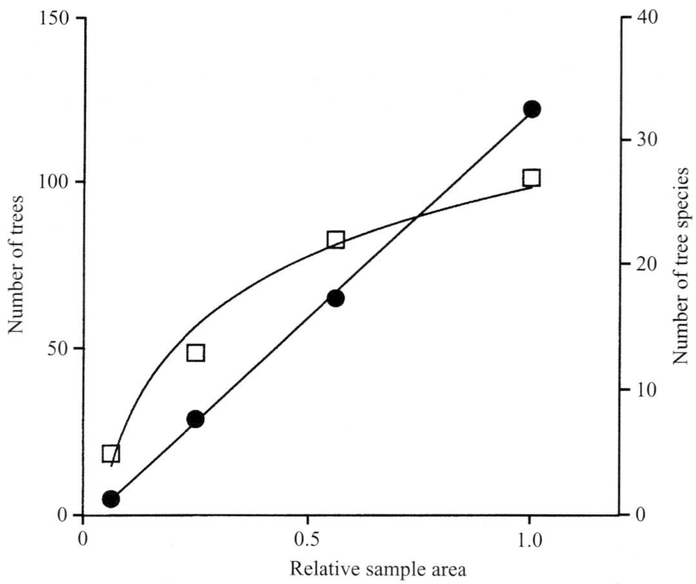

Figure 10.10 Relationship between relative sample area and the numbers of individuals (●) and species (□) of swamp forest trees in the Sedili Region, Malaysia. Maximum area is about 0.45 ha. (Original data in Corner, 1978.)

the number of species, but relations similar to those in Figure 10.10 still generally hold. Such species-area analysis is widely used in ecology (Rozenweig, 1995), the parameters of these curves often being an indicator of habitat and other ecogeographical features.

Further classification of ecosystems may be made in the abundance relations of the species in them (Hill, 1973; Kempton, 1979). These relations say nothing explicitly about the processes that link individual to canopy, but the fact that relations of a similar form occur across ecosystems had led ecologists to ask if there are higher order states that are in some way independent of the constituent forms (May, 1976; Kempton, 1979; Pielou, 1975). This must be the case to some extent for canopies, if for no other reason than the physical boundaries on plant structure referred to earlier. However, the cause and origin of a particular assemblage must ultimately lie in the traits and functions of the component individuals. From a simply practical point of view, species-area and species-abundance relations emphasise that the size of the sample or system examined will influence the measure of diversity.

10.4.2 Diversity and canopy properties

Words like stability, resilience and reliability are sometimes used to describe the way an ecosystem responds to stimuli. They are loaded with implications about what is good for (in this instance) a canopy of vegetation. For instance, a 'stable' canopy is not such a good thing for individuals that are unable to enter and reproduce in it. So, to avoid judging what is advantageous or not for a canopy, the arguments (as in the first section of the chapter) are qualified by reference to properties that cause the canopy to be in some specified state. If the state is maximum capture of resource over a long period, then the typical properties include the ability to intercept most of the sunlight, to evaporate excess energy by transpiration, to prevent heavy rain compacting or washing away the soil and to reduce large diurnal and annual fluctuations in temperature. It is argued, moreover, that diversity of form has the consequence of not only maximising some property of the canopy but also potentially allowing a system to recover from disruptions (Naeem and Li, 1997).

In agricultural systems, the coexistence of forms with different trade-offs arguably brings about greater stability of output. Returning to the hypothetical ecosystem, the millets would provide food and stabilise the soil in drought; the palms would still provide shelter and high-energy products if the land were flooded 2 m deep in water. Neither form would seem primarily to act 'for' the system, but heterogeneity in the environment allows the two forms to coexist and it is this coexistence that lends the stability.

10.4.2.1 The effect of increasing the number of forms

The argument can be extended to more species or forms. If forms within a canopy are ranked with respect to a particular function at any time, and their contributions accumulated in descending order, then curves of the type shown in Figure 10.11 generally result (Loreau, 1998). If capture of light is the function in question, then the first form intercepts so much, the second form less then the first, the third less than the second, and so on. The cumulative result can then be taken to be the property of the canopy (e.g. interception of sunlight).

In these hypothetical examples, the individual contributions are normalised with respect to the maximum value in each canopy, but canopies could also differ in the absolute maximum for the property. The two curves are shown as they are because they illustrate a difference that typically occurs between

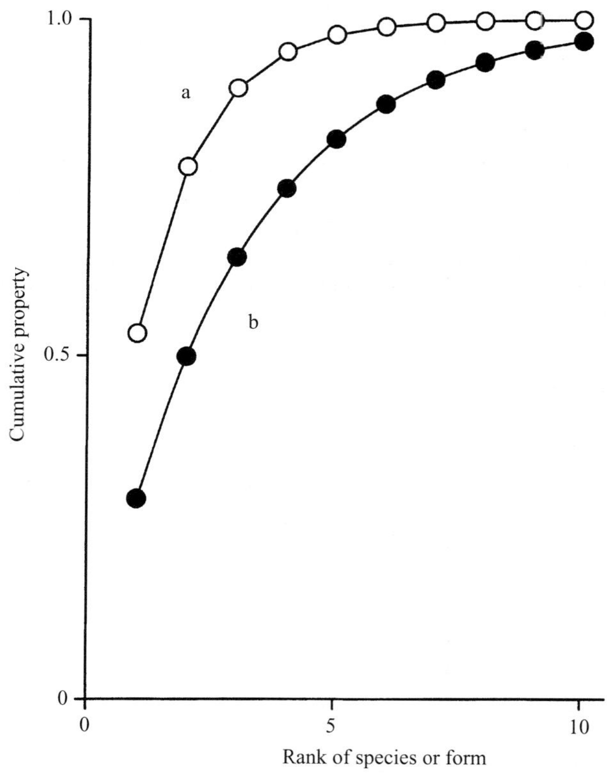

Figure 10.11 Hypothetical relationship showing the way the forms in a canopy can be ranked to show their cumulative contribution to a property of the canopy. Curves a and b represent different canopies (see text), where the maximum values of the property are normalised to unity.

systems, in that the same maximum value of the property is realised but the relative contribution of the different forms is not the same. The highly ranked forms in curve a are more able to contribute to the property of the canopy than the equivalent forms in curve b; forms after about the third ranking in curve a hardly contribute, while forms up to at least the sixth ranking in curve b still make a noticeable contribution. In both curves, many forms of low rank hardly contribute. These latter forms could number several hundred in a tropical forest or about ten in an acidic grassland. In managed agricultural ecosystems, the aim is to move in direction from curve b to a, and in the most intense systems to have just one form intercepting all the sunlight. This almost happens in some perennial plantation or grass crops supported by high agrochemical inputs. However, well-managed systems of only three to four forms (as in tropical mixed cropping) can intercept most of the radiation.

The same reasoning can be applied to a second property of the canopy, such as resistance to being blown over. Given that there are trade-offs between traits, then the forms that contributed most to the first property might contribute less to the second, while other forms, even insignificant ones in the first instance, might rise to greater importance. If Figure 10.11 could then be envisaged in three-dimensional space, with the additional axis representing another property, the resulting surface would be more convoluted and the dominants less dominant. The more properties considered, the less likely is it that any single form would dominate. A further effect, implicit in Figure 10.11, is that some of the minor forms might exist only because the dominant forms are there. For instance, tall trees in a forest canopy add much to the properties of height and shelter, which commensals, lianas and other dependants exploit.

10.4.2.2 Recovery from disturbance

These hypothetical examples illustrate a general case, alluded to in the Introduction, and relevant to both managed and more natural vegetation, that the trade-offs between functions of a particular form influence a property at a higher order in the system. The line of argument can be continued to consider a property such as resilience of the canopy. If fire or ploughing, for instance, were to severely disrupt a system, maybe to the extent of obliterating some species, it is argued that the system's resilience would depend on the number and type of species still present or able to enter it. Again, a word such as resilience (or reliability) has to be qualified with respect to a property of the canopy. Resilience might be taken to mean, for example, that the canopy recovers so that certain of its main properties (e.g. energy capture, mass) achieve those values which they had previously. Following disruption, some of the species in Figure 10.11 that contributed little to the property, might come to contribute more if other conditions changed either to favour them or to disadvantage the previous dominants.

A mundane but didactic example of the link between diversity of form and the maintenance of a higher order property is shown by a field experiment, which examined the rate at which an arable weed community diverged between different treatments imposed after ploughing was stopped. The arable community and environment were of the general types described by Squire and co-workers (in press). As in many instances of this type of study, large amounts of data on the presence and abundance of species or forms needed first to be condensed. In this instance, a principal coordinate (PCO) analysis applied to data allowed a comparison of the assemblages in terms of a simple map of PCO-space. Briefly, two communities or canopies having a similar species complement lie close together in PCO-space; those differing in many species lie farther apart. The configuration of species or forms on the PCO map is a summary of the diversity at the site.

At the beginning of the experiment, the field had been in arable rotation for many years, and was divided into plots on which either swards would be sown or the vegetation would be allowed to regenerate naturally from the seedbank, which typically contains about 30 species. The original seedbank was dominated by the annuals, *Poa* sp. and *Chenopodium album*. There was, nevertheless, substantial variation between plots (Figure 10.12A) as a result of the spatially variable weed community in this field, but the plots destined for natural regeneration and those destined to have sown swards on them were well dispersed in PCO-space. Three years later, the species complements in the natural regeneration and sown ryegrass had mostly diverged, despite overlap remaining in a few plots (Figure 10.12B).

The original cereal crops, with a few annual weed species, gave way to two different canopies in the sown and natural regeneration areas. All three canopies covered the ground for most of the year, intercepted most of the sunlight and prevented rain from hitting the soil directly. However, forms that made a very small contribution to the property of the annually disturbed weed canopy—they were among the very minor species in a graph of the type shown in Figure 10.11—rapidly assumed a greater role in the changed environment. For instance, the perennial, *Rumex acetosella*, which ranked 12th and in very low abundance in the original seedbank, rose to 5th most abundant in the sown cover and 3rd at very high abundance in natural regeneration. Diversity of form, including forms almost redundant in the original seedbank and weed flora, allowed different canopies to develop that maximised the stated property of the canopy. As in previous examples, none of the taxa was able to exploit all conditions. 'Millet' forms were better in the disturbed system and 'palm' forms in the undisturbed. A trade-off that mitigated against the individual enhanced the property of the canopy.

Several authors have given examples of how trade-offs might operate to

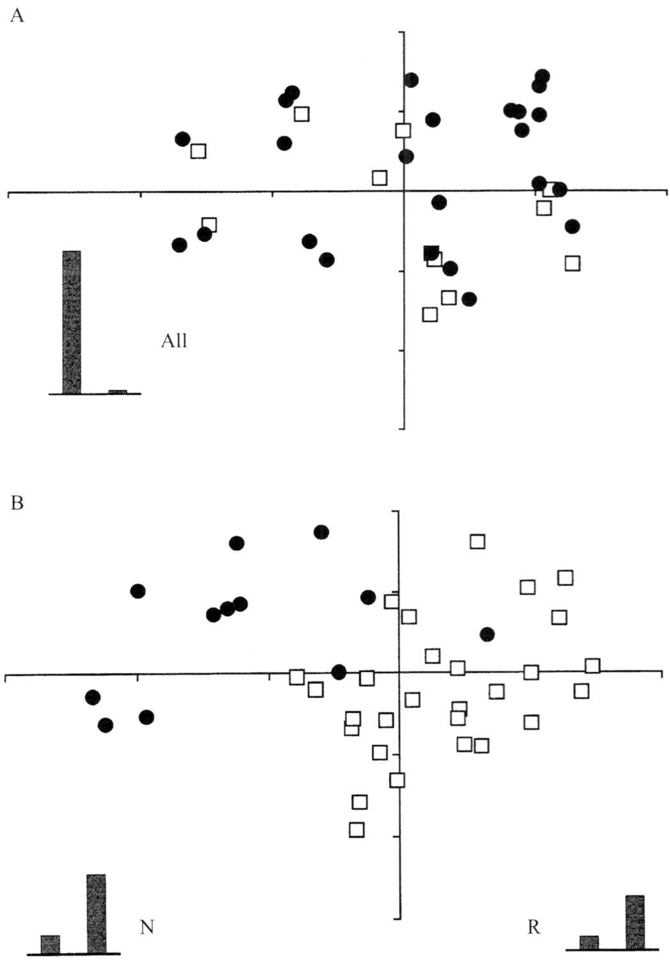

Figure 10.12 Principal coordinate analysis based on the presence and absence of 30 species in an arable seedbank: A) at the start of the experiment when future plots were defined; and B) 3 yrs later in plots sown with ryegrass covers (open squares) or allowed to regenerate naturally (closed circles). Each bar chart shows the mean proportion of (left column) *Chenopodium album* and (right column) *Rumex acetosella* in all plots at the start (All) and after 3 yrs in natural regeneration (N) and sown ryegrass (R).

maximise properties and endow stability in more complex ecosystems, especially where the canopy and root system are considered (Aerts, 1999; Tilman, 1994; Wardle *et al.*, 1998; Van der Meijden *et al.*, 1988; Sumida *et al.*, 1997).

10.5 Concluding remarks

The present chapter begins with general considerations of diversity, examines function on the scale of the stand or species, then moves on to show the potential for work at the individual level to reveal more about leaf and canopy diversity than can be gleaned from measurements only at the higher orders. The balance of effort in the field is still weighted to the scales of the bulk canopy and the species. Indeed, many of the descriptions applied to a growing collection of leaves and stems are aimed at defining higher-order structures rather than the individuals. The theme preferred here is that the behaviour of the individual is equally important, and that progress in understanding canopies will only come from a more rigorous look at the way trajectories interact and combine to give higher order properties.

There is a limit, however, to how many individual trajectories can be measured and indeed how many it is worth measuring. A general framework is now required for analysis of the individual in the system and, in this respect, the following points should be noted:

- In variable populations, sampling to derive a mean and variance over successive time intervals will be misleading if trade-offs operate within the population. This will be specially so if the means are converted to properties of the resource capture by the canopy (leaf area index, light interception), which are then used to infer behaviour in the reproductive state. If individuals die after capturing some resource or else fail to reproduce, then mean reproductive effort will be uncoupled from mean or bulk vegetative production.
- Individuals and populations lose the 'memory' of their previous states, as when a group of trajectories twist and contort as a result of local influences. A much greater understanding of 'memory' is required for assessing present and past states and for directing or conserving the evolution of a canopy.
- A methodology is needed to represent and quantify the evolution of multi-dimensional trait space. It must incorporate the interplay between individual trajectories and canopy properties both within and between generations. An approach linking the modular concepts of morphology (Porter, 1989) to functions such as resource capture, dispersal and defence would present considerable opportunity for synthesis. Mathematical approaches offer a means of a exploring trait space, helping to eject what is superfluous and directing experiments to concentrate on salient features.

Finally, it is proposed that placing emphasis on individuals within phenotypic landscapes will encourage tighter links between the molecular and physiological studies described in much of this volume and the implications of the traits for canopy properties.

Acknowledgements

This article was written while the author was funded by the Scottish Office Agriculture, Environment and Fisheries Department. Gratitude is extended to Drs Sayed Azam Ali, Sarah Collinson and Abu Sesay at the University of Nottingham for access to their data on bambara groundnut, and to colleagues at the Scottish Crop Research Institute from whose work some of the examples have been selected, notably wild raspberry (Julie Graham, Bruce Marshall), arable seedbanks and weeds (Gladys Wright. Sheena Rodger and Paula Hebden) and seasonal light interception (Gordon Dunlop). Many of the ideas developed through discussion with colleagues, John Crawford and Bruce Marshall (the latter of whom also provided helpful guidance during drafting the paper).

References

Aerts, R. (1999) Interspecific competition in natural plant communities: mechanisms, trade-offs and plant–soil feedbacks. *Journal of Experimental Botany*, **50** 29–37.

Anon (1996) *Proceedings of the International Bambara Groundnut Symposium*. University of Nottingham, Sutton Bonington, LE12 5RD, UK.

Aphalo, P.J. and Ballaré, C.L. (1995) On the importance of information-acquiring systems in plant–plant interactions. *Functional Ecology*, **9** 5–14.

Bachmann, K. (1994) Molecular markers in plant ecology. *New Phytologist*, **126** 403–418.

Ballaré, C.L., Scopel, A.L., Jordan, E.T. and Vierstra, R.D. (1994) Signalling among neighbouring plants and the development of size inequalities in plant populations. *Proceedings of the National Academy of Sciences of the United States of America*, **91** 10094–10098.

Bradshaw, A.D. (1965) Evolutionary significance of phenotypic plasticity in plants. *Advances in Genetics*, **13** 115–55.

Chapin, F.S., Autumn, K. and Pugnaire, S. (1993) Evolution of suites of traits in response to environmental stress. *American Naturalist*, **142** 78–92.

Corley, R.H.V. (1973) Effects of plant density on growth and yield of oil palm. *Experimental Agriculture*, **9** 169–80.

Corner, E.J.H. (1978) The freshwater swamp-forest of South Johore and Singapore. *Gardens' Bulletin Supplement No. 1.*, Botanic Gardens Parks and Recreation Department, Singapore.

Edmeades, G.O. and Daynard, T.B. (1979) The development of plant-to-plant variability in maize at different planting densities. *Canadian Journal of Plant Science*, **59** 561–576.

Givnish, T.J. (1999) On the causes of gradients in tropical tree diversity. *Journal of Ecology*, **87** 193–210.

Graham, J., Squire, G.R., Marshall, B. and Harrison, R.E. (1997) Spatially dependent genetic diversity within and between colonies of wild raspberry, *Rubus ideaus*, detected using RAPD markers. *Molecular Ecology*, **6** 1001–1008.

Grime, J.P. (1965) Comparative experiments as a key to the ecology of flowering plants. *Ecology*, **45** 413–415.

Grime, J.P., Hodgson, J.G. and Hunt, R. (1988) *Comparative Plant Ecology: A Functional Approach to Common British Species*. Unwin Hyman, London, UK.

Grubb, P.J. (1977) The maintenance of species richness in plant communities. *Biological Reviews*, **52** 107–145.

Harper, J.L. (1977) *Population biology of plants*. Academic Press, London, UK.

Harper, J.L. (1989) Canopies as populations, in *Plant Canopies, Their Growth, Form and Function*, (eds. G. Russell, B. Marshall and P.G. Jarvis), Cambridge University Press, Cambridge, UK, pp. 105–128.

Hay, R.K.M. and Kirby, E.J.M. (1991) Convergence and synchrony: a review of coordination and development in wheat. *Australian Journal of Agricultural Research*, **42** 661–700.

Hill, M.O. (1973) Diversity and evenness: a unifying notation and its consequences. *Ecology*, **54** 427–432.

Husband, B.C. and Barrett, S.C.H. (1996) A metapopulation perspective in plant population biology. *Journal of Ecology*, **84** 461–469.

Huston, M.J. (1979) A general hypothesis of species diversity. *The American Naturalist*, **113** 81–101.

Keddy, P.A. (1992) A pragmatic approach to functional ecology. *Functional Ecology*, **6** 621–626.

Kempton, R.A. (1979) The structure of species abundance and measurement of diversity. *Biometrics*, **35** 307–321.

Law, R., Herben, T. and Dieckmann, U. (1997) Non-manipulative estimates of competition coefficients in a montane grassland community. *Journal of Ecology*, **85** 505–517.

Loreau, M. (1998) Biodiversity and ecosystem functioning: a mechanistic model. *Proceedings of the National Academy of Sciences of the United States of America*, **95** 5632–5636.

May, R.M. (1976) Patterns in multi-species communities, in *Theoretical Ecology* (ed. R.M. May), Blackwell, Oxford, UK.

McRoberts, N., Finch, R.P., Sinclair, W., Meikle, A., Marshall, G., McNicol, J. and Squire, G.R. (1999) Assessing the ecological significance of molecular diversity in natural plant populations. *Journal of Experimental Botany*, (in press).

Moss, J.R.J. (1992) Measuring light interception and the efficiency of light utilisation by the coconut palm (*Cocos nucifera*). *Experimental Agriculture*, **28** 273–285.

Naeem, S. and Li, S. (1997) Biodiversity enhances ecosystem reliability. *Nature*, **390** 507–509.

Niklas, K.J. (1999) Evolutionary walks through a land plant morphospace. *Journal of Experimental Botany*, **50** 39–52.

Pielou, E.C. (1975) *Ecological Diversity*, John Wiley and Sons, New York, USA.

Porter, J.R. (1989) Modules, models and meristems in plant architecture, in *Plant Canopies, Their Growth, Form and Function*, (eds. G. Russell, B. Marshall and P.G. Jarvis), Cambridge University Press, Cambridge, UK, pp. 143–159.

Powell, W., Morgante, M., Andre, C., Hanafey, M., Vogel, J., Tingey, S. and Rafalski, A. (1996) The comparison of RFLP, RAPD, AFLP and SSR (microsatellite) markers for germplasm analysis. *Molecular Breeding*, **2** 225–238.

Rosenzweig, M.L. (1995) *Species Diversity in Space and Time*, Cambridge University Press, Cambridge, UK.

Schlichting, C.D. (1986) The evolution of phenotypic plasticity in plants. *Annual Review of Ecology and Systematics*, **17** 667–693.

Squire, G.R. (1990) *The Physiology of Tropical Crop Production*, CAB International, Wallingford, UK.

Squire, G.R. and Crawford, J.W. (1996) A combined theoretical and experimental approach to functional and genetic diversity. *Annual Report of the Scottish Crop Research Institute*, 1996 pp. 121–124.

Squire, G.R., Rodger, S. and Wright, G. (2000) Community-scale seedbank response to less intense rotation and reduced herbicide input at three sites. *Annals of Applied Biology*, (in press).

Sumida, A., Ito, H. and Isagi, Y. (1997) Trade-off between height growth and stem diameter growth for an evergreen oak, *Quercus glauca*, in a mixed hardwood forest. *Functional Ecology*, **11** 300–309.

Tilman, D. (1994) Competition and biodiversity in spatially structured habitats. *Ecology*, **75** 2–16.

Tilman, D. and Pacala, S. (1993) The maintenance of species richness in ecological communities, in *Species Diversity in Ecological Communities* (eds. R.E. Ricklefs and B. Schluter), University of Chicago Press, Chicago, USA, pp. 13–25.

Van der Meijden, E., Wijn, M. and Verkaar, H.G. (1988) Defense and regrowth: alternative plant strategies and the struggle against herbivores. *Oikos*, **51** 355–363.

Wardle, D.A., Barker, G.M., Bonner, K.I. and Nicholson, K.S. (1998) Can comparative approaches based on plant ecophysiological traits predict the nature of biotic interactions and individual plant species effects in ecosystems? *Journal of Ecology*, **86** 405–420.

Warming, E. (1909) *The Oecology of Plants*, Clarendon Press, Oxford, UK.

West-Eberhard, M.J. (1989) Phenotypic plasticity and the origins of diversity. *Annual Review of Ecology and Systematics*, **20** 249–278.

White, G. and Powell, W. (1997) Isolation and characterisation of microsatellite loci in *Swietania humilis* (Meliaceae): an endangered tropical hardwood species. *Molecular Ecology*, **6** 851–860.

Wright, G. and Squire, G.R. (1998) Fractal lattice design for studying plant aggregation and neighbourhood in the field. *Journal of Experimental Botany*, **49** (supplement), 43–44.

Index